The PHYSICAL WORLD

STATIC FIELDS AND POTENTIALS

University

The Physical World Course Team

Course Team Chair	Robert Lambourne
Academic Editors	John Bolton, Alan Durrant, Robert Lambourne, Joy Manners, Andrew Norton
Authors	David Broadhurst, Derek Capper, Dan Dubin, Tony Evans, Ian Halliday, Carole Haswell, Keith Higgins, Keith Hodgkinson, Mark Jones, Sally Jordan, Ray Mackintosh, David Martin, John Perring, Michael de Podesta, Sean Ryan, Ian Saunders, Richard Skelding, Tony Sudbery, Stan Zochowski
Consultants	Alan Cayless, Melvyn Davies, Graham Farmelo, Stuart Freake, Gloria Medina, Kerry Parker, Alice Peasgood, Graham Read, Russell Stannard, Chris Wigglesworth
Course Managers	Gillian Knight, Michael Watkins
Course Secretaries	Tracey Moore, Tracey Woodcraft
BBC	Deborah Cohen, Tessa Coombs, Steve Evanson, Lisa Hinton, Michael Peet, Jane Roberts
Editors	Gerry Bearman, Rebecca Graham, Ian Nuttall, Peter Twomey
Graphic Designers	Javid Ahmad, Mandy Anton, Steve Best, Sue Dobson, Sarah Hofton, Pam Owen, Andy Whitehead
Centre for Educational Software staff	Geoff Austin, Andrew Bertie, Canan Blake, Jane Bromley, Philip Butcher, Chris Denham, Nicky Heath, Will Rawes, Jon Rosewell, Andy Sutton, Fiona Thomson, Rufus Wondre
Course Assessor	Roger Blin-Stoyle
Picture Researcher	Lydia K. Eaton

The Course Team wishes to thank Carole Haswell, Michael de Podesta, Keith Higgins and Steve Swithenby for their contributions to this book. The book made use of material originally prepared for the S271 Course Team by Joy Manners, Keith Meek and Shelagh Ross. The multimedia package *Forces, fields and potentials* was written by Joy Manners and programmed by Fiona Thomson.

The Open University, Walton Hall, Milton Keynes MK7 6AA

First published 2000

Written, edited, designed and typeset by the Open University.

Published by Institute of Physics Publishing, wholly owned by The Institute of Physics, London.
IoP Publishing, Dirac House, Temple Back, Bristol BS1 6BE, UK.

US Office: Institute of Physics Publishing, The Public Ledger Building, Suite 1035, 150 South Independence Mall West, Philadelphia, PA 19106, USA.

Printed and bound in the United Kingdom by the Alden Group, Oxford.

ISBN 0 7503 0718 8

Library of Congress Cataloging-in-Publication Data are available.

This text forms part of an Open University course, S207 *The Physical World*. The complete list of texts that make up this course can be found on the back cover. Details of this and other Open University courses can be obtained from the Course Reservations Centre, PO Box 724, The Open University, Milton Keynes MK7 6ZS, United Kingdom: tel. +44 (0) 1908 653231; e-mail ces-gen@open.ac.uk

Alternatively, you may visit the Open University website at http://www.open.ac.uk where you can learn more about the wide range of courses and packs offered at all levels by the Open University.

To purchase other books in the series *The Physical World*, contact IoP Publishing, Dirac House, Temple Back, Bristol BS1 6BE, UK: tel. +44 (0) 117 925 1942, fax +44 (0) 117 930 1186; website http://www.iop.org

1.1

s207book5i1.1

STATIC FIELDS AND POTENTIALS

Introduction

In the earlier books of *The Physical World* you have learnt a great deal about motion: how to describe motion and how to predict it. You have encountered some of the forces that influence motion and what the result of that influence is. In this book you will meet an entirely new interaction, namely the *electromagnetic* interaction. As the name suggests, electric and magnetic effects are intertwined at a fundamental level, and you will discover the nature of that interdependence in Chapter 4.

To start with, however, we look at the electrostatic interaction between charged particles and draw many comparisons with the gravitational interaction between masses, with which you already have some familiarity. This is the subject of Chapter 1. Here you will learn how gravitational and electrostatic influences can be described in terms of *fields*.

Chapter 2 further explores gravity and electrostatics, building on the field concept by introducing gravitational and electric *potential*, and the interrelationships between field, force, potential and potential energy.

In Chapter 3 we are concerned with *electric currents*: the movement of free charges in response to electrical influences. In the modern world we are totally dependent on devices that operate using electric currents, usually in metal wires or *circuits*. The basic rules applying to simple circuits are explored here.

Chapter 4 reveals the magnetic effects created and experienced by moving charges and the variety of ways in which these effects manifest themselves in the real world, for example as permanent magnets.

Open University students should be aware that the video for this book (*Magnetic fields in space*) is associated with Chapter 4. There is a prompt to view it at the end of that chapter.

Chapter 1 Gravitational and electric forces and fields

1 The story of planetary motion

On what do we base our scientific theories? When we look up into the sky, how have we learned what we know about the motion of the planets? The understanding of the phenomena of the physical world usually follows from observation and successive attempts at explanation, frequently spread over a period of time and with many setbacks on the way. A very interesting example of this is the explanation of the motion of the planets.

The movement of celestial bodies had, of course, been noted over thousands of years, the planets being distinguished mainly by their relatively rapid motion in the night sky. In fact, the word *planet* is derived from the Greek word for wanderer. This motion is illustrated in Figure 1.1. In the ancient world, the Earth was accepted as the centre of the Universe, and the model proposed by the Greek astronomer Claudius Ptolemy in the second century AD was the accepted theory for the next 14 centuries. However, in 1543, the Polish astronomer Nicolaus Copernicus (1473–1543) proposed that the Earth executed a circular orbit centred on the Sun — a *helio*centric model replacing a *geo*centric model. This now relegated the Earth to the same status as the other planets, which were also held to be orbiting about the Sun. This hypothesis was most certainly not welcomed and was subjected both to ridicule and suppression. This is not uncommon with the breaking of totally new theoretical ground.

Figure 1.1 The night sky, showing the track of the Planet Mars over a period of about nine months.

The next stage in the story arose from accurate measurement (accurate, of course, in the context of that era). Over a period of more than 30 years, using a sextant and compass, the Danish astronomer Tycho Brahe (1546–1601) made accurate measurements of all the astronomical objects he could see with the naked eye in the night sky (the telescope had not been invented). His successor, the German astronomer Johannes Kepler (1571–1630), analysed Tycho's measurements and, after laborious calculations extending over about fifteen years, discovered that the observations, initially for Mars but later for the other known planets, could be explained by a heliocentric motion that followed three laws. The first of these empirical laws required the motion to be not circular but elliptical, with the Sun at one focus of the ellipse (see Figure 1.2).

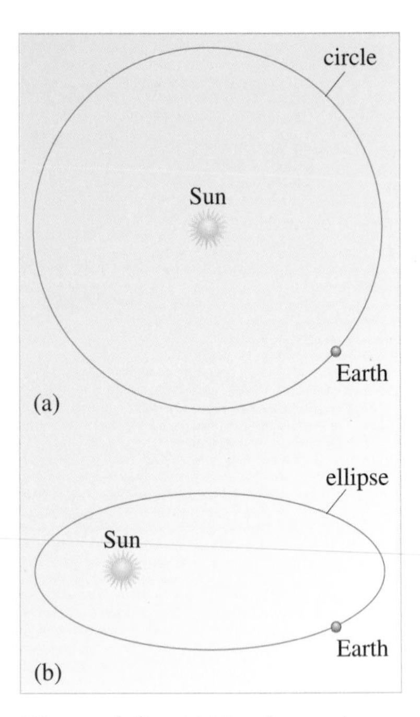

Figure 1.2 (a) The Copernican circular motion of the Earth. (b) The Keplerian elliptical motion of the Earth. (Neither diagram to scale.)

Tycho and Kepler

Tycho Brahe (Figure 1.3) and Johannes Kepler (Figure 1.4), on whose coming together the history of physics so crucially depended, were, each in his own very different ways, extraordinary characters. Tycho was a Danish nobleman of colourful and often overbearing nature, who, as a youth, had observed a partial eclipse of the Sun, which had inspired in him a lifelong passion for accurate astronomical observation. Kepler came from a degenerate family in Weil der Stadt. His father took up the life of a mercenary and narrowly avoided being hanged; his mother was brought up by an aunt who was burnt at the stake for witchcraft and herself nearly suffered the same fate. So while Tycho grew up in luxury and pursued his science much as a hobby, Kepler had a dreadful and sickly childhood, from which he escaped with the help of the educational opportunities available at the time. He went from school to seminary and on to university at Tübingen. There he continued to study for nearly four years after graduating until he was, rather unexpectedly, offered the post of 'Mathematicus of the province' at Graz in Austria. Whilst in this post he developed his ingenious (though erroneous) theory that the planetary orbits were spaced according to the sizes of spheres that could be inscribed between the five platonic solids (Figure 1.5). His book *The Mysterium*, elucidating this hypothesis, was published in 1597 and met with considerable interest.

Figure 1.3 Tycho Brahe (1546–1601).

Figure 1.4 Johannes Kepler (1571–1630).

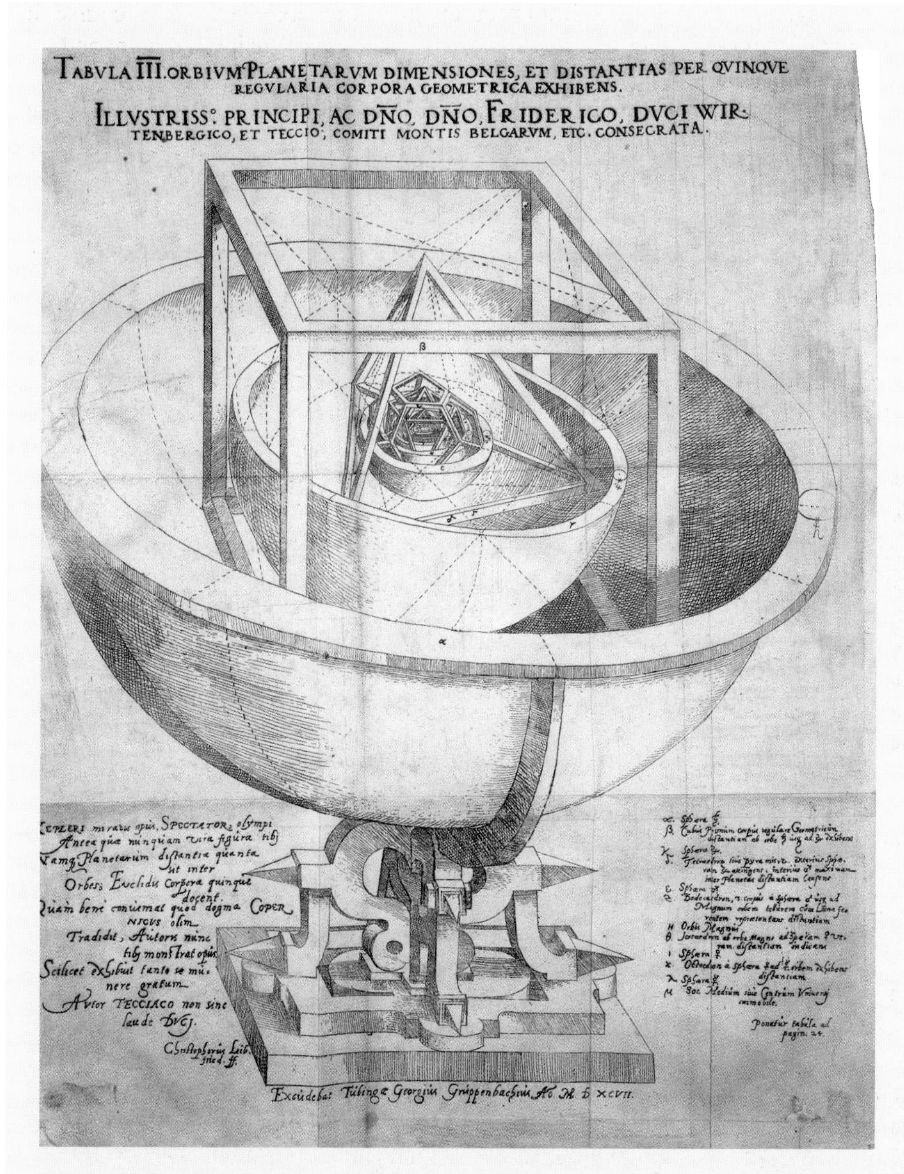

Figure 1.5 The platonic spacing of planetary orbits, according to Kepler.

Meanwhile, Tycho had been living in luxury on the island of Hveen, which had been given to him, along with an immense fortune, by Frederick II of Denmark. He built himself a fantastic observatory and furnished it with instruments the like of which had never been seen before. Here he spent 20 years adding to his mass of astronomical data until, having fallen into disfavour with the king, largely owing to his appalling treatment of his tenants, he left Denmark. In 1599, he was invited to be Imperial Mathematicus in Prague, where, finally, he was to meet Kepler. Their almost ludicrously stormy collaboration was terminated after only 18 months when Tycho died and Kepler succeeded him in his position. Tycho was the last great astronomer to hang on to the geocentric view of the Universe, though he went so far as to make the compromise that the planets other than Earth were orbiting around the Sun. He needed Kepler's genius to distil his data into a plausible theory of planetary motion, just as Kepler had need of Tycho's data to do just that. Even when they had come together, Tycho would not allow Kepler access to his data; he would throw him odd titbits here and there, much to Kepler's frustration. After Tycho's death, however, Kepler took possession of Tycho's life's work and over the next few years, which were spent in lengthy and painstaking analysis and many journeys up the wrong path, he finally concluded that the planetary orbits were ellipses with the Sun at one focus. He had then formulated his three laws of planetary motion, which formed a solid basis for the future development of the classical theory of gravitation.

Figure 1.6 Isaac Newton. Newton graduated from the University of Cambridge, England in 1665, but, due to an outbreak of plague, he spent the next 18 months at his home in Woolsthorp in Lincolnshire. In this time, which he regarded as his prime, he invented integral and differential calculus and discovered the law of universal gravitation. However, on his return to Cambridge when the University reopened in 1667, due to his secretive nature, he revealed little of his achievements. It was not until later that he was persuaded to collect his ideas on mechanics into a book, the *Principia Mathematica*, which was published in 1687. This contained, amongst other things, the formulation of his now well-known three laws of motion and their application using his universal law of gravitation to derive Kepler's three laws. Characteristically, his derivations were geometrical rather than based on his calculus, doubtless to avoid controversy about the validity of his new form of mathematics.

Did this settle any arguments? Unfortunately, it did not. Kepler's three laws were empirical, and although his first law fitted the observations, so did Ptolemy's conceptions. In the Greek world, the circle was the perfect geometrical figure. If the motion of a planet was not circular, then it might be made up of a circle imposed on another circle — an epicycle — or a circle imposed on a circle imposed on a circle, and so on. If there were no rules about the number of epicycles or their radii, the Ptolemaic method could explain the observations just as accurately – but certainly not as simply – as the Keplerian ellipses. (It is difficult to appreciate how strong was the attachment to the circle as the perfect geometrical figure. Even Galileo wanted to stick to circles.) About a hundred years later, Isaac Newton (1642–1727) demonstrated that the Keplerian ellipses arose as a consequence of combining his laws of motion with a simple force law for gravitation. The resolution of the conflict between the two systems is the subject of the next section.

2 Gravitational forces and fields

2.1 Newton's law and gravitational forces between two masses

In *Predicting motion*, you learned about Newton's three laws of motion and how they may be applied to predict the motion of bodies under the influence of specified forces. Amongst his many other great achievements, Isaac Newton (Figure 1.6) also proposed the law of universal gravitation.

Newton postulated his law of universal gravitation in order to explain the motion of the Moon. Whether he came upon the idea when an apple fell on his head as he sat reading in the orchard, as legend would have it, is doubtful, although the fact that objects always fall down towards the Earth would, of course, have been well known to him and all his contemporaries. Newton's achievement was to conceive from this the general hypothesis that, if the mass of an object is attracted towards the mass of the Earth, then perhaps every mass in the Universe is attracted towards every other mass. That is to say, every body exerts an attractive gravitational force on every other body. For any pair of bodies (such as the Earth and the apple) the gravitational forces due to each on the other are equal in magnitude and opposite in direction. The apple is observed to fall to Earth rather than the Earth falling towards the apple because, being the smaller and much less massive, the gravitational force causes a much larger acceleration in the apple than it does in the Earth (remember $\boldsymbol{F} = m\boldsymbol{a}$). Newton's **law of universal gravitation** is stated:

Every particle of matter in the Universe attracts every other particle of matter with a force whose magnitude is directly proportional to the product of their masses and inversely proportional to the square of the distance between them.

In equation form, if the masses of the two particles are m_1 and m_2 and their separation is r, then the magnitude of the force F_{grav} on either particle is:

$$F_{grav} = \frac{Gm_1m_2}{r^2}. \qquad (1.1)$$

The constant of proportionality, G, is called the **universal gravitational constant**.

Because every body attracts every other body, then, for any pair of masses 1 and 2, the force $\boldsymbol{F}_{21}$ on body 2 due to body 1, must be in the opposite direction to the force $\boldsymbol{F}_{12}$ on body 1 due to body 2, as shown in Figure 1.7. The figure illustrates the fact that the gravitational force is a **vector** and $\boldsymbol{F}_{21} = -\boldsymbol{F}_{12}$. The vector is, in each case, along the line joining the two bodies. Each force has the magnitude F_{grav}, as given by Equation 1.1.

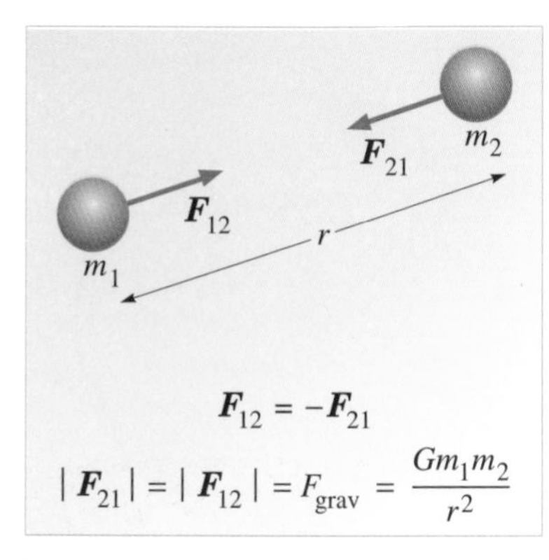

Figure 1.7 The gravitational forces exerted by two particles on each other.

The postulation of the 'inverse square' gravitational force allowed Newton to explain Kepler's laws. (In addition, in the *Principia*, he showed that, beyond its own surface, a spherically symmetric body exerts the same gravitational force on another body as would a point particle of the same mass located at the centre of the sphere. This allows considerable simplification in the mathematics.) However, the experimental study of the law and the measurement of the gravitational constant was not accomplished for another century. This achievement was reported in a paper of 1798 by another remarkable personality, the Hon. Henry Cavendish (1731–1810) (Figure 1.8). The value he obtained was very close to the modern value: $G = 6.67 \times 10^{-11}\,\mathrm{N\,m^2\,kg^{-2}}$.

Question 1.1 Show that the unit of G can also be written as $\mathrm{m^3\,s^{-2}\,kg^{-1}}$. ■

Figure 1.8 Henry Cavendish was born into an aristocratic English family. He inherited a large fortune and after leaving the University of Cambridge (without a degree), he was able to devote the remainder of his life exclusively to his scientific research which he carried out at his home in London. He avoided contact with people — especially women, to whom he would never speak — and was motivated mainly by curiosity, publishing very little of his work. His measurement of G used the principle of the torsion balance invented by his contemporary John Michell (1724–1793).

The experiment to measure G will not be discussed here. The use of the torsion balance will be discussed in Section 4 of this chapter in connection with the electrostatic force. The principle of the experiment is the same although, because of the relative weakness of the gravitational force, the dimensions of the torsion bar and the difficulty of the experiment were very different. Cavendish was the first person

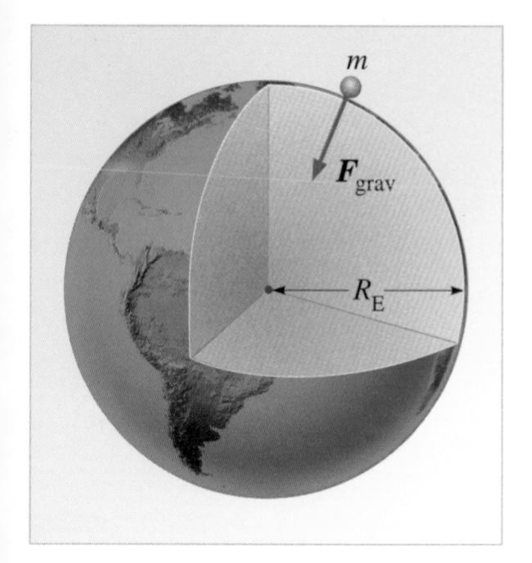

Figure 1.9 A mass m at the surface of a spherical Earth. (The Earth is shown 'cut away' to make it clear that the force is directed to the *centre* of the Earth.)

to estimate the mass of the Earth using the value of the constant he had obtained and other information known at that time, namely the magnitude of the acceleration due to gravity at the surface of the Earth, $g = 9.81\,\text{m s}^{-2}$, and the radius of the Earth, $R_E = 6400\,\text{km}$. The reasoning is as follows:

Consider a small object of mass m at the surface of the Earth, as shown in Figure 1.9. The object will experience a gravitational attraction toward the centre of the Earth and so be subject to a force of magnitude F_{grav}, given by Equation 1.1. If the mass of the Earth is M_E, then, using Newton's result that, outside its radius, a sphere exerts the same force as if all of its mass were concentrated at its centre, the magnitude of the force experienced by the object will be

$$F_{grav} = \frac{GmM_E}{R_E^2}. \tag{1.2}$$

Now, the gravitational force on the object is nothing other than the *weight* of the object, which has magnitude mg, so we can write

$$F_{grav} = mg \tag{1.3}$$

where g is the magnitude of the acceleration due to gravity at the surface of the Earth. Comparing this expression for F_{grav} with Equation 1.2, we can cancel the ms and, after some rearrangement, obtain

$$M_E = \frac{gR_E^2}{G}. \tag{1.4}$$

So, substituting the known values for the constants on the right-hand side of Equation 1.4, we find

$$M_E = \frac{(9.81\,\text{m s}^{-2}) \times (6.37 \times 10^6\,\text{m})^2}{6.67 \times 10^{-11}\,\text{m}^3\,\text{s}^{-2}\,\text{kg}^{-1}} = 6.0 \times 10^{24}\,\text{kg}.$$

Question 1.2 Two small, highly dense, bodies, each of mass 1 kg, are separated by a distance of 1 cm. (a) Calculate the gravitational force between them. (b) What proportion is this of the weight of either mass at the surface of the Earth? ■

2.2 Addition of forces: more than two masses

So far, we have considered only the gravitational force between two isolated point masses. However, if three or more massive bodies are present, we can still use Equation 1.1 to calculate the forces between each pair of masses. To find the resultant force acting on a particular mass, we can calculate the forces on it due to each of the other masses separately, and then add them together.

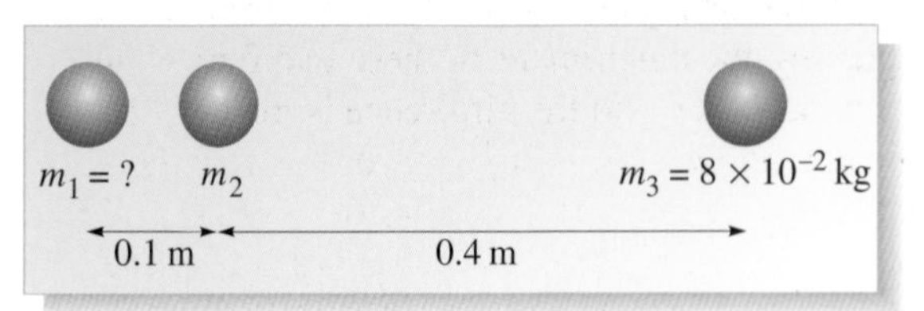

Figure 1.10 Three masses in a straight line, for use with Question 1.3.

Question 1.3 Figure 1.10 shows three masses arranged in a straight line. Assume that m_1 and m_3 are held fixed in position. Calculate the value of the mass m_1 that would be required to hold m_2 in equilibrium in the position shown. ■

The answer to Question 1.3 illustrates how, simply by using Newton's law of gravitation and knowing that the gravitational force is always attractive, it is possible to work out the direction of the resultant force on a mass due to two or more other masses. A more formal way of describing the operation you carried out in answering Question 1.3 would be to say that you calculated the total force on m_2 by taking the vector sum of the forces due to m_1 and m_3 separately. Equation 1.1 gives the magnitude of the gravitational force between two masses, but it does not give the complete picture because it does not give any information about the direction of that force.

The direction of the gravitational force

To give a full description of the gravitational interaction, we therefore need to rewrite Equation 1.1 as a vector equation. How can this be done?

We noted above that the force between two point masses always acts along the line joining the masses. If mass m_1 is taken to be at the origin, and the position of m_2 is defined by the position vector $\boldsymbol{r}$, as shown in Figure 1.11, then the force acts anti-parallel (i.e. parallel but in the opposite direction) to $\boldsymbol{r}$.

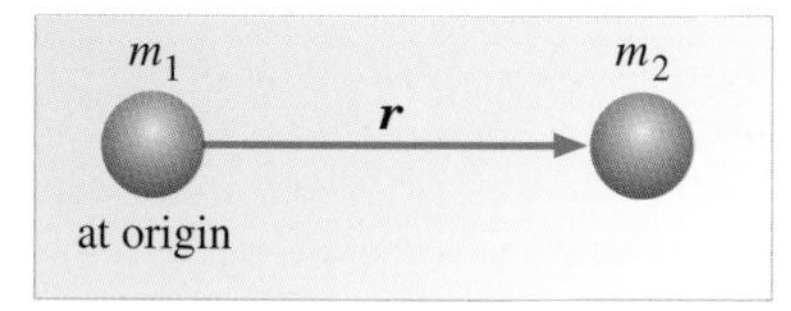

Figure 1.11 If mass m_1 is taken to be at the origin, then the position of mass m_2 can be defined by the position vector $\boldsymbol{r}$. The force on m_2 due to m_1 is always attractive and so is in the opposite direction to $\boldsymbol{r}$.

The simplest way of expressing the direction of the force is to define a **unit vector** denoted by the symbol $\hat{\boldsymbol{r}}$, (pronounced 'r hat'), which is a vector of magnitude 1 in the direction of the vector $\boldsymbol{r}$. Remember, $\boldsymbol{r}$ is the position vector of the mass for which the force is being calculated, from the other mass (which we take to be situated at the origin). In order to obtain a vector of unit magnitude from the vector $\boldsymbol{r}$, we simply divide it by its magnitude, r, hence

$$\hat{\boldsymbol{r}} = \frac{\boldsymbol{r}}{r}. \tag{1.5}$$

Note that $\hat{\boldsymbol{r}}$ is dimensionless: the operation described by the right-hand side of Equation 1.5 involves dividing a length by a length.

The equation that describes fully the gravitational force between two point masses is therefore

$$\boldsymbol{F}_{\text{grav}} = -\frac{Gm_1m_2}{r^2}\hat{\boldsymbol{r}}. \tag{1.6}$$

Note that, since the force is always attractive, the force is always toward the origin, that is, in the opposite direction to $\hat{\boldsymbol{r}}$. This is the reason for the negative sign in Equation 1.6.

One of the main problems with using Equation 1.6 lies in ensuring that $\hat{\boldsymbol{r}}$ is chosen to point in the correct direction. As it stands, Equation 1.6 is valid for calculating both the force on m_1 due to m_2 and the force on m_2 due to m_1. It is clear from the form of Equation 1.6 that, for a given m_1 and m_2, the *magnitudes* of these two forces must be equal. Their directions, however, are opposite, and the difference is manifest in the direction of $\hat{\boldsymbol{r}}$, or equivalently, of the position vector, $\boldsymbol{r}$. If you want to know the force on m_1 due to m_2 , then the position vector must be that of m_1 from m_2. It is not easy to remember in which direction to draw $\hat{\boldsymbol{r}}$, and the best safeguard is to check the direction of the force by using the rule that masses always attract each other.

Question 1.4 Imagine three point masses of equal magnitude ($m_A = m_B = m_C$) lying at the corners of an equilateral triangle, as shown in Figure 1.12. Draw arrows on the figure to represent the magnitude and direction of the gravitational forces acting on mass m_A due to each of the masses m_B and m_C. Now draw a third arrow representing the magnitude and direction of the resultant force on the mass m_A. ■

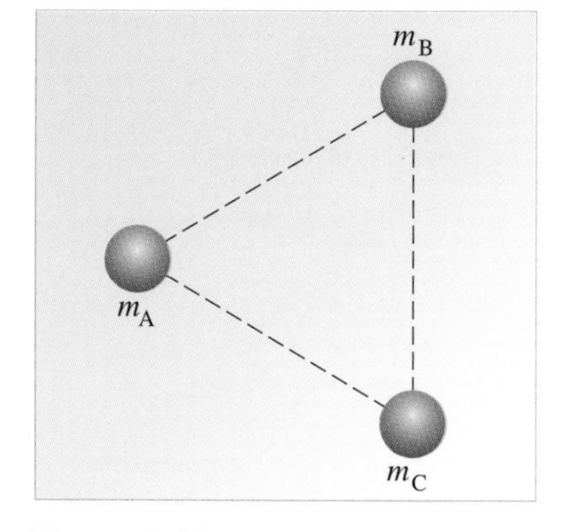

Figure 1.12 Identical masses arranged to form an equilateral triangle, for use with Question 1.4.

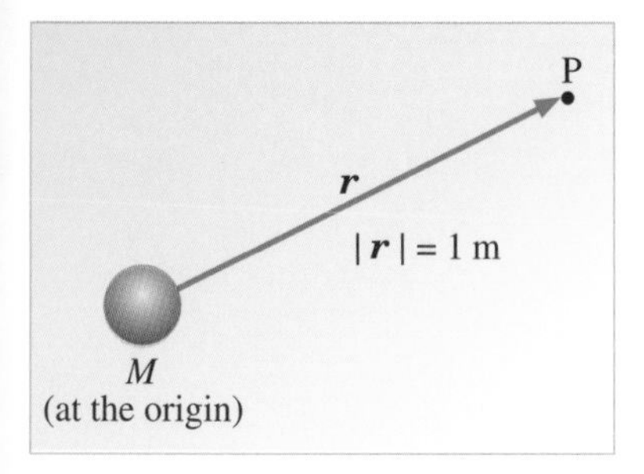

Figure 1.13 What is the force on a mass at P due to another mass, M, at the origin? For use with Question 1.5.

2.3 Introduction to the gravitational field

The concept of a *field* is extremely important in theories of the physical world. This section will begin to lay the groundwork by introducing the concept and describing, in particular, the gravitational field.

Question 1.5 A small body with mass $M = 10\,\text{kg}$ is placed at a point designated as the origin (Figure 1.13). (a) Evaluate the magnitude and direction of the gravitational force on a point mass $m_\text{P} = 1\,\text{kg}$ placed at the point P which is 1 m away from the origin and described by the position vector $\boldsymbol{r}$. (b) Using the result of part (a), evaluate the force (magnitude and direction) on a mass of 2 kg placed at the point P. ■

The important point to note about the case treated in Question 1.5 is that, once you have calculated the force on the 1 kg mass at P, you do not need to use the value of M to work out the force on a mass of 2 kg at P. This is a very important conclusion, because it means that, given the force on a known mass at a particular point, we can calculate the gravitational force on *any* mass placed at that point. Quite simply, the gravitational force is proportional to the mass on which it is acting. One way of thinking about this situation is to imagine some vector, $\boldsymbol{g}$, associated with the point P, that determines the gravitational force that would act on any mass placed at that point. This vector is called the *gravitational field* at P. A similar vector may be defined at every point in space, by dividing the gravitational force that would act on an object of mass m at any particular point by the value of m. In this way, the gravitational field can be defined at every point in space, even where there is no mass to experience a gravitational force. More precisely:

> The gravitational field is a vector quantity, defined at all points in space. Its value at any particular point is given by the gravitational force per unit 'test' mass at that point.

By a 'test' mass we mean a mass that experiences the effect of the gravitational field, but we specifically exclude the possibility that the test mass will cause those masses that are responsible for creating the field to change their positions and thereby change the gravitational field that is being investigated.

Because it is so important in physics, we will now spend some time on the field concept *in general*.

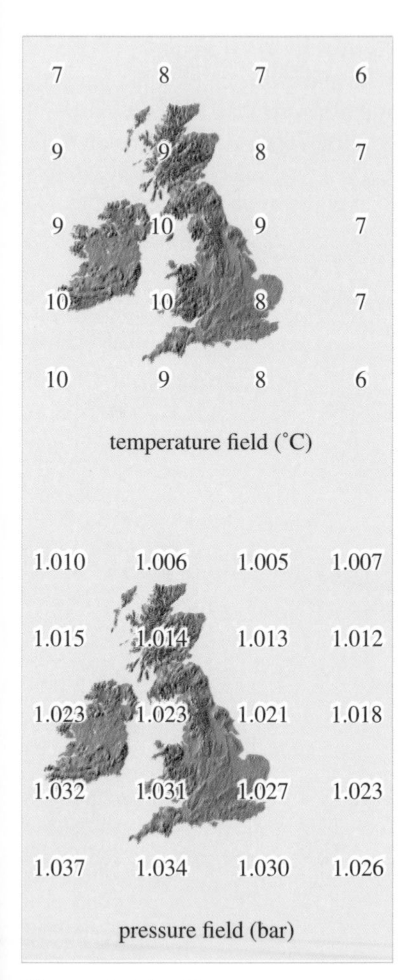

Figure 1.14 The temperature and pressure fields observed across the British Isles one day in January 1981.

2.4 Vector and scalar fields

The concept of a **field** is a very important and much-used idea in physicists' toolkits. Put into more precise language:

> A field is a physical quantity to which a definite value can be ascribed at every point throughout some region of space.

In the case of a gravitational field, the quantity is the gravitational force per unit test mass. Because force is a vector quantity, a gravitational field is represented by a vector at every point, and we therefore call it a **vector field**. However, the general definition of a field simply refers to 'a physical quantity', and that quantity need not necessarily be a vector. A **scalar** quantity might occur as a field too — only in that case it will be a **scalar field**.

You may not have thought of it in quite these terms before, but if you watch the TV weather forecasts you are in fact already quite familiar with examples of both scalar and vector fields. Figure 1.14 shows one representation of temperature and pressure fields. The temperature and pressure clearly fulfil the requirements of the definition in that they do have a value at every point in the region considered (the British Isles). It would be impractical to measure them at every point, but one could certainly obtain values for them at any point one cared to choose within the region. In Figure 1.14, the temperature and pressure values were measured at regularly spaced intervals, and written out as a grid. Figure 1.15 shows another representation of the fields, this time with 'isotherms' (lines connecting all points of equal temperature) and 'isobars' (lines of equal pressure). This representation is much more graphic, and easier to interpret. Figures 1.14 and 1.15 are examples of *scalar* fields: temperature and atmospheric pressure are scalar quanitities that are fully specified by a number and a unit. Wind velocity, on the other hand, is a vector, and the wind velocity field is therefore a vector field: a direction as well as a magnitude has to be assigned to every point in the field. The representation of the vector field is a little more difficult: in Figure 1.16 it has been achieved using arrows.

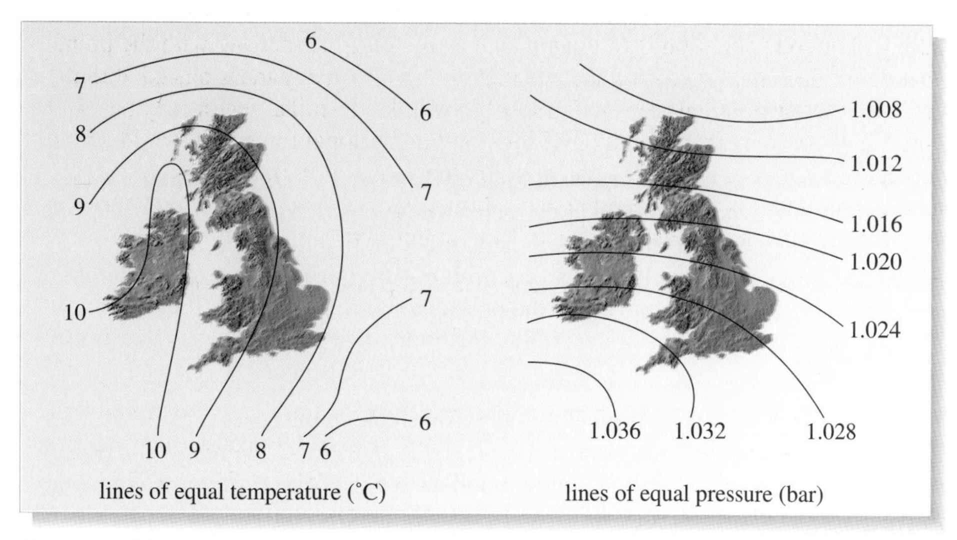

Figure 1.15 Isotherms and isobars, plotted using the same data as Figure 1.14.

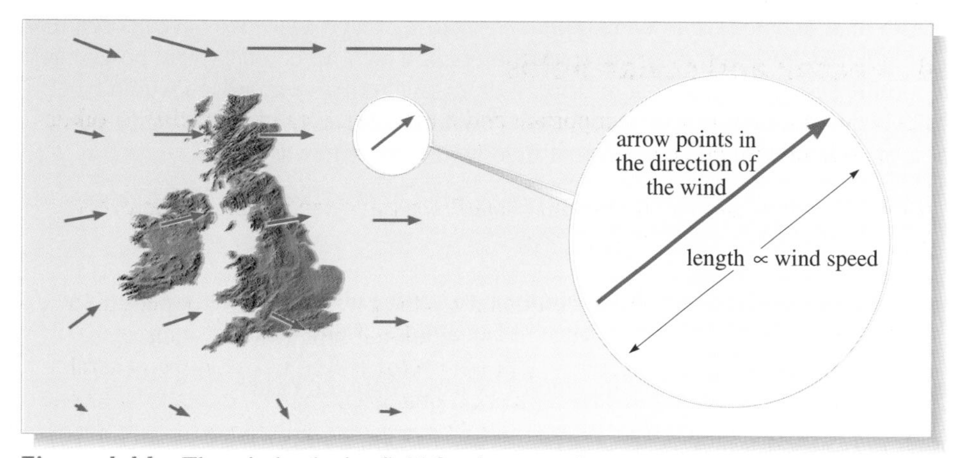

Figure 1.16 The wind velocity field for the same day.

2.5 Defining the gravitational field

We have already defined the gravitational field at a point as the force per unit mass at that point, and noted that it must be a vector. This can be expressed more formally by defining the point by its position vector $\boldsymbol{r}$, and writing:

$$\begin{pmatrix}\text{gravitational field at a point}\\ \text{defined by position vector } \boldsymbol{r}\end{pmatrix} = \frac{\boldsymbol{F}_{\text{grav}}(\text{on } m \text{ at point } \boldsymbol{r})}{m}.$$

Note that this is a vector equation, so it contains information about both the magnitude and the direction of the gravitational field. Thus:

$$\begin{pmatrix}\text{magnitude of the gravitational}\\ \text{field at point } \boldsymbol{r}\end{pmatrix} = \frac{(\text{magnitude } F_{\text{grav}} \text{ of force on } m \text{ at } \boldsymbol{r})}{m}$$

and

$$\begin{pmatrix}\text{direction of the gravitational}\\ \text{field at the point } \boldsymbol{r}\end{pmatrix} = \begin{pmatrix}\text{direction of } \boldsymbol{F}_{\text{grav}} \text{ on}\\ \text{a test mass at point } \boldsymbol{r}\end{pmatrix}.$$

All of this information can be neatly contained in the following general definition:

$$\boldsymbol{g}(\boldsymbol{r}) = \frac{\boldsymbol{F}_{\text{grav}}(\text{on } m \text{ at } \boldsymbol{r})}{m}. \tag{1.7}$$

Equation 1.7 contains a very important kind of notational shorthand. The notation $\boldsymbol{g}(\boldsymbol{r})$ means the gravitational field at the point whose position vector is $\boldsymbol{r}$. It does *not* mean $\boldsymbol{g}$ multiplied by $\boldsymbol{r}$. The reason for using notation like this is because, in general, the gravitational field $\boldsymbol{g}$ will vary with position. If we move to a different point with a different position vector, the gravitational field will usually have a different value. Another way of describing this type of situation is to say that $\boldsymbol{g}$ is a function of $\boldsymbol{r}$.

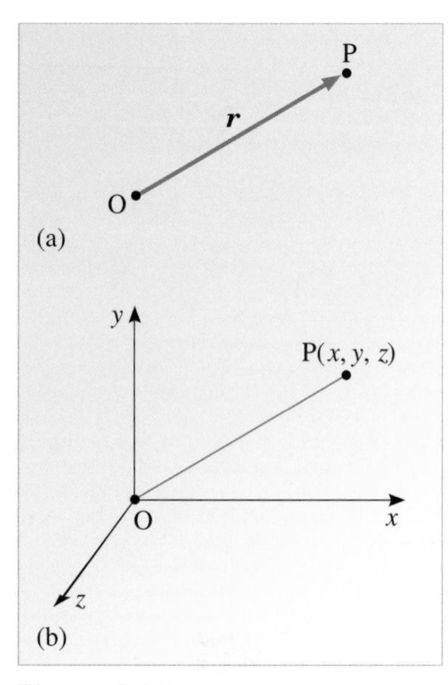

Figure 1.17 Alternative, but equivalent, ways of defining the position of point P: (a) by the position vector $\boldsymbol{r}$ and (b) by its coordinates (x, y, z). Note that $r^2 = (x^2 + y^2 + z^2)$.

The position vector notation is an extremely useful one in the context of fields. Figure 1.17a is essentially the same as Figure 1.13, and shows again how the point P referred to in Question 1.5 may be defined in terms of a position vector. To explain how to get somewhere, the first thing to do is to define the starting place. The starting place is effectively already defined by putting mass M at the origin O. The gravitational field at P is due to the presence of M, so it is entirely reasonable that the site of M should be chosen to define our starting place. Once we have chosen the fixed point O, however, there are several ways in which we could explain how to get to point P. One way is to use a position vector. An alternative approach would be to make O the origin of a set of coordinates (Figure 1.17b) and describe P as the point (x, y, z). In that case, the gravitational field would be written as

$$\boldsymbol{g}(x, y, z) = \frac{\boldsymbol{F}_{\text{grav}}(\text{on } m \text{ at } (x, y, z))}{m}. \tag{1.8}$$

In the rest of this section, we shall continue to use the position vector notation, but you should appreciate the equivalence of Equations 1.7 and 1.8.

- What is the SI unit for the magnitude of a gravitational field?

❍ From Equation 1.7, the unit for the magnitude of $\boldsymbol{g}(\boldsymbol{r})$ must be that of force divided by that of mass, that is, $\mathrm{N\,kg^{-1}}$. This is equivalent to the unit of acceleration (remember Newton's second law of motion) that is, $\mathrm{m\,s^{-2}}$. The magnitude of the gravitational field is the magnitude of the *local* acceleration due to gravity. But remember that the gravitational field varies from place to place *and* it is a vector. The frequently quoted $g = 9.81\ \mathrm{m\,s^{-2}}$ is the magnitude *at the surface of the Earth only*. Even for places with the same magnitude for g, the direction will be different (see Figure 1.18). ■

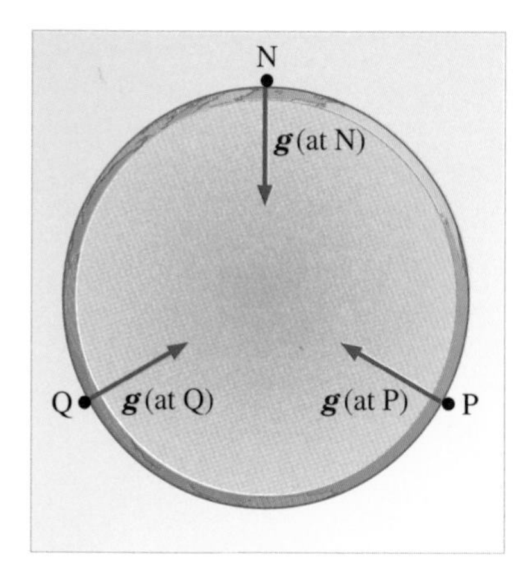

Figure 1.18 The direction of the gravitational field at the surface of a supposedly spherical Earth. The magnitudes are assumed to be the same, but note the different directions (toward the centre of the Earth in each case).

Let us now apply the general definition of a gravitational field, contained in Equation 1.7, to a particular situation, the one first described in Question 1.5 and illustrated in Figure 1.13. What is the gravitational field at the point P? Well, in Question 1.5 you calculated that the gravitational force on the 1 kg mass at the point P has magnitude 6.67×10^{-10} N and is pointing along the direction of $-\boldsymbol{r}$. The shorthand way of writing this is:

$$\boldsymbol{F}_{\text{grav}} = (-6.67 \times 10^{-10}\ \text{N})\hat{\boldsymbol{r}}$$

since $\hat{\boldsymbol{r}}$ is, as you will remember, a unit vector in the $\boldsymbol{r}$-direction. Equation 1.7 then gives

$$\boldsymbol{g}(\boldsymbol{r}) = \frac{\boldsymbol{F}_{\text{grav}}(\text{on } m \text{ at } \boldsymbol{r})}{m} = \frac{-6.67 \times 10^{-10}\ \text{N}}{1\,\text{kg}}\hat{\boldsymbol{r}} = (-6.67 \times 10^{-10}\ \text{N}\,\text{kg}^{-1})\hat{\boldsymbol{r}}.$$

Thus, the gravitational field at P, due to the mass M, is of magnitude $6.67 \times 10^{-10}\ \mathrm{N\,kg^{-1}}$ and is in the direction of $-\boldsymbol{r}$.

In Question 1.5 and subsequently, the notation has allowed us to make a distinction between a mass that we wish to consider to be the source of a gravitational field (labelled by an upper case or capital letter M) and a mass that we wish to consider to be 'feeling the effects of' the gravitational field (given the symbol lower case m). You should be aware that, although this distinction can sometimes be useful, it is actually quite artificial. Two masses m and M, separated by a distance r, will each exert an attractive force on the other, the magnitude of which, in either case, is given by

$$F_{\text{grav}} = \frac{GMm}{r^2}.$$

2.6 The force on a particle in a gravitational field

The gravitational field is defined as the gravitational force per unit mass at any point in space. This leads to the following very simple expression for the gravitational force experienced by a particle of any mass, m, at any point $\boldsymbol{r}$ in a known gravitational field $\boldsymbol{g}(\boldsymbol{r})$:

$$\boldsymbol{F}_{\text{grav}}(\text{on } m \text{ at } \boldsymbol{r}) = m\boldsymbol{g}(\boldsymbol{r}). \tag{1.9}$$

So, to find the force on a mass m at position $\boldsymbol{r}$, one simply multiplies the value of the gravitational field at position $\boldsymbol{r}$ by the mass in question. Notice that the direction of the force is completely accounted for in this equation by the direction of the gravitational field, since m is a positive quantity.

Question 1.6 Consider a mass of 0.2 kg placed in a region where there is a gravitational field of $10\,\mathrm{N\,kg^{-1}}$ in the x-direction. What would be the force on this mass? ■

2.7 The gravitational field due to a point particle

We are now in a position to work out the gravitational field $\boldsymbol{g}(\boldsymbol{r})$ at an arbitrary displacement $\boldsymbol{r}$ from a point mass M. Suppose we were to use a 'test' particle of mass m to measure the gravitational force due to the fixed mass M. Then, from Newton's law of universal gravitation,

$$\boldsymbol{F}_{\text{grav}} = \left(\frac{-GmM}{r^2}\right)\hat{\boldsymbol{r}} = m\left(\frac{-GM}{r^2}\right)\hat{\boldsymbol{r}}.$$

But, according to Equation 1.9,

$$\boldsymbol{F}_{\text{grav}}\,(\text{on } m \text{ at } \boldsymbol{r}) = m\boldsymbol{g}(\boldsymbol{r}).$$

Comparing the two expressions for $\boldsymbol{F}_{\text{grav}}$, we see that the gravitational field due to a point mass M at the origin is:

$$\boldsymbol{g}(\boldsymbol{r}) = \frac{-GM}{r^2}\hat{\boldsymbol{r}}. \tag{1.10}$$

This tells us that the magnitude of the gravitational field at displacement $\boldsymbol{r}$ from a point mass M is

$$g(\boldsymbol{r}) = \frac{GM}{r^2}. \tag{1.10a}$$

Note also that, although g represents a magnitude, and must therefore be a positive quantity, the gravitational force is always attractive, so the field will point in the *negative* $\boldsymbol{r}$-direction. However, you should be very careful in interpreting this. Remember that Equation 1.10 describes the gravitational field at a position $\boldsymbol{r}$ measured from the mass M that is the source of the gravitational field. It is only because we chose, for convenience, to measure position vectors from the position of mass M that $\boldsymbol{g}(\boldsymbol{r})$ is in the negative $\boldsymbol{r}$-direction. Had we chosen to measure positions from some other origin, then $\boldsymbol{g}(\boldsymbol{r})$ would not have been in the *negative* $\boldsymbol{r}$-direction. It is very important to remember that $\boldsymbol{g}(\boldsymbol{r})$ means the gravitational field vector at position $\boldsymbol{r}$ with respect to some specified origin. In general, the direction of $\boldsymbol{g}(\boldsymbol{r})$ is not the direction of $-\boldsymbol{r}$ (or $\boldsymbol{r}$).

Remember that, when the field is produced by a spherically symmetric body, then the gravitational effect outside the body is the same as if all the mass of the body were concentrated at a point at the centre of the sphere. So Equation 1.10 will also describe the gravitational field due to the Earth for points at, and above, the Earth's surface:

$$\boldsymbol{g}_{\text{E}}(\boldsymbol{r}) = \frac{-GM_{\text{E}}}{r^2}\hat{\boldsymbol{r}}.$$

This may be visualized by drawing the vector $\boldsymbol{g}_{\text{E}}$ at various representative points above the surface (see the box on the Earth's gravitational field).

The Earth's gravitational field

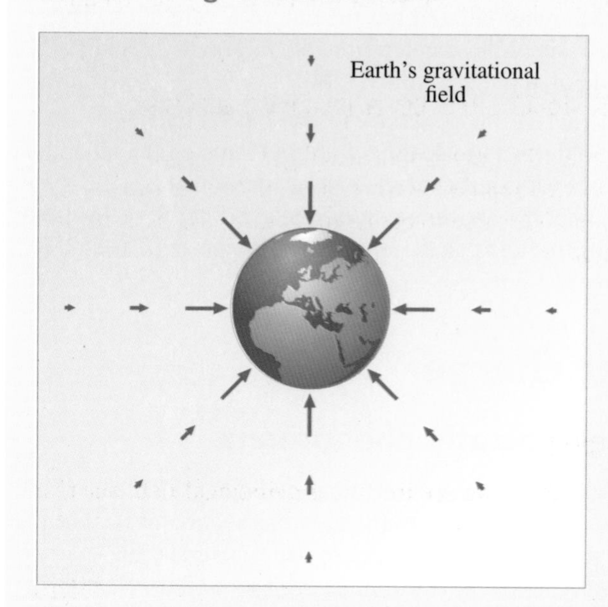

Figure 1.19 The Earth's gravitational field vector $\boldsymbol{g}_{\mathrm{E}}$ at various points above the Earth's surface.

Figure 1.19 shows how the Earth's gravitational field falls off rapidly with height above the surface. You might wonder why we are not aware of this as we climb mountains and such like. This is because, even in the highest mountain ranges, altitude changes are so small as to be invisible on the scale of this diagram. Even at the altitude of the peak of Everest, the value of r (the distance from the centre of the Earth) has increased by only 0.15% from its sea-level value, and the magnitude of g has fallen only to 99.7% of its value at sea-level. Nevertheless, small variations in the value of g over the Earth's surface are measurable and are due not only to variations in altitude but also to factors such as types of rock and their distribution and to changes in latitude. The weight of an object at the Equator is about 99.5% of its weight at the poles. This variation with latitude occurs for two reasons. First, the Earth is not exactly spherical in shape but bulges slightly round the Equator (or you could say that it is flattened at the poles). This is due to the Earth's rotation, and the effect is that places at sea-level near the Equator are farther from the centre of the Earth (larger value of r) than places at sea-level near the poles. Secondly, the rotation of the Earth itself causes objects to weigh slightly less than they would if the Earth were not rotating and this effect is most apparent at the Equator. At Quito in Ecuador, which lies on the Equator, the measured value of g is just $9.780\,\mathrm{m\,s^{-2}}$, whereas in London it is $9.811\,\mathrm{m\,s^{-2}}$ and at the North Pole $9.832\,\mathrm{m\,s^{-2}}$.

The space shuttle orbits typically at a height of 400 km above the Earth's surface. The value of g has fallen to $8.7\,\mathrm{m\,s^{-2}}$ at this height (a reduction of about 11%). This lower value of g is, however, in no sense responsible for the weightlessness of the astronauts in the space shuttle. This is due to the fact that the shuttle is orbiting at such a speed that the force of gravity provides exactly the correct centripetal acceleration for an orbit at that speed and altitude. This acceleration is the same for the shuttle, the astronauts and everything else inside it which are all effectively performing the orbit independently of each other.

Figure 1.20 A uniform field is constant in both magnitude and direction.

- What is the magnitude of $\boldsymbol{g}_E(R_E)$ at the surface of the Earth?

- Because the field is a force per unit mass, it has the unit of acceleration. The value of $g_E(R_E)$ is the magnitude of the acceleration due to gravity (usually denoted g) with a magnitude of about $9.81\,\mathrm{m\,s^{-2}}$. ■

When solving mechanics problems, we can generally ignore the vector field properties of the gravitational field if we are dealing with problems on the laboratory or similar local scale. In this case, we can regard $\boldsymbol{g}_E(\boldsymbol{r})$ as a vector that has the same magnitude and direction throughout the entire region of interest. Such a vector field is called a **uniform field**. A uniform field can be represented by the same vector at every point, as illustrated in Figure 1.20.

3 A first look at charge

3.1 Some simple electrostatic phenomena

In spite of its name, the *electrostatic* force exists between moving charges as well as stationary ones. In later sections you will discover that there are forces that *only* apply to *moving* charges, which is the reason for using the term electro*static* in *this* case.

In *Predicting motion*, you discovered the effects of forces on bodies, effects summarized in Newton's three laws of motion. In the previous section of this chapter you have learned about one of the *fundamental* forces of nature, namely the gravitational force, which not only holds the Earth in its orbit around the Sun but also holds us fairly firmly on the Earth's surface. In this section, we shall look at another force, the **electrostatic force**. This is the force responsible for binding atomic electrons to nuclei and for holding atoms together to form microscopic molecules and macroscopic solids.

Although electrostatic forces are, in general, not as apparent in everyday life as gravitational forces, it is quite easy to perform elementary electrostatic experiments. In fact, electrostatic phenomena are the basis of some well-known effects. Clingfilm sticks to itself and to containers because of electrostatic forces (Figure 1.21). You will almost certainly have seen a balloon stuck to the ceiling of a dry room after it has been rubbed on woollen clothing or something similar. In this situation, the electrostatic attraction between the balloon and ceiling is sufficiently large that it exceeds the gravitational force on the balloon and so the balloon does not fall. A similar experiment involves rubbing a piece of plastic, such as a ruler, on a woollen cloth. The ruler may then be used to lift small pieces of tissue paper, rather as a magnet lifts a steel nail. However, unlike the magnet, the ruler cannot continue to provide the lifting force indefinitely, and eventually the tissue will drop from the ruler. Another example of the action of electrostatic forces can be seen when brushing your hair in front of a mirror. If your hair is dry and you brush vigorously, you will notice that the hair is attracted towards the brush. The same 'static cling' effect causes synthetic clothing to stick to you (and, incidentally, to crackle) in dry weather.

Figure 1.21 It is electrostatic forces that cause clingfilm to cling.

In all these instances, the electrostatic force produced is comparatively large. Later in this chapter, you will discover why electrostatic forces of such strength are not often observed in everyday life. But before we can study the force itself, we must find out under what circumstances bodies like balloons, rulers and hair experience electrostatic forces. Clearly, bodies do not always experience these electrostatic forces — your hair doesn't always stand on end! Something must have happened in each case during the rubbing process to make the bodies experience an electrostatic force. This fact is conveniently expressed in the statement that bodies that experience electrostatic forces are *charged with electricity*. In this section, you will discover some of the properties of this **electric charge** and consider how electrostatic charging takes place.

Children playing in a nylon tent become highly charged!

3.2 Electric charge

Figure 1.22 Charles Augustin Coulomb (1736–1806) discovered the electrostatic force law almost exactly 100 years after Newton published his formulation of the gravitational force law.

During the 18th century, a number of distinguished and ingenious scientists carried out simple experiments, similar to those described above, which clarified the nature of electric charge. These experiments culminated in the work of Charles Augustin Coulomb (Figure 1.22), who deduced the form of the electrostatic force law. The results of these early experiments, together with their modern interpretation, can be summarized as follows. (You should check that each of these statements agrees with your general experience.)

1 *The detection of charge.* The presence of electric charge on a body can be detected only by the forces that the electric charge causes the body to produce or experience. In other words, you cannot tell simply by looking at a particular object whether it is charged or not: the only way to be sure is to see whether it can produce or experience an electrostatic force.

2 *Types of charge.* There are only two types of electric charge. Bodies carrying the *same* kind of charge *repel* one another, whereas bodies carrying *different* types of charge *attract* one another. A body carrying one type of charge can become electrically neutral (i.e. exert no electric forces) by absorbing an equal quantity of the other type of charge. This property of charge cancellation has led to the two types of charge being labelled *positive* (+) and *negative* (−), because the sum of an amount of positive charge and an equal amount of negative charge is zero.

3 *Source of electric charge.* All normal matter contains electric charge. Atoms consist of a nucleus, which is made up of protons and neutrons, and electrons, which orbit the nucleus. Electrons carry a negative charge and protons a positive charge. (The neutron, as its name implies, is electrically neutral.) When two different materials are brought into contact (and rubbing may increase the number of points of contact), electrons may be transferred from one material to the other. The direction of transfer of the electrons depends on the properties of the materials concerned and is always the same for any two materials. A plastic ruler rubbed with a woollen cloth (Figure 1.23) acquires a net negative charge, whereas a glass rod rubbed with a piece of silk ends up with a net positive charge.

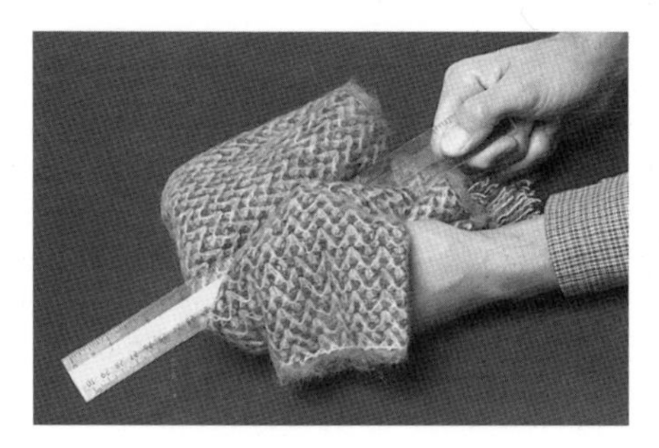

Figure 1.23 Charging by rubbing. When a plastic ruler is rubbed with a woollen cloth, electrons flow from the wool to the plastic. This process will leave an excess of electrons on the plastic so that it carries a net negative charge whereas the wool, with a deficit of electrons, carries a positive charge of equal magnitude.

4 *Conservation of charge.* In the rubbing process illustrated in Figure 1.23, *no charge has been created*: existing charges have simply been redistributed between the wool and the plastic. In fact, the *total* amount of charge in any isolated system is always constant. In other words:

The term *isolated* is used to describe a system from and into which no matter and, therefore, no charge can pass.

> In an isolated system, the total electric charge is always conserved.

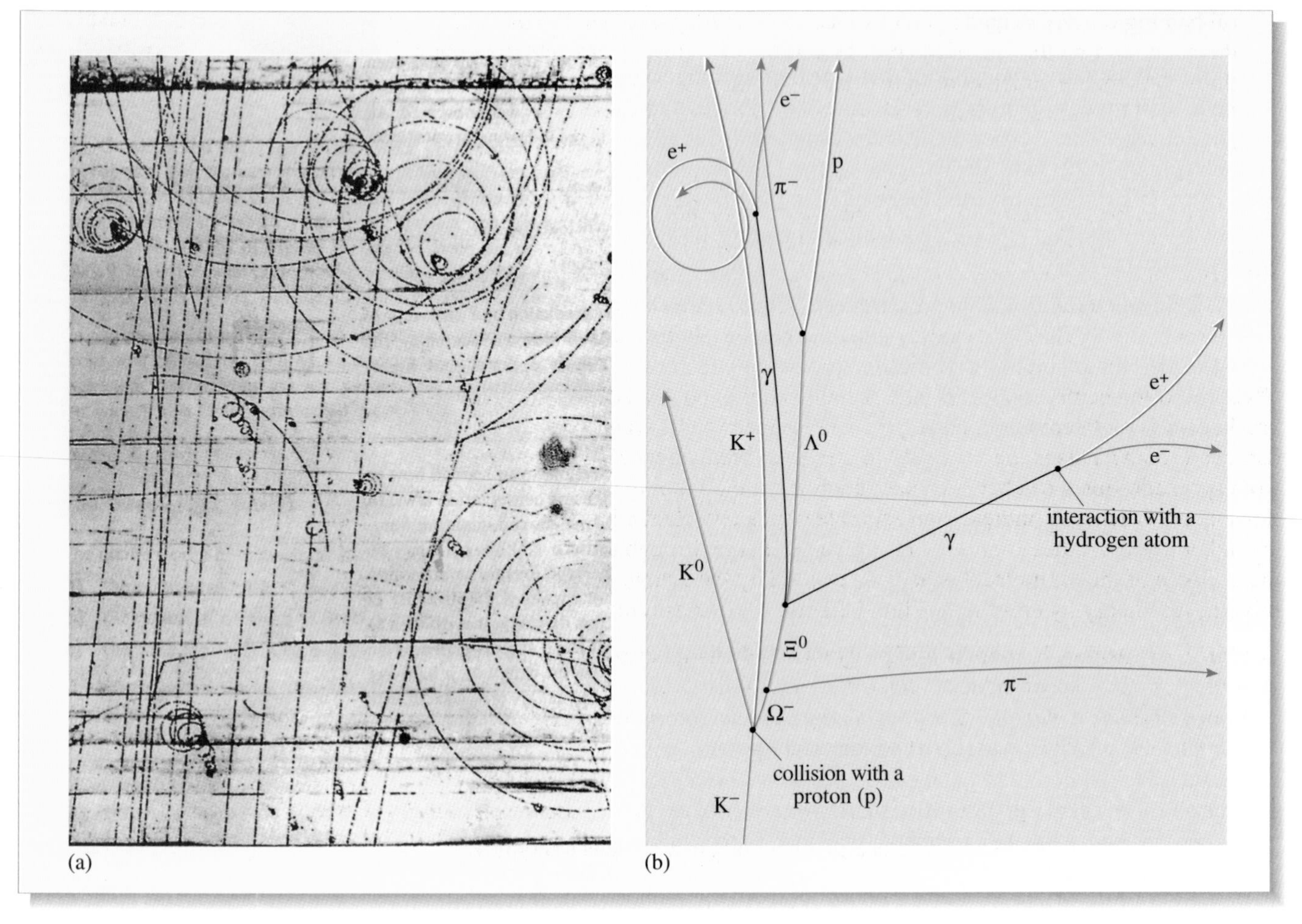

Figure 1.24 Examples of charge conservation in high energy physics. (a) A famous bubble chamber photograph that confirmed the existence of the Ω^--particle. (b) Interpretation of the tracks: the various particles are identified by letters. The tracks of positive and negative particles curve in opposite directions (the reason for this will become clear once you have studied Chapter 4). Particles with superscript 0 are neutral and do not leave tracks, so they are identified by their decay products. Notice particularly the decay of the neutral Λ^0 into a proton (p) and a negatively charged pion (π^-): total charge is conserved because equal amounts of positive and negative charge have been created.

The law of **conservation of charge** takes its place alongside the laws of conservation of linear momentum, angular momentum and energy as one of the fundamental laws of physics. The conservation of charge does *not* imply that charges can never be created or destroyed, but it *does* imply that for any positive charge created, an equal amount of negative charge must also appear. An example illustrating the law of conservation of charge is shown in Figure 1.24.

5 *Conduction of charge.* Because bodies carrying unlike charges attract one another, any body that has a deficit of electrons (i.e. any body that carries a net positive charge) will not only attract negatively charged macroscopic bodies in the vicinity, but will also attract any electrons that are close by. If these electrons are free to move, they will flow towards the positively charged body and neutralize it. It is this process of charge redistribution that prevents us detecting the electric charge on a hand-held rod of copper when it is rubbed. When copper is rubbed, it *does* acquire a charge. However, copper has the property that electric charge can flow easily through it and so the imbalance of charge in the

rubbed region can be made good by electrons flowing through the rod to or from the hand holding the end of the rod. Materials that allow electric charge to flow through them are called **conductors** and those that do not are called **insulators**. Glass and plastic are examples of insulators, whereas metals and seawater are conductors.

Certain materials behave like perfect insulators for small electrostatic forces only. When the electrostatic force increases, there comes a point when electrons can be detached from the atoms of the material and these electrons can then move a short distance. This movement of electrons is equivalent to a charge flow through the material. The material is said to have suffered electrical breakdown and become conducting. This behaviour can theoretically occur in all insulators. In most solid insulators, however, the force required to produce breakdown is so large that the effect can often be neglected. In gases, on the other hand, the occurrence of electrical breakdown is quite common. Have you ever noticed sparks when you undress in a darkened room? This is the result of the electrical breakdown of air. Clothes made from synthetic fibres can become charged through the rubbing involved in everyday movement. When you undress, you separate oppositely charged regions. When breakdown occurs, these regions can exchange charge through the air that separates them. Some of the air molecules become excited by the charge flow and then lose this surplus energy by emitting visible radiation. This causes the observed spark.

Section 3.3 considers in more detail the question of charge distribution and the way in which spontaneous or induced redistribution of charge takes place. Before moving on to these topics, make sure that you can list the main properties of electric charge. The box below provides a checklist of important points.

Box 1.1 The main properties of electric charge

1. The presence of a net electric charge on a body can be detected only via electrostatic forces.
2. There are only two types of charge: these are labelled as positive and negative. Like charges repel and unlike charges attract one another.
3. Charge is a conserved quantity.
4. All normal matter contains charge, and in certain circumstances this can be transferred from one body to another.
5. Charge can flow easily through certain substances; such materials are called conductors. Other substances, through which charge cannot easily flow, are called insulators.

Question 1.7 Two identical brushes are charged by vigorously brushing a dry moulting dog. (a) Will the hairs that fall out attract or repel one another? (b) Will the hairs be attracted to the brush or repelled from it? (c) Will the two brushes attract or repel each other? ■

3.3 The distribution of electric charge

After reading Sections 3.1 and 3.2, you may be wondering why the electrostatic force is not more apparent in everyday life. Matter has two basic properties that cause it to exert forces: one of these properties is mass, which produces gravitational forces, and the other is charge, which produces electrostatic forces. Yet, although we are aware of the gravitational force as soon as we fall out of bed in the morning, we usually have to perform an experiment before we become conscious of the electrostatic force. One of the reasons for this apparent paradox arises from the fact that there are two types of charge, positive and negative, but only one type of mass. The attraction between two equally but oppositely charged objects, and the resulting charge cancellation when they coalesce, produces a single electrically neutral body that does not exert a net electrostatic force. This explains why most lumps of matter contain precisely equal amounts of positive and negative charge and are therefore electrically neutral. However, because there is only one kind of mass, which always gives rise to an attractive gravitational force, two objects coalescing will simply form a more massive body that will exert an even greater gravitational force on other masses. As you will shortly see, the gravitational force is extraordinarily weak, but this is compensated by the very large mass of bodies such as the Earth, that exert significant gravitational forces. Although electrostatic forces are not often immediately apparent in nature, they can, under certain circumstances, produce very dramatic effects. Below we describe the conditions that give rise to two hazardous manifestations of electrostatic forces: lightning storms and the occasional explosion of oil tankers during cleaning.

Lightning storms

On hot humid days, water vapour is forced upwards through the atmosphere by rising air currents. As it rises in the atmosphere, it cools – at first condensing to form small droplets of water and then freezing to form hailstones. These hailstones grow in size as additional droplets condense on them, and eventually they become so heavy that they begin to fall under gravity. As they fall, their temperature rises and they usually melt and fall as heavy rain. This type of downpour is often accompanied by thunder and lightning. Lightning flashes develop from a zone near the base of a cloud, and are caused by the separation of charges within the cloud (Figure 1.25).

The mechanism whereby positive and negative charges are separated in a thundercloud is still controversial. Experiments are very hard to perform and often dangerous, but one fact is undisputed: the electrical activity is centred at an altitude where the temperature is between 0 °C and −10 °C. This is the only temperature range in which both hailstones and droplets of water can exist simultaneously. (Liquids that have been cooled to below their normal freezing temperature, but have not solidified, are said to be *supercooled*.) Most of the proposed mechanisms of charge separation are based on charge transfer between the rising water droplets and the falling hailstones, leaving the water droplets with a net positive charge and the hailstones with a net negative charge. The positively charged droplets are carried up by air currents, while the negatively charged hailstones are pulled down by gravity. This process will result in a net excess positive charge appearing at the top of the cloud and net negative charge appearing near the bottom, as shown in Figure 1.25. This process can continue and the amounts of charge can increase, until the electrostatic forces are so great that either:

In 1753, a scientist called Richmann was killed during a lightning experiment. His contribution to science did not end with his death, however, for his body was dissected to discover the effects of his last experiment on his vital organs.

- the vapour in the cloud undergoes electrical breakdown, allowing the electrons to flow up through the cloud in a giant spark of lightning ('cloud flash'), or
- the air beneath the cloud suffers electrical breakdown and the negative charge at the bottom of the cloud flows to the ground via a flash of forked lightning as shown in Figure 1.26 (ground flash).

The process of recharging can then begin again.

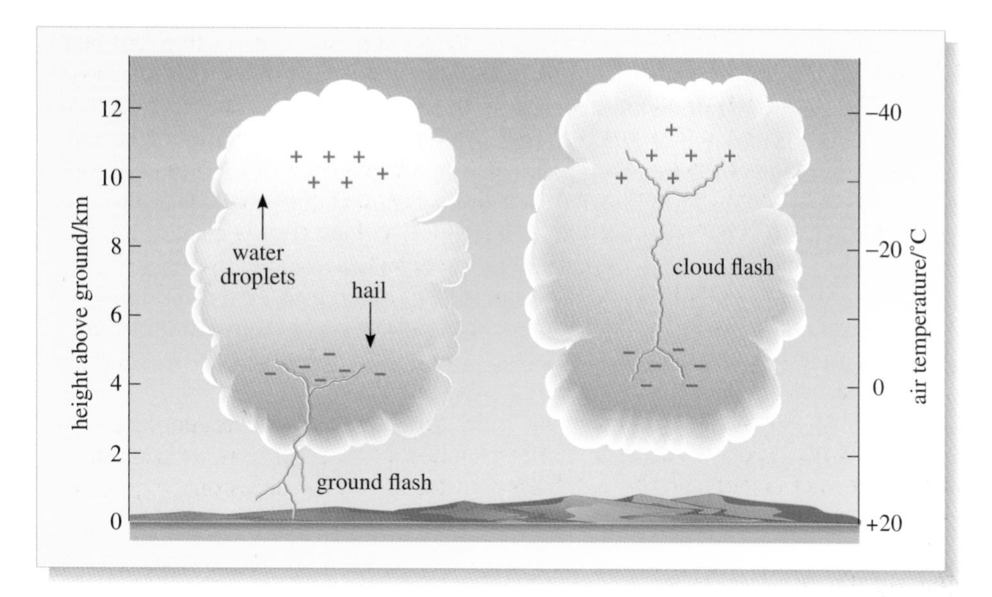

Figure 1.25 Schematic diagram of the charge distribution in a typical fully developed thundercloud. Note the thickness of the cloud, and the temperature difference across it. Two separate discharge processes are shown.

Figure 1.26 A lightning discharge. Although the visual effect is dramatic, the amount of energy liberated in an average lightning flash is modest – about the equivalent of the energy liberated in the explosion of a gallon of oil.

Interestingly, cloud flashes are much more common in tropical and equatorial regions than at higher latitudes. For many years, the accepted explanation for this was based on just two assumptions:

- Whatever the charging mechanism in clouds, it involves hailstones, and therefore the process of charge separation is confined to regions of the cloud above the altitude at which the air temperature is 0 °C (the so-called 0 °C isotherm);
- The closer the negatively charged region of the cloud is to the ground, the more ground flashes are favoured over cloud flashes.

The average air temperature close to the ground increases with decreasing latitude (i.e. from poles to Equator), so the accepted argument was that the 0 °C isotherm was farther from the ground in the tropics than at higher latitudes and this situation favoured cloud flashes over ground flashes. More recent research, however, suggests that the depth of the cloud may also be a very important factor. Typical cloud depth increases systematically from high to low latitudes, and the deeper the charging region within a cloud the more likely it is to support a cloud flash. Quantitative values enhance the plausibility of this argument. The ratio of cloud to ground flashes is about 2 at latitude 60° and around 6 at the Equator. The depth of the sub-freezing part of thunderclouds can be three times greater at the Equator than at latitude 60°, but the height of the 0 °C isotherm only changes by about 25% between these latitudes.

Oil tanker explosions

Oil tanker explosions tend to occur during cleaning, when seawater is flowing through pipes into the tanks. On the way, the water acquires an excess negative charge by rubbing on the pipe walls, while the pipe is left with an excess positive charge (Figure 1.27). These charges can build up and may cause the vapour in the tank to suffer electrical breakdown. When this happens, electrons will flow through the vapour in a spark, which may ignite the flammable oil vapour remaining in the tank. In this event, an explosion is almost inevitable because of the rapid heating of the oil vapour in the confines of the tank.

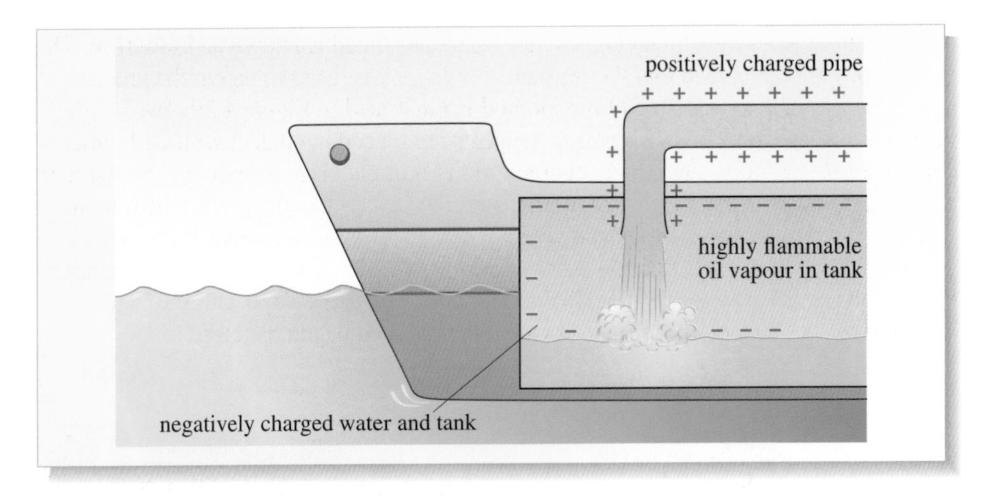

Figure 1.27 Charging by rubbing takes place as liquids flow through pipes. The charge separation between a pipe and its contents can have disastrous results in an oil tanker because of the highly explosive nature of oil vapour.

Charging by sharing and induction

Lightning and oil tanker explosions are striking illustrations of charge redistribution. Neither phenomenon is totally understood, and work is still being done to clarify the mechanisms of charge separation. However, we shall now turn our attention to a pair of well-understood methods by which charges may be redistributed. These methods were of great importance historically, because in the early days of electrostatic experimentation there was no absolute measure of quantity of charge. By using these two techniques, it was possible to give bodies identical charges in one case and opposite charges of equal magnitude in the other. This allowed Coulomb to perform quantitative experiments on the forces between two charged bodies without knowing the absolute charge on either body.

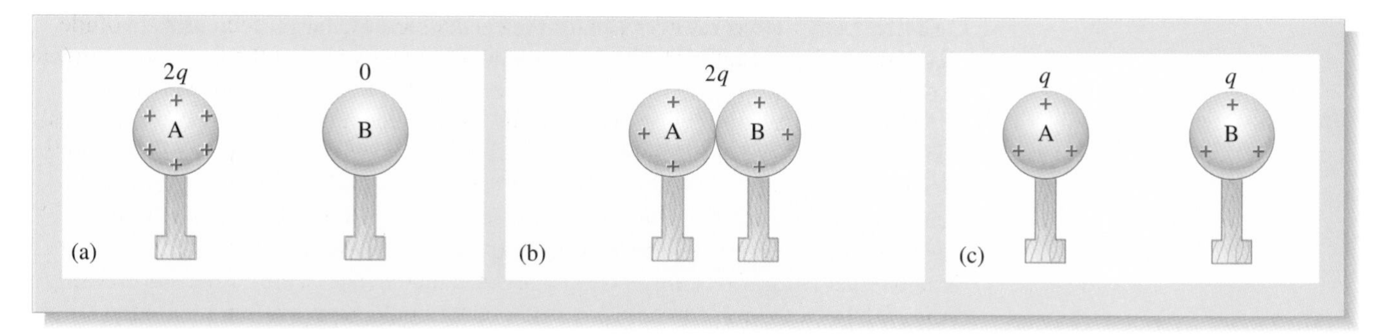

Figure 1.28 Charge sharing between conducting spheres. (a) Sphere A is initially given a positive charge $2q$, and sphere B is neutral. (b) The charge spreads over both spheres when they are brought into contact. (c) When A and B are separated, the charge left on each of them is half that originally given to A.

The first process we shall look at is **charge sharing** between two identical bodies, say two spheres A and B made of some conducting material and mounted on insulating stands (Figure 1.28a). Sphere A is charged, perhaps by rubbing it on some cloth (remember this can only be done provided the metal is not touched by the hand). Sphere A is then brought up to touch the initially neutral sphere B. Since the excess charges on A repel one another and since they can flow freely through the conducting material of the spheres, B will immediately acquire a charge (Figure 1.28b). When the spheres are separated again (Figure 1.28c), the charges will redistribute themselves uniformly, with the original charge shared equally between the two spheres.

The second process by which charges may be redistributed is known as **induction**. This method has the advantage that the original charge may be used to repeat the process again and again. The principle of the method is illustrated in Figure 1.29. Two uncharged metal spheres on insulating supports are placed in contact with each other (Figure 1.29a). When a negatively charged rod is brought close to sphere C but without touching it, some of the electrons within C are repelled by the charges on the rod and flow to sphere D, leaving a net positive charge on C and a net negative charge on D. The resulting charge distribution is shown in Figure 1.29b. The excess positive charge on the sphere closer to the rod is called the induced positive charge, and the excess negative charge on the other sphere is called the induced negative charge.

Figure 1.29 Charging by induction. In this case, both spheres are initially neutral (a) and it is a third body that is charged by rubbing and brought close to them (b). When the spheres are separated and the third body removed, the charge left on one of the spheres is opposite in sign but equal in magnitude to that left on the other.

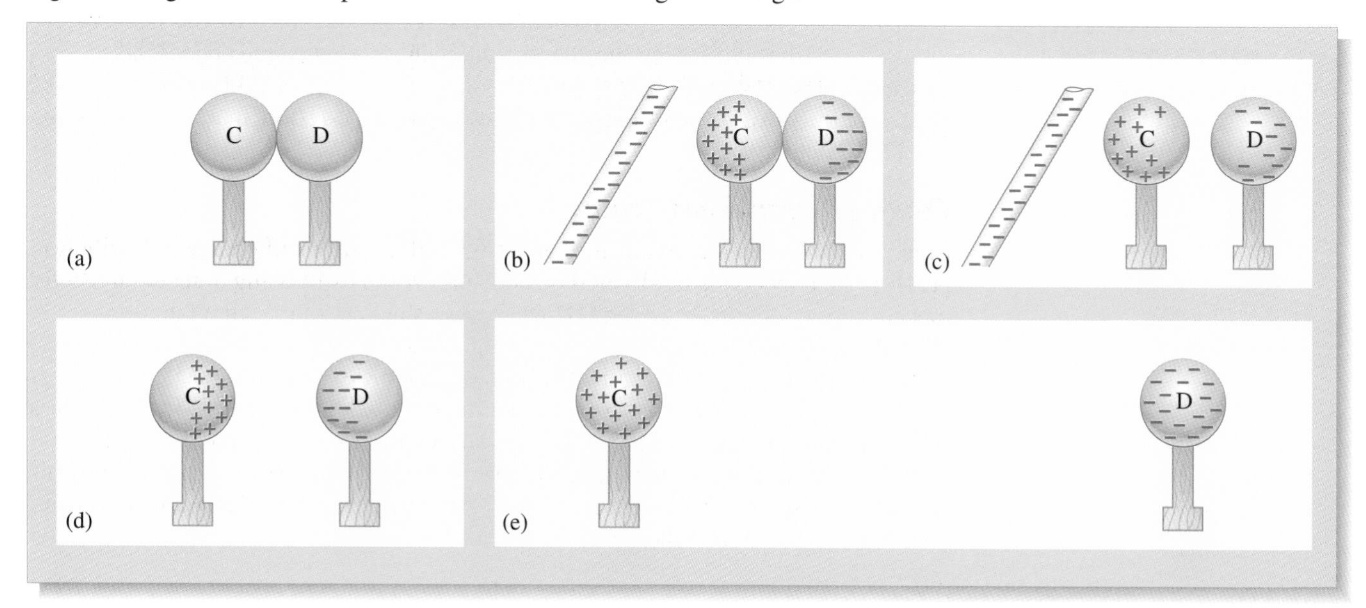

Question 1.8 Give a description of the final stages of charging by induction as illustrated in Figure 1.29c, 1.29d and 1.29e. ■

As evidenced by the ruler and tissue-paper experiment, insulators may also experience electrostatic forces due to the effects of induction. This phenomenon will be examined in more detail in Section 4.2.

4 Electrostatic forces and fields

So far, we have not considered the magnitude of an electrostatic force, only the circumstances under which such a force exists and whether it is attractive or repulsive. In this section, the description of the electrostatic force will be put on a quantitative footing, and the associated field will be introduced.

4.1 Coulomb's law and the electrostatic forces between two charges

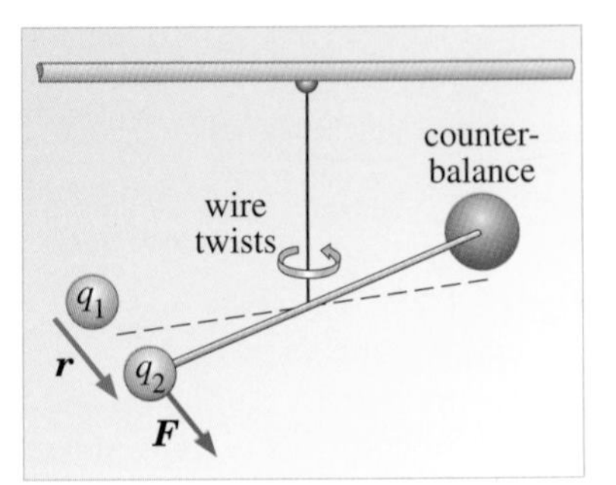

Figure 1.30 The essential details of the apparatus with which Coulomb investigated the dependence of the electrostatic force on the magnitude and separation of two 'point charges'.

The strength of the electrostatic interaction

Coulomb became, in 1785, the first person to establish the dependence of the electrostatic force on the distance between the charges. He considered the force between two very small charged bodies (i.e. bodies whose diameters were much smaller than the distance between them). This size restriction allowed him to assume that all the charge on each of the bodies was concentrated at a point. These very small charged bodies are often referred to as **point charges**. Their use allowed Coulomb to neglect the problem of how charge is distributed throughout a body. In fact, once Coulomb had established the magnitude of the force between point charges, it was a small step to extend the result to more complex charge distributions, since any charge distribution can be thought of as composed of a large number of point charges.

The experimental situation investigated by Coulomb is shown schematically in Figure 1.30. A point charge q_2 is separated from another point charge q_1 by a distance r. Because he had no means of measuring charge, Coulomb used small identical conducting spheres for the bodies so that he could successively halve the values of q_1 and q_2 by sharing the charge with identical uncharged spheres, as illustrated earlier in Figure 1.28. Using a very sensitive torsion balance (Figure 1.31) to measure the forces on the charges, Coulomb was able to show that the magnitude, F_{el}, of the electrostatic force acting on q_2 due to q_1 was:

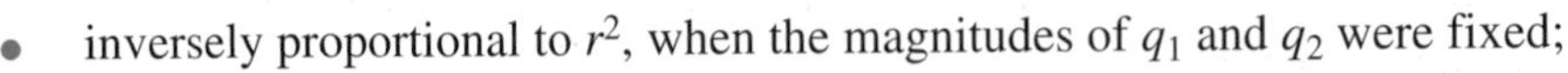

- inversely proportional to r^2, when the magnitudes of q_1 and q_2 were fixed;
- proportional to the magnitude of q_1, when q_2 and r were fixed;
- proportional to the magnitude of q_2, when q_1 and r were fixed.

Thus, in mathematical symbols: $F_{el} \propto \dfrac{1}{r^2}$, $F_{el} \propto |q_1|$ and $F_{el} \propto |q_2|$.

Figure 1.31 Coulomb's torsion balance. The force between the two small charged spheres causes a torque to act on the long horizontal beam. By twisting the wire at the suspension point, an opposite and equal torque can be applied. With zero net torque, the beam's position remains unchanged. The angle through which the suspension must be twisted to achieve this result is proportional to the electrostatic force. The whole apparatus is enclosed in a case because the forces involved are small and draughts could easily introduce appreciable errors.

These three results can be combined to give the single proportionality

$$F_{\text{el}} \propto \frac{|q_1||q_2|}{r^2}. \tag{1.11}$$

To turn this proportionality into an equation, all that is necessary is to introduce a proportionality constant. But to give a value to that constant, we need to know the unit in which q_1 and q_2 are measured. The SI unit of charge is called, very appropriately, the **coulomb** (given the symbol C). The proper definition of the coulomb depends on the magnetic force, in a way that will be discussed in Chapter 4, but, to give you some idea of the size of the unit of charge, it is worth noting that the total charge delivered by a typical lightning bolt is, very roughly, 1 coulomb.

If the charges q_1 and q_2 are located in a vacuum, the proportionality constant for the electrostatic force law is normally written in the form $\dfrac{1}{4\pi\varepsilon_0}$. The magnitude of the electrostatic force in a vacuum is thus

$$F_{\text{el}} = \frac{|q_1||q_2|}{4\pi\varepsilon_0 r^2}. \tag{1.12}$$

The quantity ε_0 (ε is the Greek letter epsilon) is known as the **permittivity of free space**. Experiments show that the value of $\dfrac{1}{4\pi\varepsilon_0}$ is $8.988 \times 10^9\,\text{N}\,\text{m}^2\,\text{C}^{-2}$.

Question 1.9 Jot down an argument to convince another student that the unit of the proportionality constant $\dfrac{1}{4\pi\varepsilon_0}$ is $\mathrm{N\,m^2\,C^{-2}}$. ■

You should notice the similarity between Equation 1.12 and the equation for the gravitational force between two masses (Equation 1.2) that you met in Section 2. In both gravitational and electrostatic interactions, the force is inversely proportional to the square of the distance between the particles. However, the two interactions have rather different strengths, as you can discover by answering the following question.

Question 1.10 What is the magnitude of the force between two charges, each of 1 coulomb, separated by 1 metre in free space? On what mass of iron would the gravitational force at the Earth's surface have this magnitude? (You may take the value for $\dfrac{1}{4\pi\varepsilon_0}$ to be $9.0 \times 10^9\,\mathrm{N\,m^2\,C^{-2}}$.) ■

The answer suggests one of two things: either that the electrostatic force is immensely strong, or that the coulomb is a very large unit of charge. In fact, there is some truth in both of these possibilities. The coulomb is, in practical terms, a very large amount of charge. The reason for choosing such a large standard unit of charge will become clear when you meet the magnetic force law in Chapter 4. In this chapter and in Chapter 2, however, you will also be concerned with the charge carried by electrons. The **electron charge** (denoted by $-e$) is very small indeed:

$$-e = -1.602 \times 10^{-19}\,\mathrm{C}.$$

Note the minus sign: the electron carries a negative charge. Protons are the positively charged constituents of the nuclei of atoms; each has charge $+e = +1.602 \times 10^{-19}\,\mathrm{C}$.

From a qualitative point of view, we would say that the electrostatic force is very much stronger than the gravitational force. However, it is not possible to make any *fundamental* quantitative comparison of the two forces because our units of charge and mass (the coulomb and the kilogram) are not fundamental in any way: they are to a large extent arbitrary. However, the following question allows you to compare the magnitudes of the electric and gravitational forces on an atomic scale.

Question 1.11 The hydrogen atom may be modelled as an electron orbiting a proton in a circular orbit of radius $5 \times 10^{-11}\,\mathrm{m}$. Treating the proton and the electron as point charges, estimate the ratio of the electrostatic force to the gravitational force between the two particles. You may take the masses of the proton and the electron to be $1.6 \times 10^{-27}\,\mathrm{kg}$ and $9.1 \times 10^{-31}\,\mathrm{kg}$, respectively. The constants are: $G = 6.7 \times 10^{-11}\,\mathrm{N\,m^2\,kg^{-2}}$ and $\dfrac{1}{4\pi\varepsilon_0} = 9.0 \times 10^9\,\mathrm{N\,m^2\,C^{-2}}$. ■

The value of e was first measured accurately by Robert A. Millikan. His method will be explained in Section 5.1. The importance of the charge on an electron is that it is generally believed to be the smallest amount of charge that can have an independent existence and be transferred from one body to another. In a rubbing experiment, you might transfer about 10^{12} electrons (i.e. a charge of about $10^{-7}\,\mathrm{C}$) to the object being rubbed. This may seem like a huge number of electrons. It is, however, only a tiny fraction, about 1 in 10^{13}, of the total number of electrons in the object. Because the electron charge is so small, in macroscopic electrostatics we can assume electric charge to be infinitely divisible and continuous and we can ignore its 'graininess'. We may, for example, imagine that a negatively charged body has the charge spread continuously over it, rather than worry about the finite number of separate electrons acting as point charges. We do exactly the same thing in dealing with the mass of an ordinary object: we imagine it to be infinitely divisible, although the atomic nature or 'graininess' of all matter implies that it is not really so.

4.2 Coulomb's law

Question 1.12 Figure 1.32 shows three charges arranged in a straight line. Charges q_1 and q_2 are held fixed in position. (a) In which direction will q_3 move if it is free to do so? (b) Calculate the magnitude of the force on q_3. ■

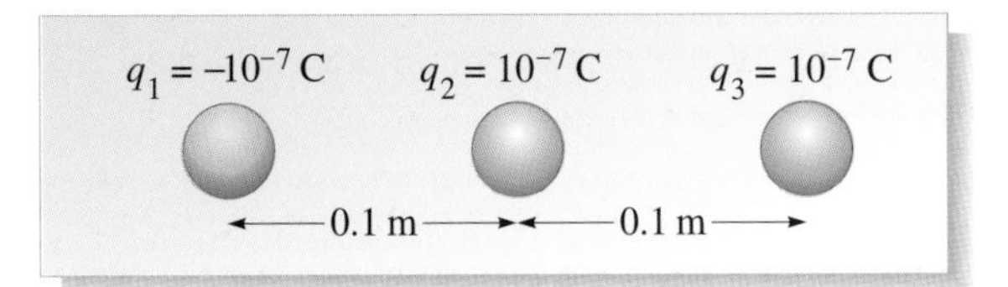

Figure 1.32 Three charges in a straight line; for use with Question 1.12.

In Question 1.12 you used the fact that 'like charges repel, unlike charges attract' to determine the resultant force on one point charge in a group of three. Of course, groups of charges will not, in general, be arranged in a straight line and so we must make the vector nature of the electrostatic force explicit, as we did for the gravitational force in Section 2.2.

If charge q_1 is taken to be at the origin, and the position of q_2 is defined by the position vector $\boldsymbol{r}$, as shown in Figure 1.33, then the force on q_2 acts either parallel or anti-parallel (i.e. parallel but in the opposite direction) to $\boldsymbol{r}$.

As we did for gravitation, we shall make use of the unit vector $\hat{\boldsymbol{r}}$. The electrostatic force between two point charges in a vacuum is therefore

$$\boldsymbol{F}_{\text{el}} = \frac{1}{4\pi\varepsilon_0}\frac{q_1 q_2}{r^2}\hat{\boldsymbol{r}}. \qquad (1.13)$$

This is usually referred to as **Coulomb's law in free space**. Remember that, because Equation 1.13 is a vector equation, the quantities appearing on the right-hand side are the true values of the charges q_1 and q_2, not their magnitudes as was the case in Equation 1.12. You can see how the signs work out by comparing Figure 1.33b and 1.33c. If the charges are of the same sign, then $\boldsymbol{F}$ is parallel to $\hat{\boldsymbol{r}}$ (i.e. repulsive); if they are of opposite sign, then $\boldsymbol{F}$ is anti-parallel to $\hat{\boldsymbol{r}}$ (i.e. attractive).

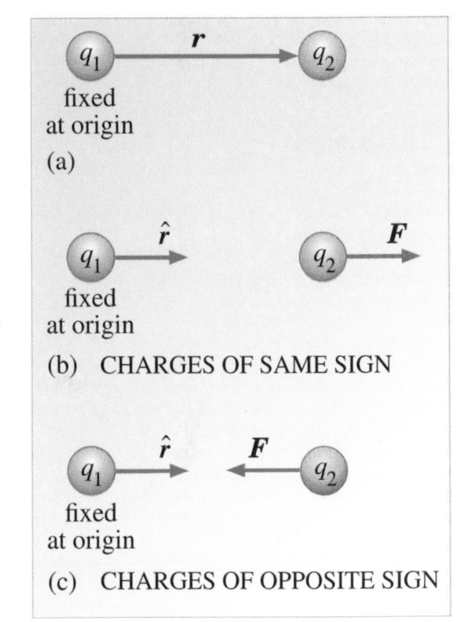

Figure 1.33 (a) If charge q_1 is taken to be at the origin, then the position of charge q_2 can be defined by the position vector $\boldsymbol{r}$. (b) Force on q_2 due to q_1 when both charges are positive, or both are negative. (c) Force on q_2 due to q_1 when the charges are of opposite sign.

As for the case of the gravitational force, we have to be careful with our choice of the unit vector $\hat{\boldsymbol{r}}$. If you are calculating the force *on q_2 due to q_1*, then $\hat{\boldsymbol{r}}$ must be the unit vector in the direction of the position vector *of q_2 from q_1*, which is taken to be at the origin.

If you compare Equation 1.6 for the gravitational force with Equation 1.13 for the electrostatic force, you will notice that the minus sign has been included explicitly in Equation 1.6 to accommodate the fact that masses are always positive and always attract each other ($\boldsymbol{F}_{\text{grav}}$ anti-parallel to $\hat{\boldsymbol{r}}$). Charges can be positive or negative and, because like charges repel ($\boldsymbol{F}_{\text{el}}$ parallel to $\hat{\boldsymbol{r}}$), there is no minus sign in Equation 1.13.

In cases where there are more than two point charges present and there is therefore no obvious choice of origin, the resultant force on any particular charge may be found by taking the vector sum of all the forces on that charge due to the other charges. The force that any other point charge exerts on the chosen charge will always be parallel or anti-parallel to the line joining those two charges.

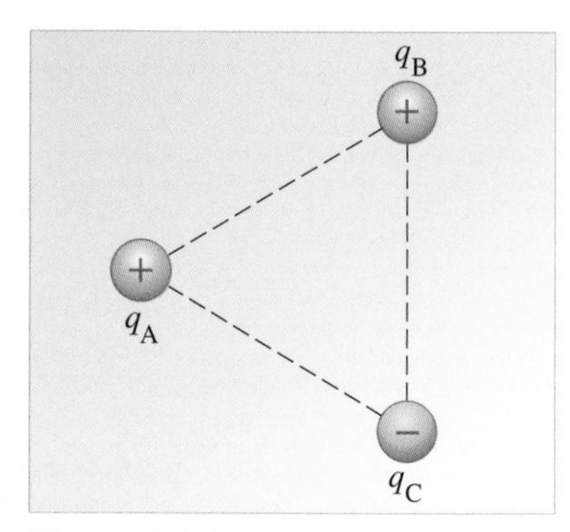

Figure 1.34 Identical charges arranged to form an equilateral triangle; for use with Question 1.13.

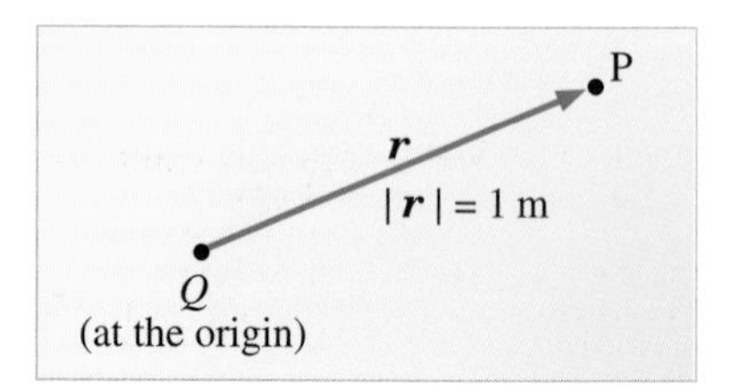

Figure 1.35 What is the force on a charge at P, due to another charge at the origin? For use with Question 1.14.

Question 1.13 Imagine three point charges of equal magnitude but with q_A and q_B positive and q_C negative, lying at the corners of an equilateral triangle, as shown in Figure 1.34. Draw on the figure arrows representing the magnitude and direction of the electrostatic forces acting on charge q_A because of the presence of charges q_B and q_C. Now draw a third arrow representing the magnitude and direction of the resultant force on the charge q_A. ■

4.3 Defining the electric field

In this section, we will introduce the concept of the *electric field*. The definition of the electric field is very similar to that of the gravitational field given in Section 2. The argument that we use to introduce it is, therefore, also similar to Section 2.

Question 1.14 A small body carrying a charge $Q = 10^{-6}$ C is placed at a point designated as the origin (Figure 1.35). (a) Evaluate the magnitude and direction of the electrostatic force on a point charge $q = 10^{-7}$ C placed at the point P, which is 1 m away from the origin and described by the position vector $\boldsymbol{r}$. (b) Using the result of part (a), evaluate the force (magnitude and direction) on a charge of 2×10^{-7} C placed at the point P. ■

From the example treated in Question 1.14 it is clear that the electrostatic force experienced by a charge q at a certain point is proportional to the charge q. Thus, we may define the electric field $\mathscr{E}$ at that point as the electrostatic force that would act on a charge placed at that point, divided by the value of the charge. In this way, the electric field can be defined at any point in space, even though there may be no charge at that point to experience an electrostatic force. More precisely:

> The **electric field** is a vector quantity defined at all points in space. Its value at any particular point is given by the electrostatic force per unit positive test charge at that point.

This can be expressed mathematically by defining the point by its position vector $\boldsymbol{r}$, and writing:

$$\begin{pmatrix}\text{electric field at a point defined} \\ \text{by position vector } \boldsymbol{r}\end{pmatrix} = \frac{\boldsymbol{F}_{\text{el}}(\text{on charge } q \text{ at } \boldsymbol{r})}{q}.$$

This is a vector equation, so it contains information about both the magnitude and the direction of the electric field. Thus:

$$\begin{pmatrix}\text{magnitude of electric} \\ \text{field at } \boldsymbol{r}\end{pmatrix} = \frac{(\text{magnitude } F_{\text{el}} \text{ of force on charge } q \text{ at } \boldsymbol{r})}{(\text{magnitude of charge } q)}$$

and

$$\begin{pmatrix}\text{direction of electric} \\ \text{field at } \boldsymbol{r}\end{pmatrix} = \begin{pmatrix}\text{direction of } \boldsymbol{F}_{\text{el}} \text{ on a} \\ \textit{positive} \text{ charge at } \boldsymbol{r}\end{pmatrix}.$$

All of this information can be expressed in the following general definition:

$$\mathscr{E}(\boldsymbol{r}) = \frac{\boldsymbol{F}_{\text{el}}(\text{on } q \text{ at } \boldsymbol{r})}{q}. \tag{1.14}$$

Here, as usual, the notation $\mathscr{E}(\boldsymbol{r})$ means the electric field at the point whose position vector is $\boldsymbol{r}$: it does *not* mean $\mathscr{E}$ multiplied by $\boldsymbol{r}$. In other words, $\mathscr{E}$ is a function of $\boldsymbol{r}$.

Question 1.15 Use Equation 1.14 to work out the SI unit for the magnitude of an electric field. ■

Figure 1.36 gives a few examples of electric fields.

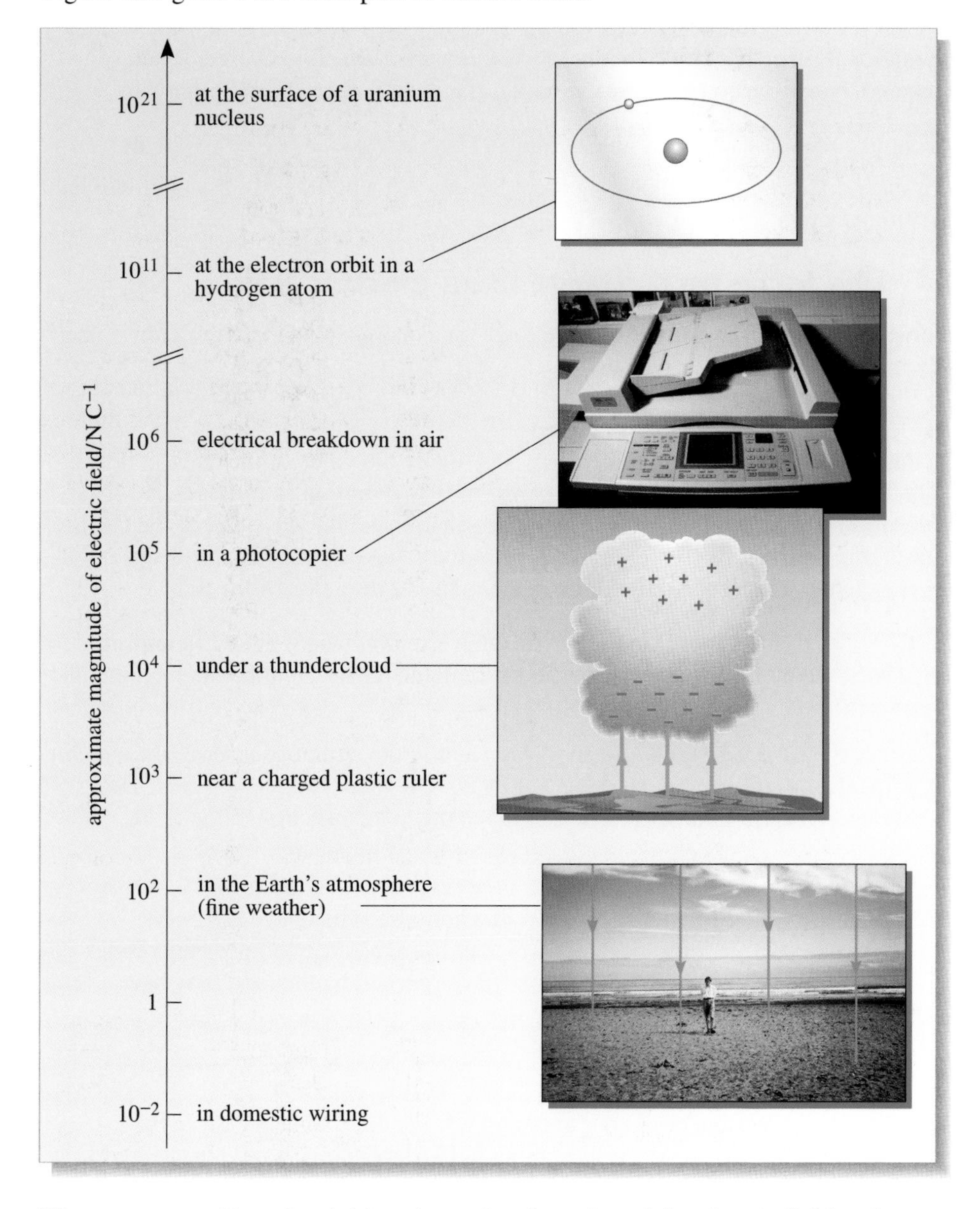

Figure 1.36 Typical magnitudes of electric fields.

We can now use Equation 1.14 to determine the value of the electric field at the point P in Figure 1.35. In Question 1.14 you calculated that the electrostatic force on the 10^{-7} C charge at point P has magnitude 9×10^{-4} N and is pointing along the direction of $\boldsymbol{r}$. The shorthand way of writing this is $\boldsymbol{F}_{\text{el}} = (9 \times 10^{-4}\,\text{N})\,\hat{\boldsymbol{r}}$, since $\hat{\boldsymbol{r}}$, as you may remember, is a unit vector in the $\boldsymbol{r}$-direction. Equation 1.14 then gives

$$\mathscr{E}(\boldsymbol{r}) = \frac{\boldsymbol{F}_{\text{el}}(\text{on } q \text{ at } \boldsymbol{r})}{q} = \frac{(9 \times 10^{-4}\,\text{N})}{(10^{-7}\,\text{C})}\hat{\boldsymbol{r}} = (9 \times 10^{3}\,\text{N}\,\text{C}^{-1})\hat{\boldsymbol{r}}.$$

Thus, the electric field at P, due to the charge Q, is of magnitude $9 \times 10^{3}\,\text{N}\,\text{C}^{-1}$ and is in the direction of $\boldsymbol{r}$.

You may have noticed that in Question 1.14 and subsequently, we have employed the same kind of distinction that we used in gravitation, namely that between a charge that we wish to consider to be the source of an electric field (labelled by an upper case or capital letter Q) and a charge that we wish to consider to be 'feeling the effects of' the electric field (given the symbol lower case q). As in the gravitational case, this useful distinction is quite artificial. Two charges q and Q separated by a distance r will each exert a force on the other, the *magnitude* of which, in either case, is given by Coulomb's law, i.e. $F_{\text{el}} = \dfrac{1}{4\pi\varepsilon_0}\dfrac{|q||Q|}{r^2}$. This notational distinction will appear again in Section 4.5 when we discuss the field due to a point charge.

4.4 The force on a charge in an electric field

In order to obtain an expression for the force on a charge in an electric field we need only rearrange Equation 1.14 thus:

$$\boldsymbol{F}_{\text{el}}\,(\text{on } q \text{ at } \boldsymbol{r}) = q\boldsymbol{\mathscr{E}}(\boldsymbol{r}) \qquad (1.15)$$

So, the force on a charge q at position $\boldsymbol{r}$, is given by the value of the electric field at position $\boldsymbol{r}$ multiplied by the charge in question. The direction of the force is determined in this equation by the *sign* of the charge and the *direction* of the electric field. If q is positive, the force on it will be in the same direction as the electric field. If q is negative, the force on it will be in the opposite direction to the field.

● Consider a charge of -3×10^{-6} C placed in a region where there is a uniform electric field of 4×10^5 N C^{-1} in the x-direction. What would be the force on this charge?

❍ Since the field is in the x-direction, we can use the component form of Equation 1.15: $(F_{\text{el}})_x = q\mathscr{E}_x$, i.e. $(F_{\text{el}})_x = (-3 \times 10^{-6}\,\text{C}) \times (4 \times 10^5\,\text{N C}^{-1}) = -1.2\,\text{N}$. The minus sign shows that the force acts in the opposite direction to the field. ■

Example 1.1

Electrons are emitted into a vacuum from a hot wire. They have negligible initial speed, and are then accelerated by a uniform electric field of magnitude 10^4 N C^{-1}. What speed will the electrons have reached by the time they have travelled 10 cm from the wire?

Solution

Preparation We can orient our axes any way we like, so let's choose the x-axis to be the electrons' direction of motion. Electrons are negatively charged, so the force on them is in the opposite direction to the field. $\boldsymbol{\mathscr{E}}$ must therefore point in the $-x$-direction.

Choosing the x-direction to be the electrons' direction of travel, then, in the usual notation, $s_x = 0.1$ m and $u_x = 0$. Electrons will be accelerated in the opposite direction to the field, so

$$\mathscr{E}_x = -10^4\,\text{N C}^{-1}.$$

We are asked to find v_x.

Can we use the constant acceleration equations? Yes, because $\boldsymbol{\mathscr{E}}$ is uniform, $\boldsymbol{F}_{\text{el}}$ is constant, and so $\boldsymbol{a}$ is constant.

Thus we can calculate the magnitude of the force on an electron by using $(F_{\rm el})_x = q\mathscr{E}_x$, and determine the resulting acceleration by using $a_x = \dfrac{(F_{\rm el})_x}{m_{\rm e}}$. We can then calculate the final speed using the constant acceleration equation $v_x^2 = u_x^2 + 2a_x s_x$.

Working The secret of solving this problem lay in devising the plan of attack. Now it is just a matter of substituting the data. The electron's charge is -1.6×10^{-19} C, so

$$(F_{\rm el})_x = q\mathscr{E}_x = (-1.6 \times 10^{-19}\,{\rm C}) \times (-10^4\,{\rm N\,C^{-1}}) = 1.6 \times 10^{-15}\,{\rm N}.$$

Therefore

$$a_x = \frac{(F_{\rm el})_x}{m_{\rm e}} = \frac{(1.6 \times 10^{-15}\,{\rm N})}{(9.11 \times 10^{-31}\,{\rm kg})} = 1.8 \times 10^{15}\,{\rm m\,s^{-2}}.$$

Now, with initial velocity $u_x = 0$, the final velocity v_x after a displacement $s_x = 0.1$ m will be given by

$$v_x^2 = u_x^2 + 2a_x s_x$$
$$= 2 \times 1.8 \times 10^{15}\,{\rm m\,s^{-2}} \times 0.1\,{\rm m}.$$

Thus $\quad v_x = \pm 2 \times 10^7\,{\rm m\,s^{-1}}$.

So the final speed is

$$v = |v_x| = 2 \times 10^7\,{\rm m\,s^{-1}}.$$

The original data were given to just one significant figure, and the final result is quoted to the same precision.

Checking The unit is correct, with v coming out in m s^{-1}, which provides a check on the algebra. As regards the actual value of v, the speed of the electron is one-tenth of the speed of light, which does not seem unreasonable.

4.5 The electric field due to a point charge

As an application of the general definition of the electric field, we can now formulate an expression for the electric field $\mathscr{E}(\boldsymbol{r})$ at an arbitrary displacement $\boldsymbol{r}$ from a point charge Q. Suppose we have a fixed point charge Q. A test charge q at $\boldsymbol{r}$ would experience an electrostatic force due to Q given by Coulomb's law:

$$\boldsymbol{F}_{\rm el}(\text{on } q \text{ at } \boldsymbol{r}) = \left(\frac{qQ}{4\pi\varepsilon_0 r^2}\right)\hat{\boldsymbol{r}} = q\left(\frac{Q}{4\pi\varepsilon_0 r^2}\right)\hat{\boldsymbol{r}}.$$

But, according to Equation 1.15

$$\boldsymbol{F}_{\rm el}(\text{on } q \text{ at } \boldsymbol{r}) = q\mathscr{E}(\boldsymbol{r}).$$

Comparing the two expressions for $\boldsymbol{F}_{\rm el}$ we see that the **electric field due to a point charge** is

$$\mathscr{E}(\boldsymbol{r}) = \left(\frac{Q}{4\pi\varepsilon_0 r^2}\right)\hat{\boldsymbol{r}}. \tag{1.16}$$

Equation 1.16 tells us two things. First, it tells us that the magnitude of the electric field at displacement $\boldsymbol{r}$ from a charge Q is

$$\mathscr{E}(\boldsymbol{r}) = \frac{|Q|}{4\pi\varepsilon_0 r^2}. \tag{1.16a}$$

Note the modulus sign around the charge Q in Equation 1.16a: this ensures that the magnitude of the electric field will always be positive even if the source charge is negative. Note also that, although $\mathscr{E}$ represents a magnitude, and is therefore not emboldened, $\boldsymbol{r}$ is still a position vector and therefore *is* emboldened. Secondly, Equation 1.16 shows that the electric field $\boldsymbol{\mathscr{E}}(\boldsymbol{r})$ is in the $+\hat{\boldsymbol{r}}$-direction if the charge Q is positive and in the $-\hat{\boldsymbol{r}}$-direction if Q is negative. Remember, however, that $\boldsymbol{\mathscr{E}}(\boldsymbol{r})$ is in the + or $-\hat{\boldsymbol{r}}$-direction only because we chose to measure position vectors from the position of charge Q. In general, the direction of $\boldsymbol{\mathscr{E}}(\boldsymbol{r})$ is not the direction of $\boldsymbol{r}$ or $-\boldsymbol{r}$.

4.6 Addition of electric fields

Any charge will always be surrounded by an electric field. Now, suppose two or more charges are placed close together so that their fields overlap. How can we work out their combined effect?

Well, a little algebra will show that adding fields is essentially the same process as adding forces. Figure 1.37 shows a situation in which two source charges Q_1 and Q_2 interact with a third test charge q placed at a point P that has position vector $\boldsymbol{r}$ (in this case, with respect to some arbitrary origin that is not marked on the diagram).

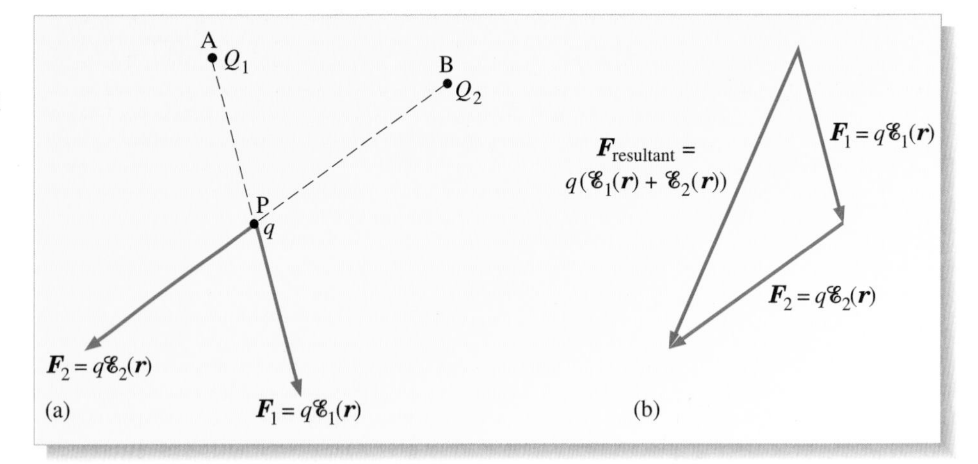

Figure 1.37 (a) The charge q at the point P (defined by position vector $\boldsymbol{r}$) experiences forces $\boldsymbol{F}_1$ and $\boldsymbol{F}_2$ due to the presence of the two charges Q_1 and Q_2. (b) The resultant force $\boldsymbol{F}_{\text{resultant}}$ on q at P is equal to the vector sum of the individual forces acting on it. This implies that the total electric field at the point P, as tested by a charge q, is equal to the vector sum of the individual electric fields due to each of the source charges separately.

Suppose that the fields produced individually by Q_1 and Q_2 are denoted, respectively, $\boldsymbol{\mathscr{E}}_1$ and $\boldsymbol{\mathscr{E}}_2$. As experienced by the test charge q, the forces $\boldsymbol{F}_1$ and $\boldsymbol{F}_2$ due to these sources are $q\boldsymbol{\mathscr{E}}_1(\boldsymbol{r})$ and $q\boldsymbol{\mathscr{E}}_2(\boldsymbol{r})$, respectively. (This follows directly from the definition of the electric field.) The resultant force, $\boldsymbol{F}_{\text{resultant}}$, on q is obtained by adding the forces $\boldsymbol{F}_1$ and $\boldsymbol{F}_2$ vectorially, as illustrated in Figure 1.37b. So

$$\begin{aligned}\boldsymbol{F}_{\text{resultant}} &= \boldsymbol{F}_1 + \boldsymbol{F}_2\\ &= q\boldsymbol{\mathscr{E}}_1(\boldsymbol{r}) + q\boldsymbol{\mathscr{E}}_2(\boldsymbol{r})\\ &= q(\boldsymbol{\mathscr{E}}_1(\boldsymbol{r}) + \boldsymbol{\mathscr{E}}_2(\boldsymbol{r})).\end{aligned}$$

But the resultant electric field $\boldsymbol{\mathscr{E}}_{\text{resultant}}(\boldsymbol{r})$ is defined by Equation 1.14 as the resultant electrostatic force per unit charge. So in this case

$$\boldsymbol{F}_{\text{resultant}} = q\boldsymbol{\mathscr{E}}_{\text{resultant}}(\boldsymbol{r}).$$

Comparing equations, we must have

$$\mathscr{E}_{\text{resultant}}(\boldsymbol{r}) = \mathscr{E}_1(\boldsymbol{r}) + \mathscr{E}_2(\boldsymbol{r}) \tag{1.17}$$

Thus, the resultant electric field is the vector sum of the individual electric fields produced by the two sources. In other words, electric fields add vectorially just as forces do. Notice that, in this case, the electric field, $\mathscr{E}_{\text{resultant}}(\boldsymbol{r})$, is highly unlikely to be in the direction of $\boldsymbol{r}$.

Question 1.16 A positive and a negative charge of equal magnitude are placed at a distance s from each other on the x-axis as shown in Figure 1.38. Determine the direction of the electric field at point P, which is equidistant from both charges. ■

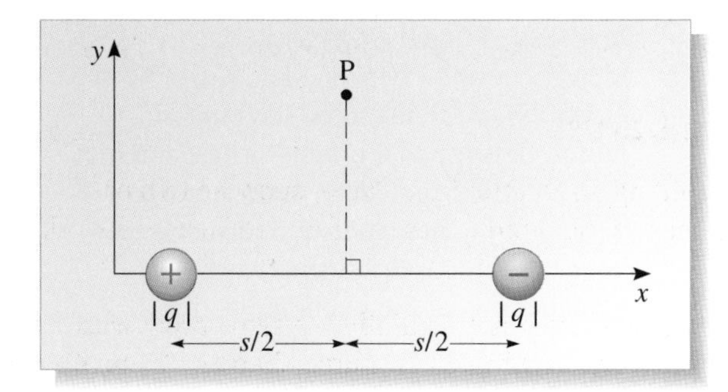

Figure 1.38 What is the direction of the electric field at P? For use with Question 1.16.

Question 1.17 Figure 1.39 shows two charges $Q_1 = 5.0 \times 10^{-6}$ C and $Q_2 = -3.0 \times 10^{-6}$ C, placed such that if Q_1 is taken to be at the origin of coordinates then Q_2 is at the position $x = -4$ m, $y = -3$ m. Calculate the magnitude and direction of the electric field at the point P, which has coordinates $x = 0$ m, $y = -3$ m. ■

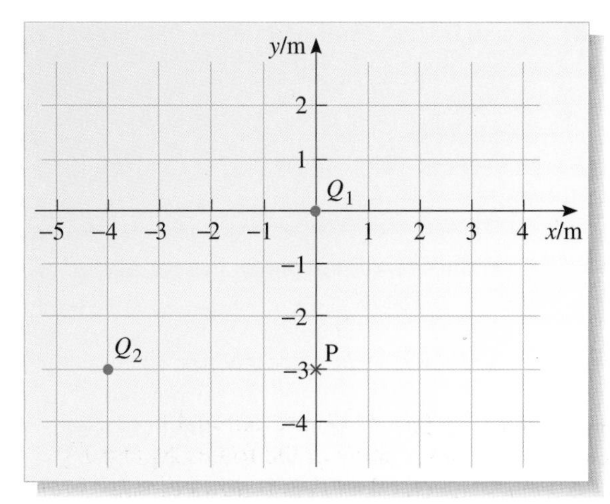

Figure 1.39 What is the electric field at P? For use with Question 1.17.

4.7 Representations of electric fields

So far in this chapter, vectors have been represented by arrows. The tail of the arrow is placed at the point where the vector is acting, the length of the arrow represents the magnitude of the vector and the direction of the arrow represents the direction of the vector. Vector fields can be represented by a suitable array of such arrows. So, for example, the gravitational field of the Earth was illustrated in Figure 1.19 by an array of arrows placed at various distances above the Earth's surface. There is, however, another useful way of illustrating vector fields and that is by using **field lines.**

Michael Faraday

Michael Faraday (Figure 1.40) was born in Newington, Surrey, in 1791, one of ten children of a blacksmith. The family was poor and, after they moved to London, the 14 year old Faraday was apprenticed to a bookbinder. There he studied voraciously. In 1812 he was given tickets to attend the lectures of Humphrey Davy at the Royal Institution. He took careful notes, illustrated them and sent them first to the President of the Royal Society and then to Davy hoping to impress them sufficiently to be given a job as a scientific assistant. In 1813 he began work as Davy's assistant.

Figure 1.40 Michael Faraday (1791-1867).

Faraday's numerous discoveries included methods of liquefying gases, the identification of benzene and the laws of electrolysis. He invented the electric motor, developed the transformer and the first continuous electric generator. He was not a mathematician, and seems to have had little interest in atomic explanations of the phenomena he investigated. His great gifts were the ability to choose the right experiment, to implement it and to realize the significance of unexpected observations. His interpretations were pictorial.

Faraday declined many honours but, in 1824, he did accept fellowship of the Royal Society, an honour that is granted to distinguished scientists by the invitation of the fellows themselves. In 1833, Faraday succeeded Davy as professor of chemistry and, between the years of 1825 and 1862, he gave many very popular lectures at the Royal Institution. He remained an independent thinker, respected by, but at some distance from, the establishment. He declined to assist the government in the production of poisonous gas for use in the Crimean War. He died at Hampton Court in 1867.

Field lines were invented by Michael Faraday (Figure 1.40), for the purpose of illustrating electric fields. He imagined the space around an array of point charges, for example, to be filled with directional field lines with the following two properties:

- The tangent to the field line is parallel to the electric field at that point;
- The lines are closer together where the field is strong and farther apart where it is weaker.

Thus the electric field round a point positive charge is represented by a set of radial lines pointing away from the charge as in Figure 1.41a. If the charge is negative (Figure 1.41b) then the lines point towards the charge. Thus we have another rule for electric field lines:

- Field lines emerge from positive charges and disappear into negative charges.

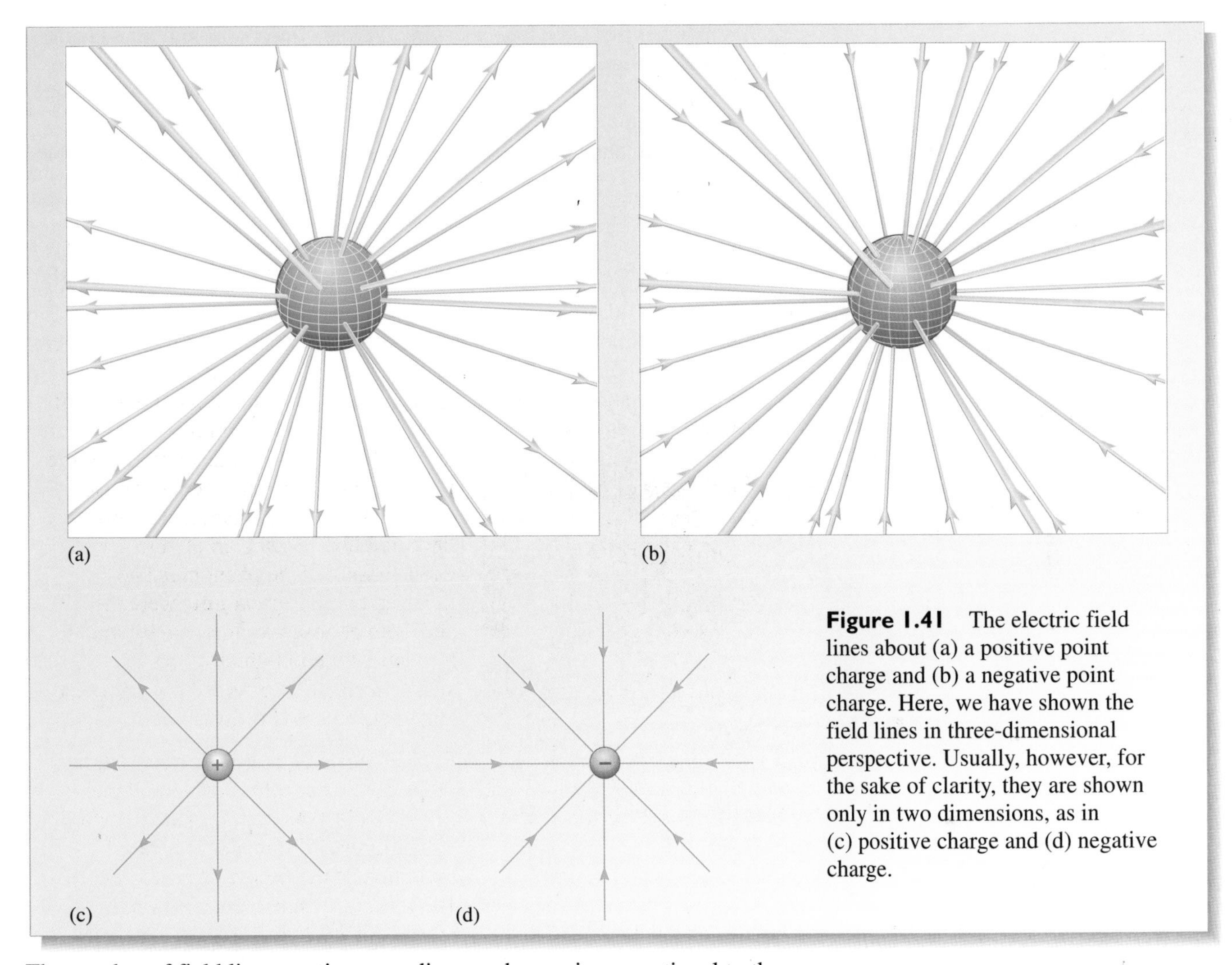

Figure 1.41 The electric field lines about (a) a positive point charge and (b) a negative point charge. Here, we have shown the field lines in three-dimensional perspective. Usually, however, for the sake of clarity, they are shown only in two dimensions, as in (c) positive charge and (d) negative charge.

The number of field lines starting or ending on charges is proportional to the magnitude of the charge. In any one picture you can choose the number of field lines per unit charge, but the chosen number must remain consistent within that picture. Thus, the field around the two charges shown in Figure 1.42 has eight field lines emanating from the 4 C charge and four ending on the −2 C charge. It is

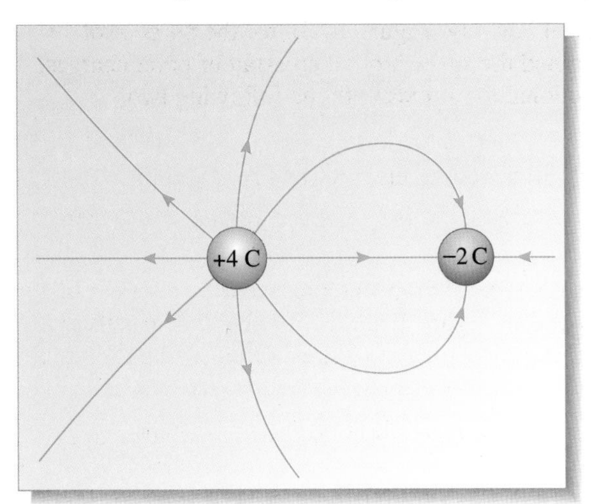

Figure 1.42 The number of field lines starting or ending on a charge is proportional to the magnitude of the charge. You should remember that it is not always possible to get this accurately correct in a two-dimensional representation of a three-dimensional field.

important to remember that field lines exist in all three dimensions,and the spacings between them, and the number per unit charge, are accurately correct only in three dimensions. On paper, we can generally show the field lines only in two dimensions, which involves taking some kind of cross-section through the three-dimensional picture. Inevitably, this sometimes results in uneven spacings between the field lines.

Figure 1.43 gives you an opportunity to practise constructing the field lines around a simple array of charge. This is an important skill, because applications of electrostatic phenomena frequently feature fields that have been tailored by control of the distribution of charge. In Figure 1.43a some of the electric field vectors have been drawn in to help you. Remember that the number of field lines starting or ending on charges is proportional to the magnitude of the charge. It is also helpful to note that, very close to point charges, the electric field will be dominated by the field due to that charge, so the field lines starting and ending on the charge will be very close to a set of equally spaced radial lines. This is illustrated in Figure 1.43b where we have shown ten field lines starting on the positive charge. How many field lines will end on each of the negative charges?

Question 1.18 Draw in the field lines on Figure 1.43. ■

Figure 1.43 For Question 1.18. Draw in the field lines on these two diagrams. In (a) the two charges have the same magnitude and some of the electric field vectors in the region have been shown to help you. In (b) the positive charge has twice the magnitude of each of the two negative charges, and you have been given the beginnings of the field lines on the positive charge.

A uniform field (one that has a constant magnitude and direction over the region of interest) is represented by a series of equally spaced, parallel field lines. One way of creating a uniform electric field is by placing equal and opposite charges on each of two large, closely spaced parallel conducting plates (Figure 1.44). Then the region between the plates contains a uniform electric field.

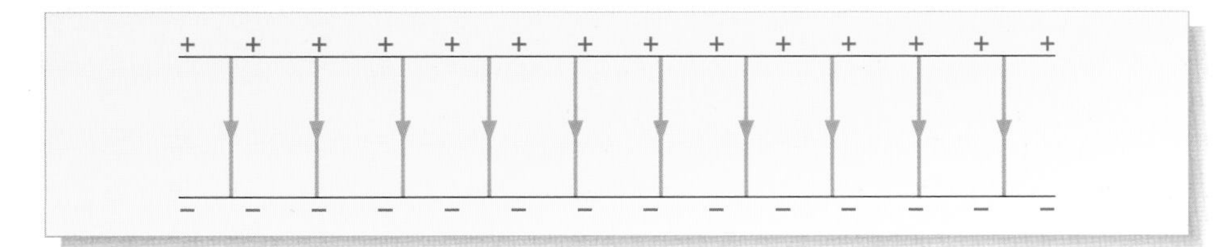

Figure 1.44 A uniform electric field between two oppositely charged, parallel plates.

There is one other important feature of field lines that you may have noticed: field lines never intersect. In other words, they never cross each other.

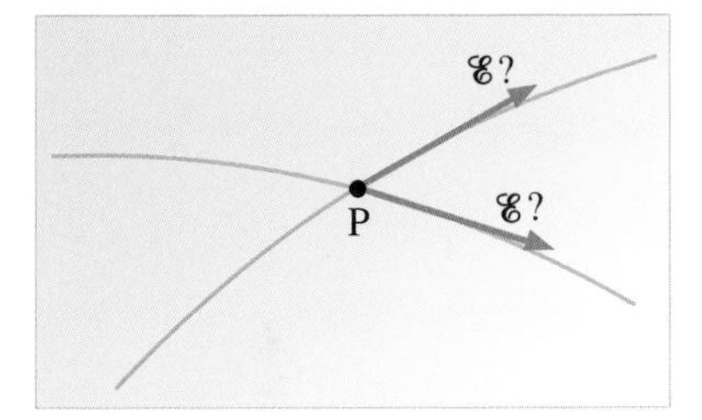

Figure 1.45 The situation if field lines were to intersect.

● Why do field lines not intersect?

❍ Suppose for a moment this did occur (see Figure 1.45). At the intersection point P, the field would have to be simultaneously parallel to two lines that point in different directions. This is clearly a contradiction – the field has a definite direction (not a choice of two!). The supposition that field lines may intersect cannot be true. ■

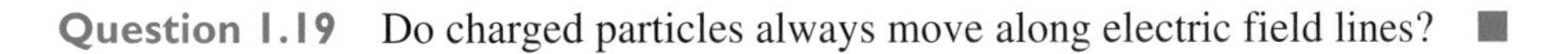

Question 1.19 Do charged particles always move along electric field lines? ■

4.8 Field lines in general

The use of field lines is not restricted to the representation of electric fields. You will have noticed from the similarity of the discussions in Section 2 and Section 4 that the electrostatic and gravitational interactions have a great deal in common. The similarity extends to the fact that gravitational fields can equally well be represented by field lines. Thus the gravitational field of the Earth can be shown as in Figure 1.46.

Figure 1.46 The gravitational field lines in the vicinity of the Earth.

Because the gravitational interaction is always attractive, the field lines always point towards and terminate on masses.

● Where do gravitational field lines begin?

❍ Because gravitational field lines terminate on masses, but never begin on them, the only place that gravitational field lines can begin is at infinity. ■

Question 1.20 Figure 1.47 shows the field lines in a region containing two objects, both of which are charged and have mass.

(a) With no more information than this, can you tell whether the field shown is the gravitational field between the two objects or the electric field?

(b) Given the additional information that the two objects carry charges of the opposite sign, can you tell which type of field is shown? ■

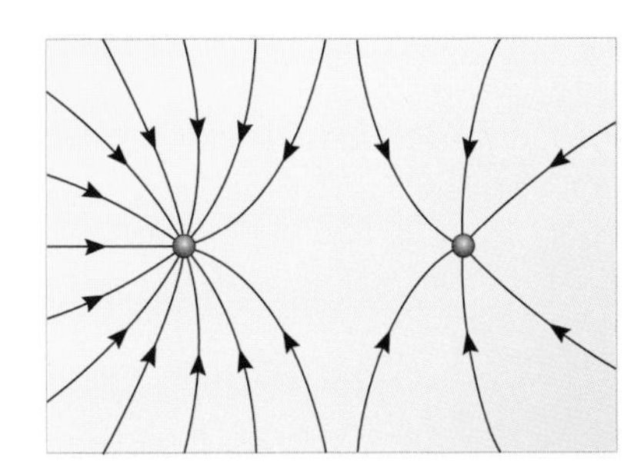

Figure 1.47 A gravitational field or an electric field? For use with Question 1.20.

Open University students should leave the text at this point and use the multimedia package *Forces, fields and potentials: part 1*, which provides more practice with constructing and visualizing fields. The activity will take about two hours.

5 Applications of electric fields

Electric fields and electric forces are used very widely. The applications described in this sectión illustrate some of the basic principles on which many electrostatic devices are based. The first example is the Millikan oil drop experiment, which was the first method used for the accurate determination of the charge on the electron.

5.1 Millikan's oil drop experiment

Before Robert A. Millikan, an American physicist, began his series of experiments in 1909, the existence of the electron was uncertain. A number of experiments suggested that electric charge was atomic in nature, i.e. it existed only in small indivisible amounts, but no one had succeeded in measuring the charge on one of these 'atoms of electricity'. In 1897, J.J. Thomson had managed to measure the **charge to mass ratio of the electron**, $-e/m_e$, but his experiments provided only a rough estimate of the amount of charge carried by one electron.

- Why is it easier to measure the charge to mass ratio than to measure the charge itself?

○ Well, the reason is straightforward. The acceleration $\boldsymbol{a}$ of an electron in a uniform electric field $\mathscr{E}$ is given by

$$\boldsymbol{a} = \frac{\boldsymbol{F}_{el}}{m_e} = \frac{-e\boldsymbol{\mathscr{E}}}{m_e} \tag{1.18}$$

Thus, acceleration, the easily measurable parameter, yields only the charge to mass ratio. The problem lies in finding a way of eliminating the electron mass. ■

The idea behind Millikan's experiment to determine the charge on the electron was to attach a number of electrons to an oil drop. In this way, the mass in Equation 1.18 became the mass of the oil drop plus the mass of the electrons. Because the mass of the oil drop is much larger than that of an electron, it can be more easily measured. In the experiment, small droplets of oil were allowed to fall into a region of uniform electric field $\boldsymbol{\mathscr{E}}$ between two oppositely charged plates (Figure 1.48). The drops were charged by friction as they were formed, and their charge could be changed by ionizing the air around them with X-rays or a radioactive source. The field was directed so that the electric force on the negatively charged drops was in the opposite direction to the gravitational force (Figure 1.48b). Millikan looked at the drops through a microscope, selected one drop (which we shall suppose had mass m_{drop} and carried charge q), and adjusted the electric field until the electric and gravitational forces balanced, and the drop was stationary, i.e. neither falling nor rising. Thus,

magnitude of upward force = magnitude of downward force

$$|q|\mathscr{E} = m_{drop}g \tag{1.19}$$

where g is the magnitude of the acceleration due to gravity.

Millikan wanted to measure q. He could determine $\mathscr{E}$ by some other means, and he knew g, but he did not know the mass of the oil drop. To determine this mass, he devised a very ingenious procedure. He switched off the electric field and allowed the drop to fall under gravity until it reached its terminal speed, i.e. the speed at which the magnitudes of the forces on the oil drop due to the air resistance and gravity are exactly equal. (The force due to air resistance is discussed in *Classical physics of matter*.) Millikan knew that this terminal speed depended on the cross-sectional area of the oil drop and the air resistance. From the terminal speed of the drop, and using the known value of the air resistance, he was therefore able to calculate the radius and hence (knowing the density of the oil) the mass of the oil drop.

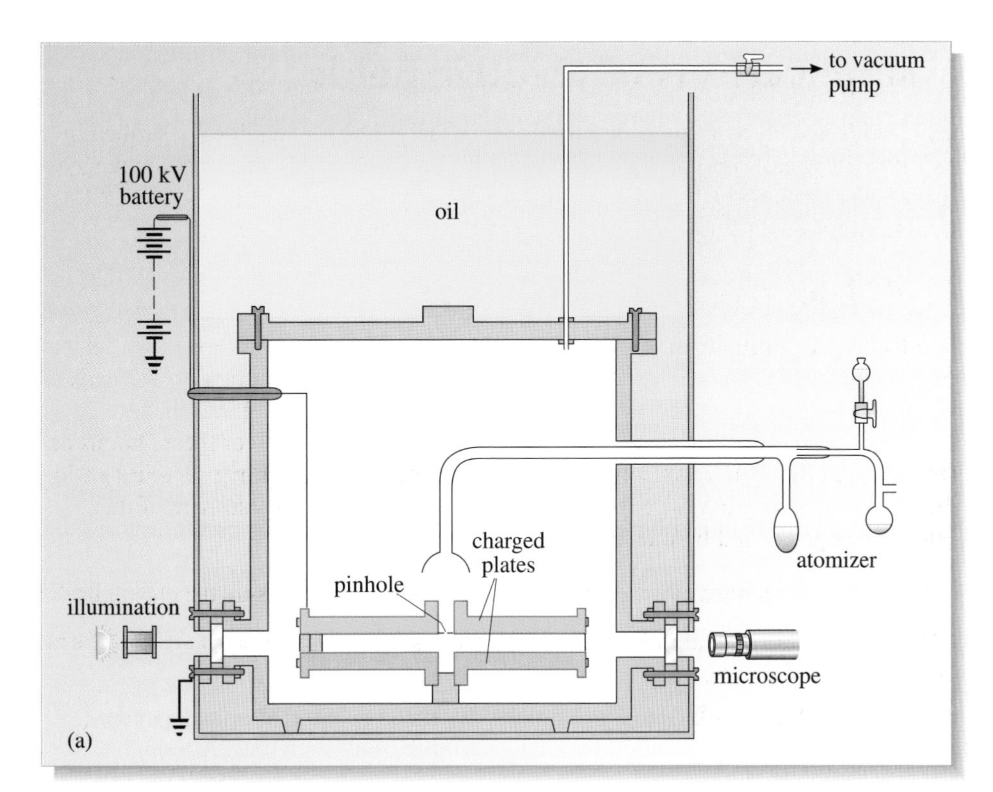

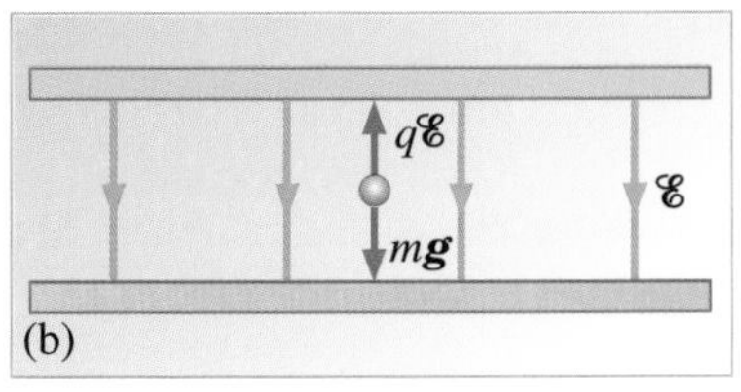

Figure 1.48 (a) A diagram of Millikan's original apparatus. A fine spray 'atomizes' the oil, dispersing it into tiny drops. The electric field is applied by connecting a battery across the plates. The area between the plates is brightly lit and the falling drops can then be observed through a travelling microscope. The downward speed of a drop can be determined by measuring the time it takes to fall through a known distance. (b) Schematic view of an oil drop between the charged plates. The upward electrostatic force $q\mathscr{E}$ and the downward gravitational force $m\mathbf{g}$ must exactly balance when the drop is stationary. Note that q is negative so the electrostatic force is in the opposite direction to the electric field.

Having determined the mass of an individual drop, it was a simple matter to work out the charge carried by the drop directly from Equation 1.19.

A very small selection of Millikan's original data (converted into SI units) is shown in Table 1.1.

Table 1.1 A selection of Millikan's data.

Charge on drop/ -10^{-19} C	Number of electrons, n	Charge/ $-n \times 10^{-19}$ C
8.20	5	1.640
11.49	7	1.641
13.13	8	1.641
11.49	7	1.641
9.87	6	1.645
8.20	5	1.640
8.20	5	1.640
8.20	5	1.640

On examining the values of the charges obtained (the first column), Millikan noted that they were all integral multiples of a certain value, which he took to be the ultimate unit of charge: the charge on the electron itself. The second column of Table 1.1 shows the appropriate integral multiples; the corresponding values for the electron's charge are given in the final column. Millikan repeated this experiment on sixty consecutive days, and observed drops with charges ranging from -1.6×10^{-19} C to -217.6×10^{-19} C. In other words, some of his drops carried a single electron charge and one actually carried 136 electron charges. From all these results, he finally calculated the charge on an electron to be

$$-e = -(1.630\,3 \pm 0.000\,8) \times 10^{-19}\,\text{C}.$$

The currently accepted value is about 2% smaller than this, and the reason for the discrepancy is simply that the value Millikan used for the force due to air resistance was wrong: he had a systematic error in the analysis of his experiment. When his data are reworked using the current estimate of the force due to air resistance, the value Millikan would have obtained is within half a per cent of the currently accepted value, which is (in 2000):

$$-e = -(1.602\,176\,462 \pm 0.000\,000\,063) \times 10^{-19}\,\text{C}.$$

For virtually all the calculations that you make in this course, the approximation $e = 1.60 \times 10^{-19}$ C will be sufficiently accurate. Millikan's assumption that the charge he measured was the smallest amount of charge capable of an independent existence has been amply verified by a large number of experiments. Although quarks, the building blocks of the nuclear particles, are said to carry charges of $\pm\frac{e}{3}$ and $\pm\frac{2e}{3}$, it is believed that quarks cannot exist as free particles. Certainly, free particles with charges that are a fraction of e have never been definitely observed.

The significance of Millikan's result should not be underestimated. First, it confirmed the atomic nature of electricity and gave a value for the smallest amount of free charge. Secondly, because the charge to mass ratio of the electron was already known, it provided the first determination of the electron's mass. Millikan was awarded the 1923 Nobel prize in physics, in part for the oil drop experiment.

Question 1.21 In an experiment similar to Millikan's, the mass of an oil drop is calculated to be 4.31×10^{-14} kg. If an electric field of magnitude $1.15 \times 10^{5}\,\text{N}\,\text{C}^{-1}$ is required to hold it stationary, how many electron charges is the drop carrying?

Question 1.22 In Millikan's experiment (Figure 1.47), the oil drops were charged by friction before they were allowed to fall through a small hole in the top, positively charged, plate. Why weren't they all attracted to the plate before they went through the hole? ■

5.2 Ink-jet and electrostatic printing

Ink-jet printing

Today, there are many methods of putting ink onto paper, whether in a printing works or in an office, and several of the methods make use of electrostatics. Ink-jet printing is one of the most modern, and certainly the most versatile, of the processes. Japanese and Chinese characters can be printed in this way, as easily as any others.

The underlying idea is to charge droplets of ink very precisely, by using computer control. At the top of the apparatus shown in Figure 1.49 you can see the charging electrode that charges up the ink droplets. In some printers, the drops are simply sprayed from a gun. In bubble-jet printers, tiny amounts of the ink are boiled in a fine tube and ejected as minute bubbles. The amount of charge transferred to any particular drop determines how much it will be deflected as it passes through the electric field between the deflecting plates; a drop carrying a large charge will be deflected more than one with a smaller charge. Precision charging means that a number of ink droplets will be deflected in such a way that they are deposited very close to each other in a pattern, forming a character. Any uncharged droplets fall straight down and are collected in the gutter and recycled.

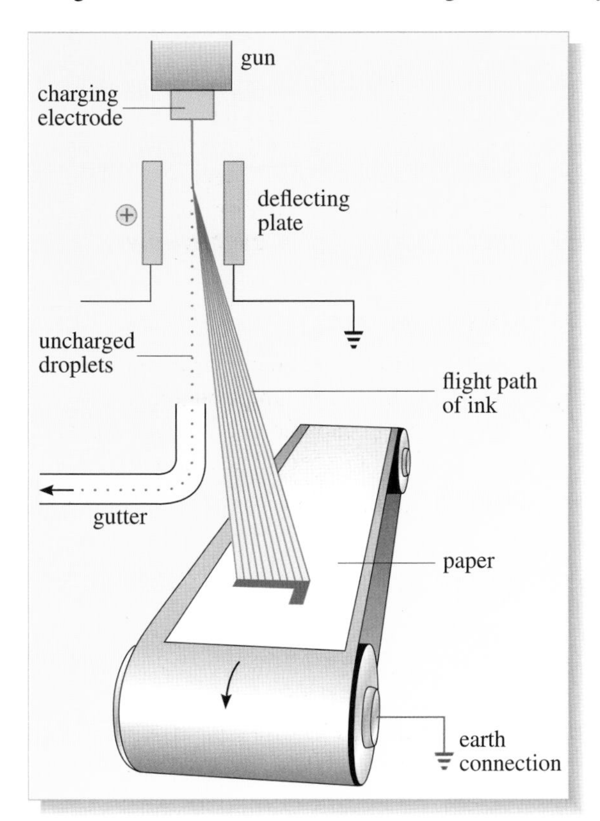

Figure 1.49 The ink-jet printer uses very precise charging, under computer control, to deflect the jet of ink onto the paper.

This method allows very fast printing, roughly 20 metres of words each second. Because there is no contact by any typeface, this process is suitable for the most sensitive of surfaces, and it is almost silent. Any change of character range needs only a change of computer program.

Electrostatic printing

The best-known electrostatic printing process is that called xerography, more commonly termed 'photocopying'. Initial developments were made by Chester Carlson and Otto Kornei in the late 1930s. A public demonstration was given of the process in 1948, and the first commercial equipment was introduced in 1950.

The key materials underlying xerography are photoconductors, i.e. materials that conduct electricity when illuminated with blue or ultraviolet light. The property of photoconduction is also exploited in some lightmeters. In early xerox copiers, the element selenium was used as the photoconductor. However, selenium is toxic, so it is nowadays seldom used, having been replaced by non-toxic compounds with similar properties. The process consists of first charging the photoconductive layer in the dark (Figure 1.50). This is accomplished using a high-voltage device that gives the layer a uniform positive charge. The electric field in the vicinity of the charged plate is about $10^5\,\mathrm{N\,C^{-1}}$.

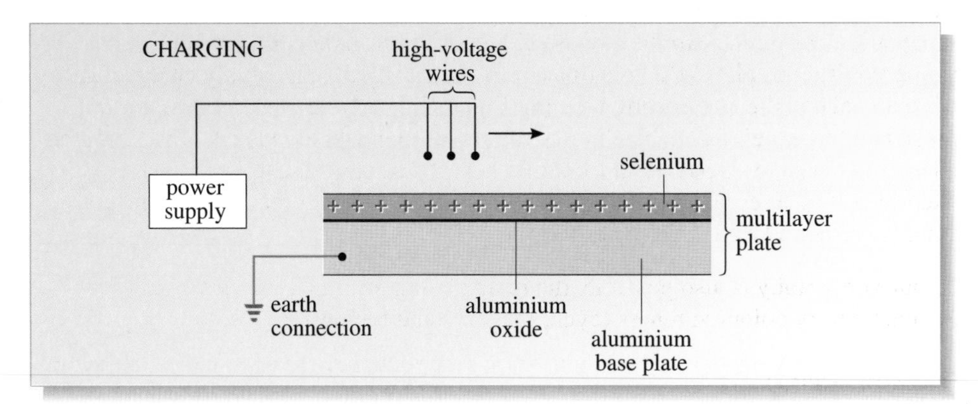

Figure 1.50 Production of a uniform positive charge on a layer of photoconductive material.

During an exposure, an image of the document to be copied is projected onto the charged layer (which may be the surface of a drum, but which for simplicity is shown on Figure 1.51 as a flat plate). The projection is usually done with the aid of both lenses and mirrors, arranged so that reduction or enlargement can be achieved; for simplicity, Figure 1.51 shows just a lens. The lens produces an inverted image which, in the case shown, is the same size as the original. Where light falls on the photoconductive layer, the layer conducts, and positive charge leaks through the aluminium oxide layer onto the earthed base plate. The positive charge is retained in the dark area.

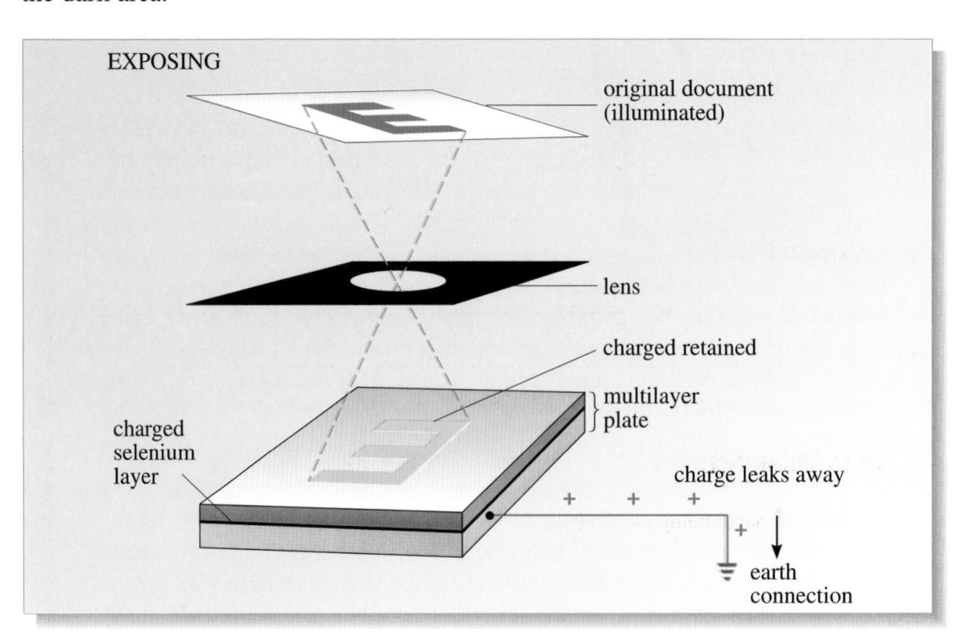

Figure 1.51 An image of the document is projected onto the photoconductive layer.

There are now positive charges on the drum or plate distributed so that they correspond to an exact image of the document. This has to be converted into an inked image. To do this, the ink, a black powder known as toner, must be given a negative charge. This can be done directly by induction, using a highly charged rod (Figure 1.52a). An alternative method uses a two-component toner, in which the powdered ink is mixed with tiny insulating glass beads. As a result of friction, the beads and toner take on opposite charges: the beads positive and the toner negative. The toner is then spread across the photoconductive plate during the developing phase (Figure 1.52b). The charged areas of the plate attract the ink so the toner adheres to the charged sections of the plate, leaving the plate selectively powdered with ink. A sheet of paper is then given a strong positive charge and placed in contact with the powdered drum or plate. The toner particles are attracted to the highly charged paper in what is called the transferring process (Figure 1.53). Note that at this stage the image is a 'positive', that is black on white rather than white on black. The final step is fixing – the paper is passed in front of a heater that melts the toner to such an extent that it adheres permanently to the paper. A photocopy comes out warm! Its charge is removed by an earthed comb or brush as it is emerging from the machine.

Colour xerography is also possible, the image being built up of three layers of complementary coloured toners (cyan, magenta and yellow).

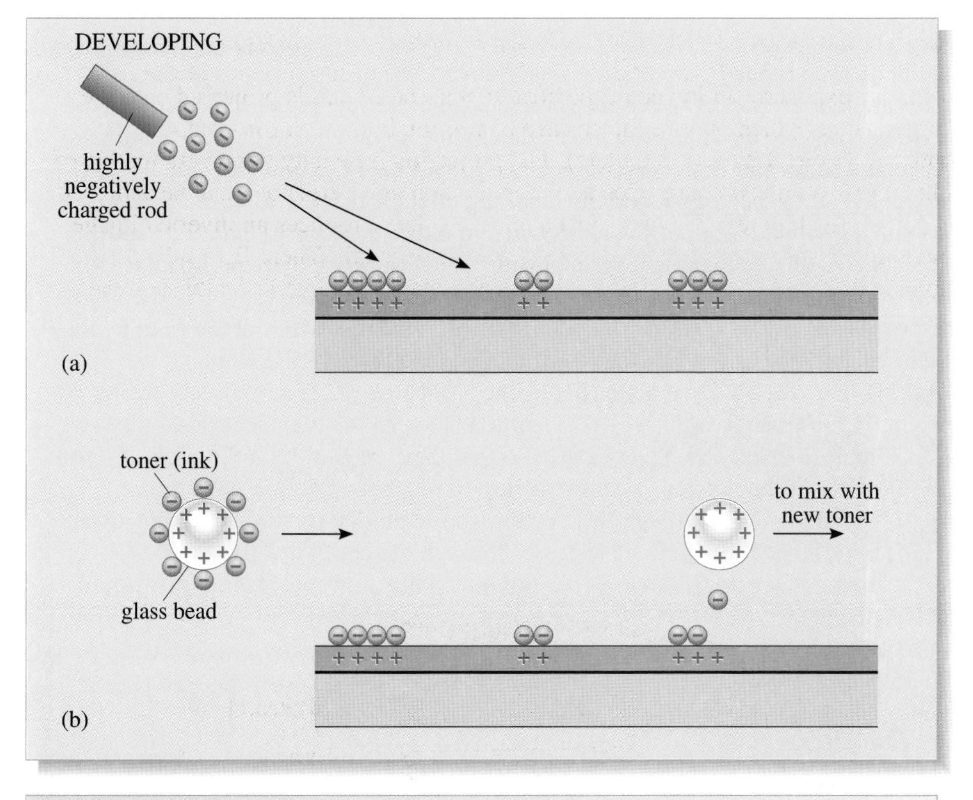

Figure 1.52 Two ways of applying negatively charged ink to the photoconductive plate.

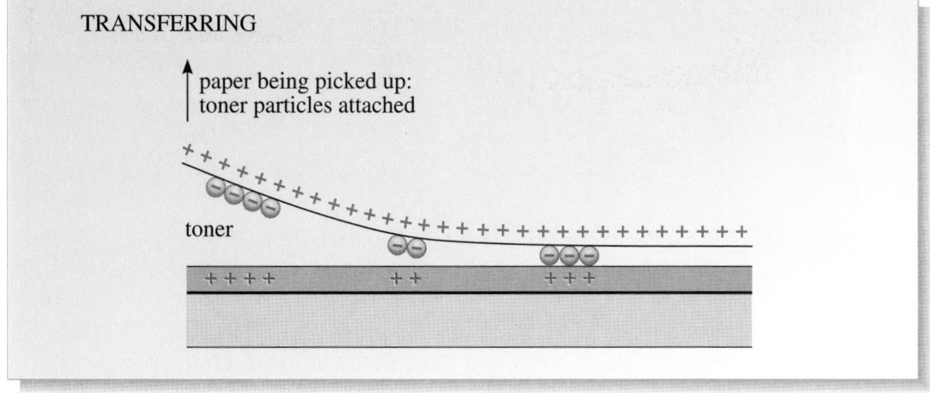

Figure 1.53 Highly charged paper pulls away the particles of toner.

5.3 Liquid crystal displays

Many modern devices such as watches, thermometers and pocket calculators employ liquid crystal displays (LCDs) as their means of delivering information to the user. The usual appearance is that of black figures against a grey background, and it is the application of a simple electric field that switches the elements of the display on and off.

Liquid crystals are fluids consisting of stiff, rod-like, organic molecules, which can form structures with some order, rather like matches in a match box do if left to themselves. Liquid crystals therefore exhibit some of the properties of crystalline solids, hence their name.

Figure 1.54 shows the construction of a simple example of a seven-element cell that would form one digit in a display. The liquid crystal substance is sandwiched in the 10 μm (10^{-6} m) space between transparent plate electrodes and two sheets of polarizer supported on glass plates. Details of one element are shown in Figure 1.55. A thin layer of rubbed polymeric material covers the inside surfaces of the transparent electrodes. The rubbing produces microscopic grooves, which tend to cause the rod-like liquid crystal molecules to line up parallel to them. The grooves above and below the liquid crystal medium are arranged so that they are perpendicular to each other, and the resulting effect is that the molecules twist through 90° from top to bottom as shown in the figure. With the liquid crystal molecules in this arrangement the plane of polarization of polarized light is also twisted through 90° as it passes through the medium. Because the top and bottom electrodes are covered by crossed polarizers, the rotating effect on the light by the liquid crystal medium means that the light can pass through the whole device resulting in a greyish, clear state (Figure 1.55a). However, if a relatively strong electric field (about 3×10^5 N C^{-1}) is applied between the two electrodes, the rod-shaped molecules tend to be reoriented so that they are parallel to the field (Figure 1.55b). Without the twisted crystal structure to rotate its plane of polarization, polarized light coming through the top electrode cannot pass through the crossed polarizer at the bottom so no light passes through the cell. This gives the characteristic black appearance of the figures on the grey background of a liquid crystal display.

Polarized light is discussed in detail in *Dynamic fields and waves*. The essential point is that a *polarizer* of the kind shown in Figure 1.54 will pass only light that is *polarized* parallel to the polarizer.

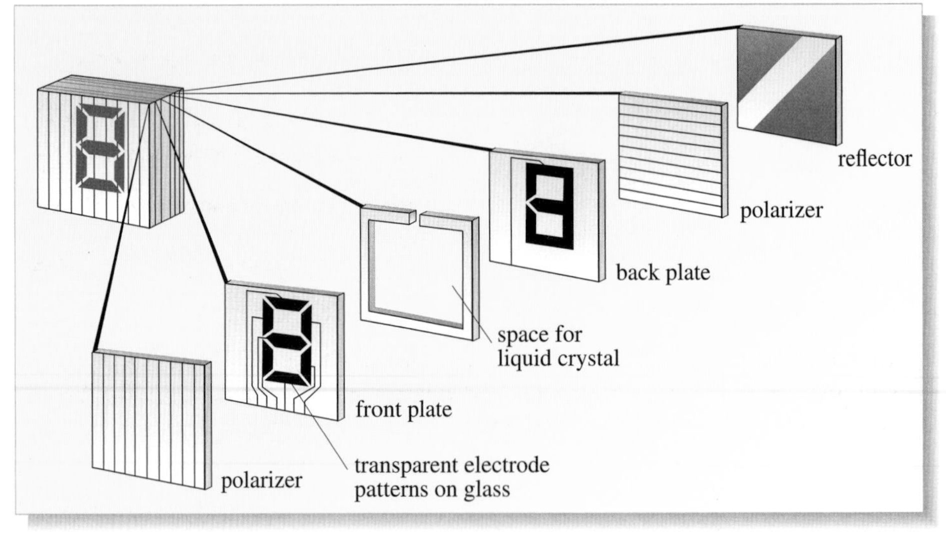

Figure 1.54 A simple type of LCD.

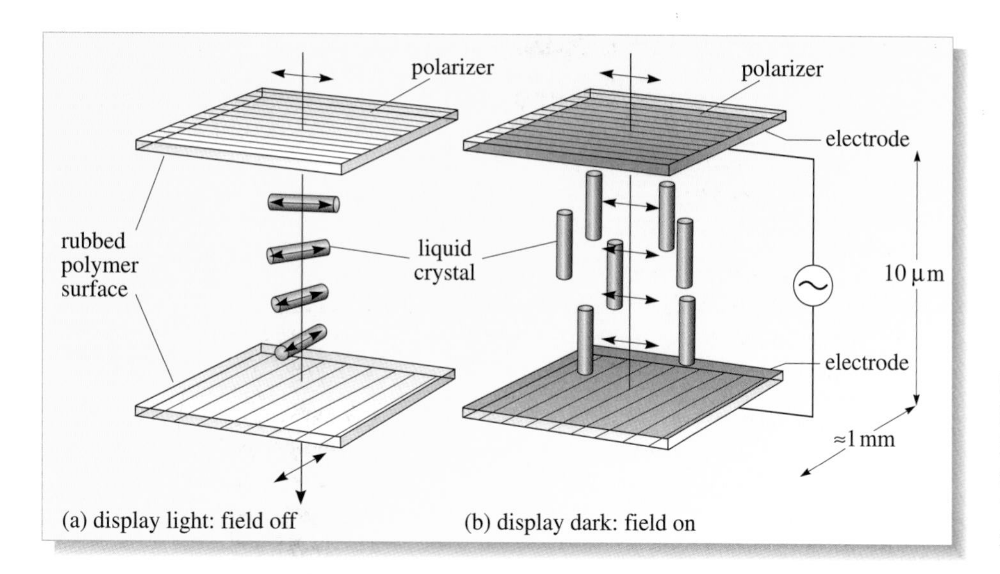

Figure 1.55 One element of an LCD (note the difference in the horizontal and vertical scales) showing (a) light transmitting or 'off' mode and (b) dark or 'on' mode.

5.4 The Earth's electric field

There is a great deal of electrical activity in the Earth's atmosphere. Thunderstorms are the most spectacular manifestation of this activity but even in calm, fine, weather there is an electric field in the Earth's atmosphere. This field (Figure 1.56) points vertically downwards and, near the surface of the Earth, has a magnitude of about $130\,\mathrm{N\,C^{-1}}$ on average. The atmospheric electric field is due to the fact that there is a net negative charge on the Earth and a net positive charge in the ionosphere, which is a layer of the atmosphere about 60 km above the Earth's surface. Thunderstorms play an important role in maintaining this charge separation because lightning sends negative charge down to the ground and positive charge up to the ionosphere. There are about 2000 thunderstorms occurring in the world at any given time, supplying charge at an average rate of 2000 coulombs every second. This charge build-up does not continue indefinitely, however, because, in regions of fine weather, the motion of ions transfers charge in the opposite direction. Because there is a greater concentration of positive ions (represented by red dots in Figure 1.56) near the Earth's surface, the electric field is greatest in this region.

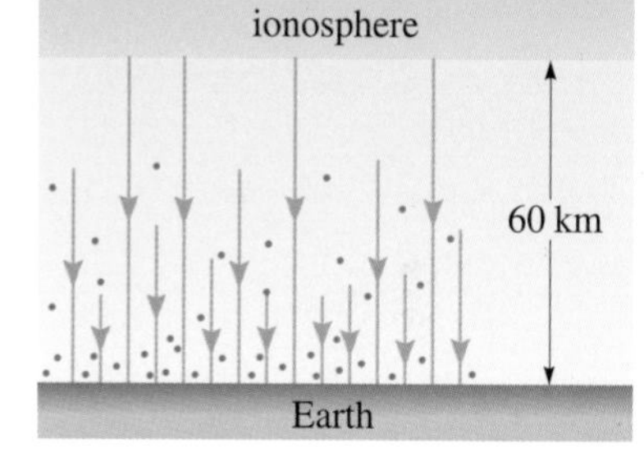

Figure 1.56 The atmospheric electric field.

Interestingly, the magnitude of the atmospheric electric field varies considerably throughout the course of a day and is greatest (on average) everywhere on Earth at the same time, about 18.00 GMT. This is due to the large number of thunderstorms occurring in the Amazon basin in the afternoon.

6 Closing items

6.1 Chapter summary

1 All matter exerts a gravitational attraction on all other matter. The attraction of point masses separated by a distance r may be summed up in the vector form of Newton's law of universal gravitation:

$$\boldsymbol{F}_{\text{grav}} = \frac{-Gm_1m_2}{r^2}\hat{\boldsymbol{r}} \tag{1.6}$$

where G is the universal gravitational constant.

2 The gravitational field $\boldsymbol{g}(\boldsymbol{r})$ is a vector quantity defined at all points in space. At any particular point $\boldsymbol{r}$ its value is given by the gravitational force per unit mass at that point. So

$$\boldsymbol{g}(\boldsymbol{r}) = \frac{\boldsymbol{F}_{\text{grav}}\text{ (on } m \text{ at } \boldsymbol{r})}{m}. \tag{1.7}$$

3 The gravitational force experienced by a point mass m at $\boldsymbol{r}$ is given by

$$\boldsymbol{F}_{\text{grav}}\text{(on } m \text{ at } \boldsymbol{r}) = m\boldsymbol{g}(\boldsymbol{r}) \tag{1.9}$$

4 The gravitational field at $\boldsymbol{r}$ due to a point mass M situated at the origin is given by

$$\boldsymbol{g}(\boldsymbol{r}) = \frac{-GM}{r^2}\hat{\boldsymbol{r}}. \tag{1.10}$$

5 Charges exert a force on each other. The properties of charge are summarized in Box 1.1 of Section 3.2.

6 Coulomb's law of force between charges situated in a vacuum may be summarized in the vector equation

$$\boldsymbol{F}_{\text{el}} = \frac{1}{4\pi\varepsilon_0}\left(\frac{q_1 q_2}{r^2}\right)\hat{\boldsymbol{r}}. \tag{1.13}$$

The constant ε_0 is called the permittivity of free space.

7 The electric field $\mathscr{E}(\boldsymbol{r})$ is a vector quantity, defined at all points in space. At any particular point $\boldsymbol{r}$ its value is given by the electrostatic force per unit charge at that point. So

$$\mathscr{E}(\boldsymbol{r}) = \frac{\boldsymbol{F}_{\text{el}}\text{(on } q \text{ at } \boldsymbol{r})}{q}. \tag{1.14}$$

8 The electrostatic force experienced by a point charge q at $\boldsymbol{r}$ is given by

$$\boldsymbol{F}_{\text{el}}\text{ (on } q \text{ at } \boldsymbol{r}) = q\mathscr{E}(\boldsymbol{r}). \tag{1.15}$$

9 The electric field due to a point charge Q at the origin is given by

$$\mathscr{E}(\boldsymbol{r}) = \left(\frac{Q}{4\pi\varepsilon_0 r^2}\right)\hat{\boldsymbol{r}}. \tag{1.16}$$

10 Fields of a given type, due to different sources, may be added vectorially using the triangle or parallelogram rule.

11 The similarities between the interaction of masses via gravitational fields and the interaction of charges via electric fields are summarized in Box 1.2 below.

12 Gravitational and electric fields can be represented pictorially by means of field lines.

13 Applications and occurrences of electric fields are widespread. Discussed in this chapter are Millikan's oil drop experiment, ink-jet and electrostatic printing, and liquid crystal displays. The Earth has a naturally occuring electric field.

Box 1.2 Gravitational and electric fields

	Gravitation	Electrostatics
Force laws for two point particles	$\boldsymbol{F}_{\text{grav}} = \dfrac{-G m_1 m_2}{r^2}\hat{\boldsymbol{r}}$	$\boldsymbol{F}_{\text{el}} = \dfrac{1}{4\pi\varepsilon_0}\dfrac{q_1 q_2}{r^2}\hat{\boldsymbol{r}}$
Force law in field	$\boldsymbol{F}_{\text{grav}} = m\boldsymbol{g}(\boldsymbol{r})$	$\boldsymbol{F}_{\text{el}} = q\boldsymbol{\mathscr{E}}(\boldsymbol{r})$
Field due to a point particle	$\boldsymbol{g}(\boldsymbol{r}) = \dfrac{-GM}{r^2}\hat{\boldsymbol{r}}$	$\boldsymbol{\mathscr{E}}(\boldsymbol{r}) = \left(\dfrac{Q}{4\pi\varepsilon_0 r^2}\right)\hat{\boldsymbol{r}}.$

6.2 Achievements

Now that you have completed this chapter, you should be able to

A1 Explain the meaning of all the newly defined (emboldened) terms introduced in this chapter.

A2 State Newton's law of gravitation and use it in simple calculations.

A3 Relate to one another the concepts of gravitational force and gravitational field, and use the relationship in simple calculations.

A4 List the properties of electrostatic charge, as summarized in Box 1.1 at the end of Section 3.2.

A5 Describe and explain the processes of charging by rubbing, by sharing and by induction.

A6 State Coulomb's law, and use it in simple calculations involving small charged objects in a vacuum.

A7 Define the terms vector field, scalar field, electric field and gravitational field, and use these definitions in simple calculations.

A8 Remember that the electric field between an infinitely wide pair of parallel and oppositely charged flat plates is confined to the region within the plates and is uniform and perpendicular to the plane of the plates.

A9 Relate to one another the concepts of electrostatic force and electric field, and use the relationship in simple calculations.

A10 Represent simple electric and gravitational fields using field lines.

A11 Describe the mechanisms of lightning storms, certain oil-tanker explosions, Millikan's oil drop experiment, ink-jet and electrostatic printing and liquid crystal displays.

A12 Draw comparisons between the gravitational and electrical interactions, in particular the form of the force laws and fields both for point particles and in the case of uniform fields.

6.3 End-of-chapter questions

Question 1.23 (a) Write down Newton's law of gravitation for two point masses m_1 and m_2.

(b) Two masses of 96 kg and 128 kg are in fixed positions at points A and B, respectively. A mass of 1 kg is at point C, and the angle CAB is a right angle. Draw a diagram showing the construction of the gravitational field at C due to the two masses.

(c) If the distance CB is 8 cm and CA is 4 cm, what is the magnitude of the resultant field at C? (Take $G = 6.67 \times 10^{-11}\,\mathrm{N\,m^2\,kg^{-2}}$.)

Question 1.24 At what height above the Earth's surface has the value of the gravitational field fallen to 81% of its value at sea-level? (Take $R_E = 6380\,\mathrm{km}$.)

Question 1.25 If a charged object is brought up to an insulated uncharged metal rod, the electrons in the metal flow to one end of the rod (Figure 1.57). What causes the flow of electrons to stop?

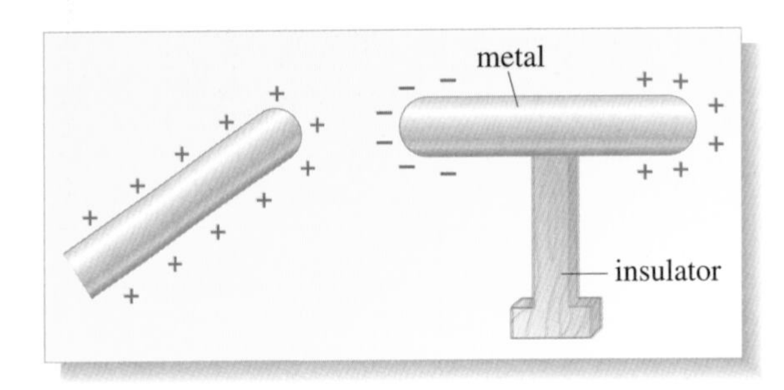

Figure 1.57 Redistribution of charge in an insulated metal rod when a positively charged object is brought up to it.

Question 1.26 (a) Two small conducting spheres are known to repel each other due to an electrostatic force. What can be said about their charges?

(b) As two small spheres are brought close together, they are found to attract each other by an electrostatic force. What can be said about their charges? (*Caution*: Part (b) is not quite as easy as it may seem at first!)

Question 1.27 If the force on an electron at a particular point is $10^{-14}\,\mathrm{N}$ in the direction of the positive y-axis, calculate the electric field at that point and the force on an α-particle at the same point. (The charge on an α-particle is $2e$.)

Question 1.28 (a) Draw a rough sketch to show how the charge distribution within a conducting, neutral, spherical object is affected by the presence nearby of a wire carrying a large negative charge.

(b) If the object with this charge distribution were placed in a uniform electric field, would it move if it were free to do so?

(c) The object is placed in a non-uniform electric field. The field variation is in the same direction as the displacement of the charge, and is such that the change in the magnitude of the field across the object, $\Delta\mathscr{E}$, is $1\,\mathrm{N\,C^{-1}}$. The force on the object is found to be of magnitude $10^{-9}\,\mathrm{N}$. Estimate the number of electrons displaced within the object.

Question 1.29 In a Millikan's oil drop type of experiment, the electric field $\mathscr{E}$ is adjusted such that a particular drop is stationary. The magnitude of the field is then momentarily increased and subsequently returned to its initial value. How will this affect the motion of the droplet? ■

Chapter 2 Gravitational and electric potential

1 Where does a star's energy come from?

When a cold, extensive cloud of gas and dust (Figure 2.1), somewhere in outer space, starts to contract under gravity, what are the changes it has to go through before it can become a shining star? What milestones does it have to reach and pass, and what are the underlying principles behind the various processes that occur during the star's formation?

Figure 2.1 The Coal Sack near the Southern Cross is an example of a cold dense gas and dust cloud of the kind from which new stars may be born.

The gas cloud (or part of it) that is eventually to become a star may have a total mass similar to that of our Sun (about 2×10^{30} kg), consist mostly of hydrogen with a number density of around 10^{11} particles m^{-3}, and have an average temperature of about 20 K. At such a low temperature the atoms are not moving very fast: the expression $\langle E_{\text{trans}} \rangle = (3/2)kT$ implies a mean speed for a hydrogen atom in the cloud of about 700 m s^{-1}. This may seem fast in everyday terms but it is very slow for a hydrogen atom!

$\langle E_{\text{trans}} \rangle$ is the average translational kinetic energy of an atom in the cloud; k is Boltzmann's constant and T is the absolute temperature of the cloud. The origin of this expression is discussed in *Classical physics of matter*, Chapter 2.

For the gas cloud to become a star and start radiating light, the hydrogen atoms have to start undergoing fusion. The first step in that process is for two hydrogen nuclei (protons) to fuse together to form deuterium. In order to do this, the two protons must be able to come into contact. Of course, there is a very powerful force preventing this from happening, namely the electrostatic repulsion between the two protons. However, if the protons are moving fast enough, they will have enough kinetic energy to overcome this repulsion and get close enough together for fusion to occur. Where can they acquire all this kinetic energy? Well, as the gas cloud contracts under its own gravitational attraction, the constituent atoms are essentially falling towards each other, and towards the centre of the cloud. As they fall, they speed up, just as any falling object near the surface of the Earth speeds up. The atoms collide with each other and share out this kinetic energy until a significant proportion of them are moving fast enough to undergo fusion. In this process, we say that gravitational potential energy is being converted into kinetic energy; and, as you will see in this chapter, as the protons approach each other and are slowed down by their mutual electrostatic repulsion, their kinetic energy is then reconverted to potential energy, but this time it is *electrostatic* potential energy. Once the protons are 'in contact', they become subject to the strong nuclear force, which is a very powerful, but short range, attractive force. The strong nuclear force overwhelms the electrical repulsion, and causes the protons effectively to 'fall' together as they fuse to form deuterium, liberating particles and energy as they do so. It is in this way that the gravitational potential energy of the cloud can contribute to the energy eventually radiated by the star.

The whole process is a bit like a roller-coaster ride where potential energy is converted into kinetic energy and then back again to potential energy several times until, eventually, you reach the bottom of the ride, and all the original potential energy has been finally converted to other forms, most notably thermal energy.

In this chapter, we will develop the concepts for the energy approach and show how these concepts can be applied to the understanding of a vast range of phenomena and applications, with specific examples of up-to-date theory and technology.

2 Potential energy

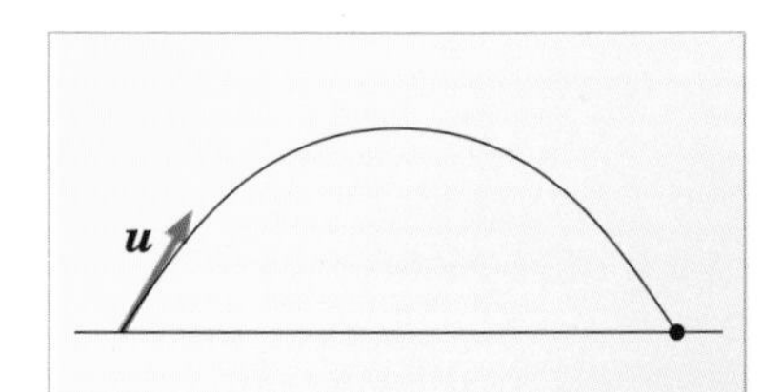

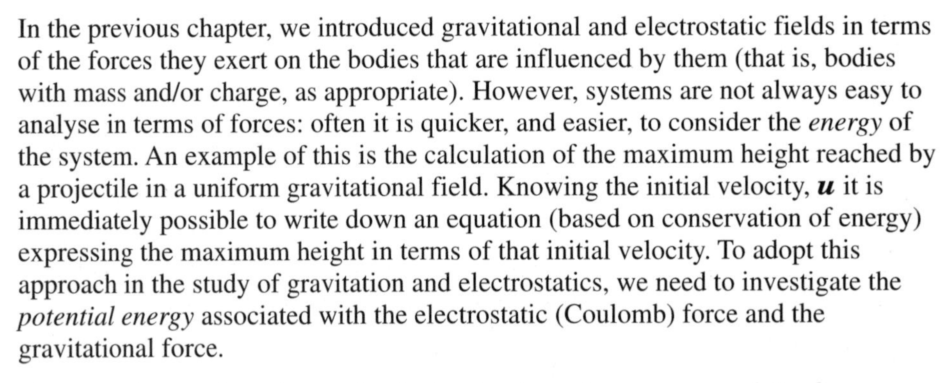

In the previous chapter, we introduced gravitational and electrostatic fields in terms of the forces they exert on the bodies that are influenced by them (that is, bodies with mass and/or charge, as appropriate). However, systems are not always easy to analyse in terms of forces: often it is quicker, and easier, to consider the *energy* of the system. An example of this is the calculation of the maximum height reached by a projectile in a uniform gravitational field. Knowing the initial velocity, $\boldsymbol{u}$ it is immediately possible to write down an equation (based on conservation of energy) expressing the maximum height in terms of that initial velocity. To adopt this approach in the study of gravitation and electrostatics, we need to investigate the *potential energy* associated with the electrostatic (Coulomb) force and the gravitational force.

You should know from the mechanics books *Describing motion* and *Predicting motion* that, when a constant force $\boldsymbol{F}$ acts on an object that moves through a displacement $\boldsymbol{s}$, the energy transferred to the object is given by $\boldsymbol{F} \cdot \boldsymbol{s}$. Any object that moves around in some kind of force field must, therefore, experience changes in its *energy*. If the force field is *conservative*, then the work done in moving from one point to another will be independent of the path followed, and it will be possible to associate a potential energy with each configuration of the system. In this way, if an object is moved against some conservative force, for example, if a mass is raised from the floor to a certain height, or if a positive charge is moved towards another positive charge, then that object's potential energy must increase. In a uniform field such as the gravitational field near the surface of the Earth, this increase is easy to calculate. When an object of mass m is raised through a height Δh, the increase in potential energy is $mg\,\Delta h$, where g is the magnitude of the acceleration due to gravity. This energy change is interpreted as a change in the **gravitational potential energy** of the object. The energy is described as *potential* energy because, if the object is subsequently released and allowed to fall freely under the influence of gravity, then the potential energy is recovered in the form of kinetic energy after the object has fallen through the vertical height Δh.

A directly analogous situation, involving **electrostatic potential energy**, occurs if we move a positive charge q through a distance Δr in a uniform electric field, $\mathscr{E}$, and in a direction exactly opposite to the direction of that field. The work done by the conservative field will be $-q\mathscr{E}\,\Delta r$ and the corresponding increase in potential energy will be the negative of this, $q\mathscr{E}\,\Delta r$. If the field is not uniform, the energy is rather more difficult to calculate (it involves an integral), but the principle is the same.

Let us now consider the potential energy function under more general conditions. You have already met (in Chapter 2 of *Predicting motion*) the important relationship between a conservative force and the related potential energy function:

$$F_x = -\frac{\mathrm{d}E_{\mathrm{pot}}}{\mathrm{d}x}. \tag{2.1a}$$

F_x is the component of the force $\boldsymbol{F}$ in the x-direction and $\mathrm{d}E_{\mathrm{pot}}/\mathrm{d}x$ is the rate of change of potential energy with the x-coordinate: in graphical terms, $\mathrm{d}E_{\mathrm{pot}}/\mathrm{d}x$ is the gradient of the graph of potential energy E_{pot} versus position x. Similar expressions apply for the components of force in the y- and z-directions:

$$F_y = -\frac{\mathrm{d}E_{\mathrm{pot}}}{\mathrm{d}y} \tag{2.1b}$$

$$F_z = -\frac{\mathrm{d}E_{\mathrm{pot}}}{\mathrm{d}z}. \tag{2.1c}$$

In general, we can say that the component of the force in any direction is minus the gradient of the graph of potential energy versus position in that direction.

Equation 2.1 implies that:

- the magnitude of the force component is greatest when the gradient of the graph of potential energy versus position is steepest, i.e. F_x has its greatest magnitude when $\dfrac{\mathrm{d}E_{\mathrm{pot}}}{\mathrm{d}x}$ has its greatest magnitude.
- the direction of the force component is the direction in which the potential energy decreases, i.e. F_x is positive if the potential energy decreases in the positive x-direction (so that $\dfrac{\mathrm{d}E_{\mathrm{pot}}}{\mathrm{d}x}$ is negative).

Remember that Equation 2.1 is a general equation connecting the force with the gradient of potential energy. The potential energy can be gravitational, electrostatic or any other form. So, how do we work out what the potential energy of a system is? Let's start with the simple system of two point particles.

The electrostatic force between two point charges and the gravitational force between two point masses both obey an inverse square law, so in either case we may make use of a unit vector, $\hat{\boldsymbol{r}}$, to write

$$\boldsymbol{F} = \frac{k}{r^2}\hat{\boldsymbol{r}}. \tag{2.2}$$

Here, $\boldsymbol{F}$ can be either the gravitational force or the electrostatic force and all the constants have been collected together as $k = q_1q_2/4\pi\varepsilon_0$ in the electrostatic case or $k = -\,Gm_1m_2$ in the gravitational case. It will help if we consider one of the particles as being fixed at the origin and consider the force on the other one, which is free to

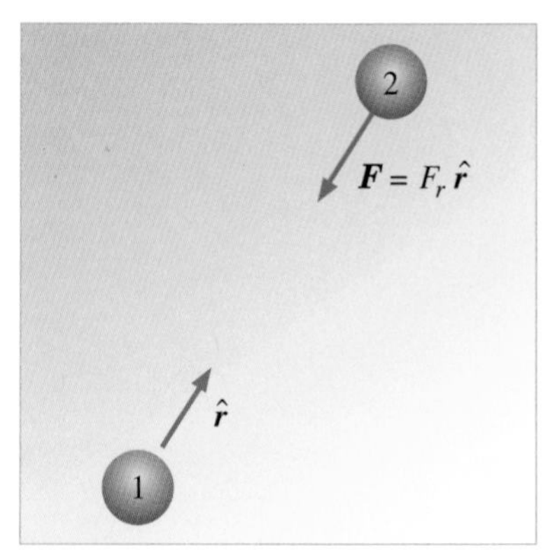

Figure 2.2 The gravitational or electrical force $\boldsymbol{F}$ on particle 2 due to particle 1 (which is fixed at the origin) is always in the direction of either $\hat{\boldsymbol{r}}$ or $-\hat{\boldsymbol{r}}$, and can therefore be fully specified by its radial component F_r so that $\boldsymbol{F} = F_r\hat{\boldsymbol{r}}$. In this case, we have shown an attractive force so F_r is negative.

move (Figure 2.2). Then, if $\boldsymbol{r}$ is the position vector of the movable particle from the fixed one, $\hat{\boldsymbol{r}} = \boldsymbol{r}/r$ points from the fixed particle to the movable one. Although $\hat{\boldsymbol{r}}$ is not in a fixed direction in space (it depends where the movable particle is), the force $\boldsymbol{F}$ in Equation 2.2 is always in the direction of $\hat{\boldsymbol{r}}$ or $-\hat{\boldsymbol{r}}$. Thus, we can fully specify the force by giving the component F_r, that is, the component of the force in the direction of $\hat{\boldsymbol{r}}$. Then $\boldsymbol{F} = F_r\hat{\boldsymbol{r}}$. If F_r is positive, the force will be in the direction of $\boldsymbol{r}$. If F_r is negative, the force will be in the $-\boldsymbol{r}$ direction. By analogy with Equation 2.1, the force component F_r is related to the potential energy by the equation

$$F_r = -\frac{\mathrm{d}E_{\text{pot}}}{\mathrm{d}r}. \tag{2.3}$$

Now F_r is obtained from Equation 2.2, that is $F_r = \frac{k}{r^2}$. Putting this together with Equation 2.3, we have

$$\frac{\mathrm{d}E_{\text{pot}}}{\mathrm{d}r} = -\frac{k}{r^2}. \tag{2.4}$$

So, we now know that the gradient, or derivative with respect to r, of the potential energy (for two point masses or charges) is given by $-\frac{k}{r^2}$.

Question 2.1 What is $\frac{\mathrm{d}E_{\text{pot}}}{\mathrm{d}r}$ in each of the following cases? (C and k are constants, independent of r.)

(a) $E_{\text{pot}} = kr + C$. (d) $E_{\text{pot}} = -\frac{k}{r} + C$.

(b) $E_{\text{pot}} = \frac{k}{r} + C$. (e) $E_{\text{pot}} = \frac{k}{r^2} + C$.

(c) $E_{\text{pot}} = \frac{k}{r} - C$. (f) $E_{\text{pot}} = \frac{k}{r^3} - C$.

In which case(s) does your expression for $\frac{\mathrm{d}E_{\text{pot}}}{\mathrm{d}r}$ correspond to Equation 2.4? ■

The answer to this question shows that the gravitational or electrostatic potential energy for two point particles, separated by a distance r, is of the form

$$E_{\text{pot}} = \frac{k}{r} + C \tag{2.5}$$

where k is the appropriate collection of constants (for gravity or electrostatics) and C is a constant that can take any value (positive or negative). To simplify matters as much as possible, it is conventional to take the potential energy of the system to be zero when the component parts of the system are at an infinite separation from one another (i.e. when the force is effectively zero). This convention is quite arbitrary, but has the convenient consequence that, when the distance r between the two particles is infinity, the first term in Equation 2.5 is zero, and so the value of C must also be zero. Thus, substituting the appropriate constants for the two cases, we have the final expressions for the **gravitational potential energy of two point masses**:

$$E_{\text{grav}} = -\frac{Gm_1m_2}{r}, \tag{2.6}$$

and the **electrostatic potential energy of two point charges** (in free space):

$$E_{\mathrm{el}} = \frac{q_1 q_2}{4\pi\varepsilon_0 r} \tag{2.7}$$

where E_{grav} and E_{el} are taken to be zero when $r = \infty$.

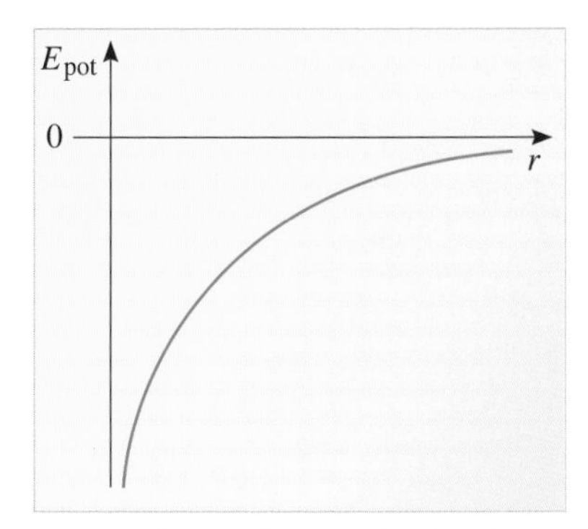

Figure 2.3 The gravitational potential energy of two masses, or the electrostatic potential energy of two charges of opposite sign, separated by a distance r.

Figure 2.3 shows the variation of the gravitational potential energy of two masses (corresponding to Equation 2.6). The gravitational potential energy of a pair of masses is always negative (if we take the zero of potential energy to be at infinite separation) because the force between masses is always attractive. The shape of the potential energy curve for two charges of the opposite sign is the same as that in Figure 2.3 since the force between them is also attractive. For two positive or two negative charges however, the shape is inverted (Figure 2.4) since the force is repulsive and any *increase* in separation corresponds to a *reduction* in potential energy.

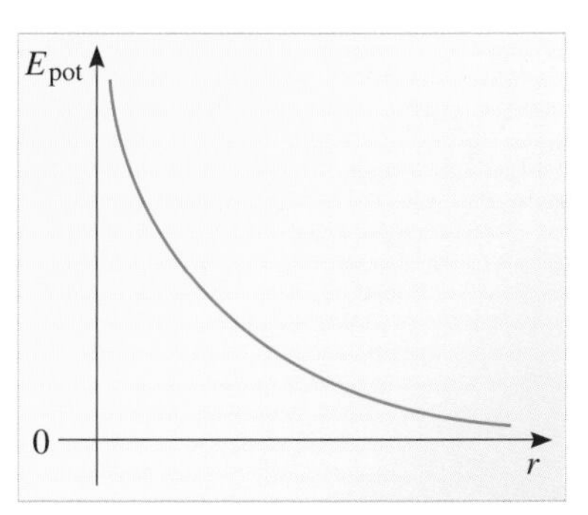

Figure 2.4 The electrostatic potential energy of two charges of the same sign, separated by a distance r.

Equations 2.6 and 2.7 are of great importance and you should remember them. Together with the Newton and Coulomb force laws, they will enable you to solve many problems in gravity and electrostatics.

Before leaving this section, there is one further subtlety of which you should be aware. It is meaningless to refer to the potential energy of an object in isolation. The equations we have derived are for the potential energy of two masses or two charges separated by a distance r. The potential energy only exists because of the forces between the two objects. In certain contexts, we might refer to the potential energy of an object without specifying the rest of the system with which the object is interacting. For example, physicists often use the phrase 'the potential energy of a body … at a height h'. In that case, the other part of the system with which the body is interacting is clearly the Earth, via the gravitational force. But, beware of allowing such loose descriptions to obscure the origins of potential energy.

2.1 Gravitational potential energy near the surface of the Earth

The expression (Equation 2.6) obtained in the previous section refers to the potential energy of two masses. It can be applied to systems as large or larger than galaxies and as small or smaller than atoms.

Question 2.2 Calculate the gravitational potential energy of the Earth–Moon system. The masses of the Earth and Moon are 6.0×10^{24} kg and 7.4×10^{22} kg and the mean distance from the Earth to the Moon is 3.8×10^{8} m.
Take $G = 6.7 \times 10^{-11}\,\mathrm{N\,m^2\,kg^{-2}}$. ■

In Question 2.2, you used Equation 2.6 for the potential energy of two masses. This varies as $1/r$ and was shown graphically in Figure 2.3. For a mass at, or near, the *surface* of the Earth, the gravitational field can be assumed to be uniform. That is, the gravitational force on an object is effectively constant up to any normally attainable height above the ground. In terms of the potential energy graph, this means that, over the range of r that this height implies, the gradient of the potential energy versus r graph is essentially constant (Figure 2.5 overleaf).

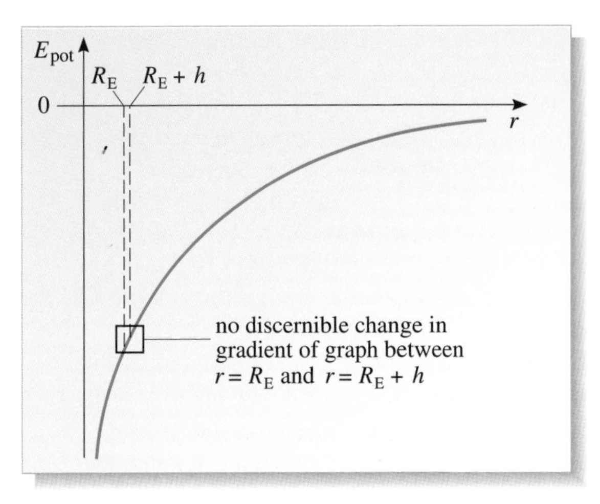

Figure 2.5 Over a small range of r, at $r = R_E$ (the radius of the Earth), the gradient of the graph of E_{pot} against r can be taken to be constant.

Near the surface of the Earth, the most obvious place to set the zero of gravitational potential energy is at ground level. Then, the gravitational potential energy at height h is $E_{grav}(h) = -W_{cons}$ where W_{cons} is the work done by the conservative gravitational field as the mass is raised from the ground to height h. The force is constant and equal to $m\boldsymbol{g}$ so $W_{cons} = -mgh$ since h increases upwards and $\boldsymbol{g}$ points downwards. Thus

$$E_{grav}(h) = mgh. \tag{2.8a}$$

It must again be emphasized that the place chosen for the zero of potential energy is arbitrary, and the most important thing to remember is that the *change* in gravitational potential energy, ΔE_{grav}, of a mass m, when it moves through a vertical distance Δh near the surface of the Earth, is

$$\Delta E_{grav} = mg\,\Delta h \tag{2.8b}$$

where Δh is positive when the mass moves upwards.

Question 2.3 (a) What is the potential energy of a 5 kg mass at a height of 20 m above the ground, assuming that the Earth's gravitational field is uniform up to that height? Take the zero of gravitational potential energy to be at ground level and assume $g = 9.8\,\mathrm{N\,kg^{-1}}$.

(b) With what speed would the mass strike the Earth's surface if released from rest at that height? (Neglect air resistance.) ■

Example 2.1

A common process in stars is the fusion of two hydrogen nuclei, each consisting of a single proton, to form a deuterium ('heavy hydrogen') nucleus consisting of a proton and a neutron bound together. Assuming the electrostatic interaction to dominate unless the protons are in contact, and taking the radius of a proton as 10^{-15} m, estimate the minimum temperature of a star in which fusion to deuterium takes place at an appreciable rate.

In answering this question, you may assume that, even for protons in contact, the particles act as though all their charge is concentrated at their centres. (This principle is discussed in more detail in the solution to Question 2.6.)

Solution

Preparation What is it that we need to calculate here? From the question, we conclude that two hydrogen nuclei will fuse if they can approach each other to within a separation of order of 10^{-15} m. But, protons repel each other electrostatically. They will only get close enough to each other to fuse if their initial translational kinetic energy exceeds the electrostatic potential energy of two protons when they are in contact. Thus, we need:

$$(\text{initial } E_{\text{trans}}) > (E_{\text{pot}} \text{ when protons are in contact}).$$

The apparent 'conversion' of a proton into a neutron does not constitute a violation of charge conservation, because the fusion reaction also results in production of a *positron* — a particle with the same mass as the electron and charge +e, together with an electrically neutral particle called a neutrino.

Working For fusion to be a common event within the star, a significant proportion of the protons must possess at least this much kinetic energy. A rough criterion would be that the mean translational kinetic energy $\langle E_{\text{trans}} \rangle$ should be equal to or greater than E_{pot} for two protons in contact. We know that

$$\langle E_{\text{trans}} \rangle \approx \frac{3kT}{2}$$

so for two colliding protons the total kinetic energy is $3kT$.

We also have

$$E_{\text{pot}} = E_{\text{el}} = \frac{e^2}{4\pi\varepsilon_0 r}.$$

This is essentially Equation 2.7, with q_1 and q_2 both equal to e, the charge on a proton. Here, we are modelling the protons as point charges separated by twice the proton radius and we are asked only to estimate the temperature so we need not use many significant figures.

Electrostatic potential energy of two protons at a separation of 2×10^{-15} m (see Figure 2.6) is

$$E_{\text{el}} = \frac{e^2}{4\pi\varepsilon_0 r} = 9 \times 10^9 \text{ N m}^2 \text{ C}^{-2} \times \frac{(1.6 \times 10^{-19} \text{ C})^2}{2 \times 10^{-15} \text{ m}} = 1.2 \times 10^{-13} \text{ J}.$$

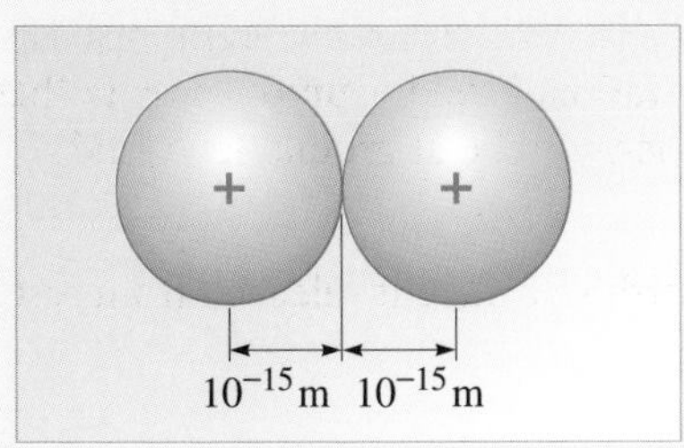

Figure 2.6 Two protons, each with radius 10^{-15} m, 'in contact'.

If the temperature is such that

$$2\langle E_{\text{trans}} \rangle = 3kT \approx 1.2 \times 10^{-13} \text{ J}$$

then a reasonable proportion of the hydrogen nuclei will have sufficient energy to fuse. Therefore

$$T \approx \frac{1.2 \times 10^{-13} \text{ J}}{(3 \times 1.4 \times 10^{-23} \text{ J K}^{-1})} \approx 3 \times 10^9 \text{ K}.$$

Checking Is the method sensible? We can reassure ourselves on that score by thinking about how a change in the variables would affect the result.

The working shows that if the particles involved were more highly charged, or had to be brought closer together, then an even higher temperature would be required for fusion to take place. This is in accordance with what one would expect. Is this a reasonable value for T? It is fairly common knowledge that fusion (whether in bombs or in stars) only takes place at high temperatures. But without further information we cannot really tell whether our estimate of roughly 10^9 K is a more sensible value than, say, 10^6 K or 10^{16} K.

Comment: *In fact, fusion occurs in stars at temperatures much lower than 10^9 K. There are two reasons for this: (a) the protons have a wide spread of energies, and a few have many times the mean translational energy; (b) an effect called quantum tunnelling allows fusion to take place more readily. You will find out more about this in Chapter 2 of* Quantum physics: an introduction.

Question 2.4 How much energy is required to bring two table tennis balls, each with a charge of 10^{-6} C, from a separation of 10 m to a separation of 10 cm? (Assume free space conditions.)

Question 2.5 The average distance between the electron and proton in a hydrogen atom is about 10^{-10} m. Estimate the electrostatic potential energy of the atom.

Question 2.6 α-decay, a type of radioactive decay, can be thought of as an α-particle (a helium nucleus consisting of two protons and two neutrons) escaping to just outside the atomic nucleus and then flying off, propelled by electrostatic repulsion. Imagine an α-particle expelled from a nucleus of radius 10^{-14} m to leave a residual nucleus containing 81 protons. Estimate the magnitude of the force with which the α-particle is expelled, and its final speed. (*Hint:* The crucial step here is setting up a model for the process. You may have difficulty with this, but, if so, think about it first and then read the Preparation stage of the solution. You should then be able to solve the problem yourself.)

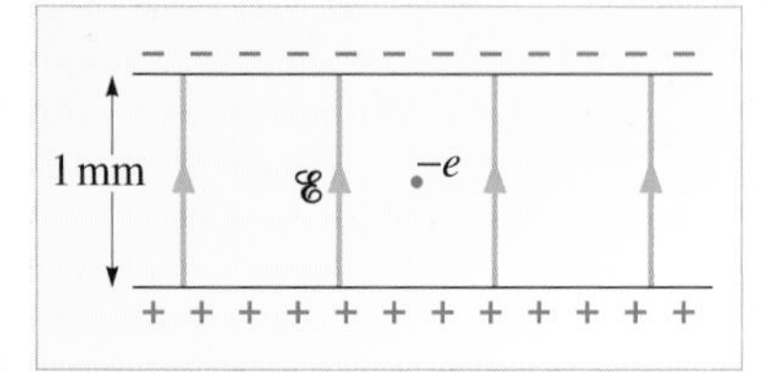

Figure 2.7 For use with Question 2.7.

Question 2.7 A uniform electric field of magnitude 10^4 N C^{-1} exists between two oppositely charged, parallel, conducting plates, which are 1 mm apart (Figure 2.7). By how much will the electrostatic potential energy of an electron change, if it is moved from the positive plate to the negative one? ■

One last important thing to realize about potential energy, and changes in potential energy, is that they scale linearly with the magnitude of the test mass or test charge. Think about the potential energy of a charge q in the electric field due to another charge Q. The potential energy is $qQ/4\pi\varepsilon_0 r$. If we were to replace q by a charge twice as large, that is, $2q$, the potential energy at any value of r would be exactly twice what it was for q. This behaviour is reminiscent of the fact that, in a given electric field, the force on a charge scales with the value of the charge in question. In the case of potential energy, it leads to the definition of *potential* which is the subject of the next section.

3 Gravitational and electric potential

3.1 Defining the electric potential

In the previous chapter, the electric field was introduced as the electrostatic force per unit charge:

$$\mathscr{E}(\boldsymbol{r}) = \left(\frac{1}{q}\right)\boldsymbol{F}_{\text{el}}\ (\text{on } q \text{ at } \boldsymbol{r})\,. \qquad \text{(Eqn 1.14)}$$

The beauty of the electric field is that it is independent of the value of the test charge q. This section introduces another concept, the **electric potential**, which is related to potential energy in exactly the same way as the electric field is related to the electrostatic force. In other words, the electric potential at a point is defined as the electrostatic potential energy per *unit* charge at that point:

$$\text{electric potential at point } \boldsymbol{r} = \frac{(\text{electrostatic potential energy of charge } q \text{ at } \boldsymbol{r})}{(\text{charge } q)}.$$

Denoting the electric potential by the letter V, we can write the following general definition:

$$V(\boldsymbol{r}) = \left(\frac{1}{q}\right)E_{\text{el}} \quad (\text{with } q \text{ at } \boldsymbol{r}). \qquad (2.9)$$

The electric potential is a *scalar* quantity: $V(\boldsymbol{r})$ means the electric potential at the point with position vector $\boldsymbol{r}$. It does *not* mean that V acts in the $\boldsymbol{r}$-direction. You should see from Equation 2.9 that the electric potential is zero where the electrostatic potential energy is zero. We can rearrange Equation 2.9 to obtain an expression for the electrostatic potential energy of a charge q at position $\boldsymbol{r}$:

$$E_{\text{el}} = qV(\boldsymbol{r}) \qquad (\text{with } q \text{ at } \boldsymbol{r})\,. \qquad (2.10)$$

Now, we shall look a little more closely at the definition of electric potential given in Equation 2.9 by considering the particular case of a point charge. Figure 2.8 shows the arrangement: q is a point charge with position vector $\boldsymbol{r}$ with respect to a fixed charge Q at the origin. In Section 2, Equation 2.7, you saw that the electric potential energy E_{el} of a pair of charges like this can be expressed by

$$E_{\text{el}}(\boldsymbol{r}) = \left(\frac{1}{4\pi\varepsilon_0}\right)\frac{qQ}{r}. \qquad (2.10\text{a})$$

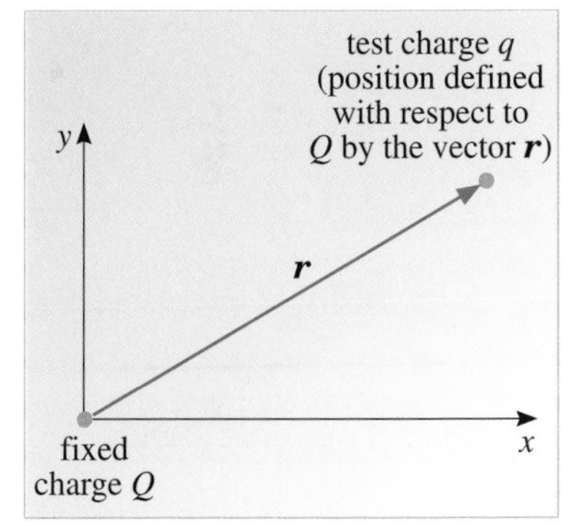

Figure 2.8 The electrostatic potential energy E_{el} of a test charge q with a displacement defined by the position vector $\boldsymbol{r}$ from a source charge Q is given by Equation 2.10a. The electric potential V at position $\boldsymbol{r}$ from the source charge is given by Equation 2.11.

Inserting this value for E_{el} into Equation 2.9 yields an expression for the electric potential due to a point charge Q:

$$V(\boldsymbol{r}) = \left(\frac{1}{4\pi\varepsilon_0}\right)\frac{Q}{r} \quad (\text{for charge } Q \text{ at origin})\,. \qquad (2.11)$$

Note that if the sign of Q is negative, then the electric potential $V(\boldsymbol{r})$ is also negative.

Thus, the electric potential due to a point charge Q is independent of the value of the test charge q and depends only on the source charge Q and the position $\boldsymbol{r}$. At each point in space, it can be specified by a number (multiplied, of course, by an appropriate unit, which will be defined later). Since the electric potential has no direction associated with it, it is a scalar field, and not a vector field. This is implicit in the notation we use — a simple italic V. It is partly because the electric potential is a scalar quantity that it is often easier to handle than the electric field. You will see evidence of this later in the chapter.

In the previous section, you saw that it is only *changes* in the potential energy that are physically significant: the useful quantity is the potential energy *difference*. Is the same true for the electric potential? Does its value have any significance or is it only differences that are important? Does it mean anything to say that the potential at a point has a given value? Well, Equation 2.10 shows that electric potential is zero when potential energy is zero, so if the zero of potential energy is arbitrary, so must be the zero of electric potential. What is important is the potential difference between two points. It is only after multiplying the potential difference by the charge being moved that the physically significant quantity, the change in potential energy, can be found.

If we always dealt with artificially simple systems, such as point charges, the choice of a position at which the electric potential was zero would be obvious. We would choose infinite separation from the source charge, i.e. $r = \infty$ in Equation 2.11. Often, in more complicated systems, a different choice is made, and it is important to realize why. Because the zero of potential is arbitrary, it is **potential difference**, the physically significant quantity, that is measured. In experiments, infinity is not accessible, and so the potential differences are conveniently measured between the point of interest and a defined, accessible point. For example, because of its moisture content, the Earth is a good conductor and so, as you will see later, is at the same potential everywhere. We might, therefore (and often do), choose the electric potential of the Earth to define our zero. Then all potential differences would be measured between the point of interest and the Earth. Since the electric potential difference is the important quantity, we often use Equation 2.9 to relate a change in electrical potential to a corresponding change in electrical potential energy. Thus:

$$\Delta V = \frac{\Delta E_{\text{el}}}{q} \qquad (2.12)$$

where the Δ symbol, as usual, represents a change in the quantity that follows it, and the $\boldsymbol{r}$ in parenthesis has been dropped for simplicity.

Figure 2.9 Alessandro Volta (1745–1827) was an Italian physicist who invented the first battery (known as the voltaic pile). This was a very important achievement as it allowed the first production of large electric currents. He also invented the electrophorus (an early electrostatic generator) and an electroscope. He was born in Como, where he was later a professor, and subsequently moved to Pavia. The unit of electric potential (and potential difference), the volt, is named after him.

Rearranging Equation 2.12 gives us the very important relation which allows us to calculate the change in potential energy of a charge q when it moves through a potential difference ΔV, namely

$$\Delta E_{\text{el}} = q\,\Delta V. \qquad (2.13)$$

This is a very widely used equation as it is the basis for energy consumption calculations in electrical devices.

It is clear from Equation 2.13 that a unit of electric potential (and potential difference) is the joule per coulomb, J C^{-1}. This unit is used so frequently, however, that it is given its own name, the **volt**, after Alessandro Volta (Figure 2.9). It is represented by the symbol V:

$$1 \text{ volt} = 1 \text{ joule per coulomb, i.e. } 1\text{ V} = 1\text{ J C}^{-1}.$$

At first sight, potential difference may appear to be rather an abstract quantity, but in fact you use the concept every time you refer to the voltage of a battery. When you say that a car has a 12 V battery, this is really a concise way of saying that, when no current flows, the potential difference between the negative terminal of the battery and the positive terminal is 12 V. Figure 2.10 shows the values of some potentials and potential differences.

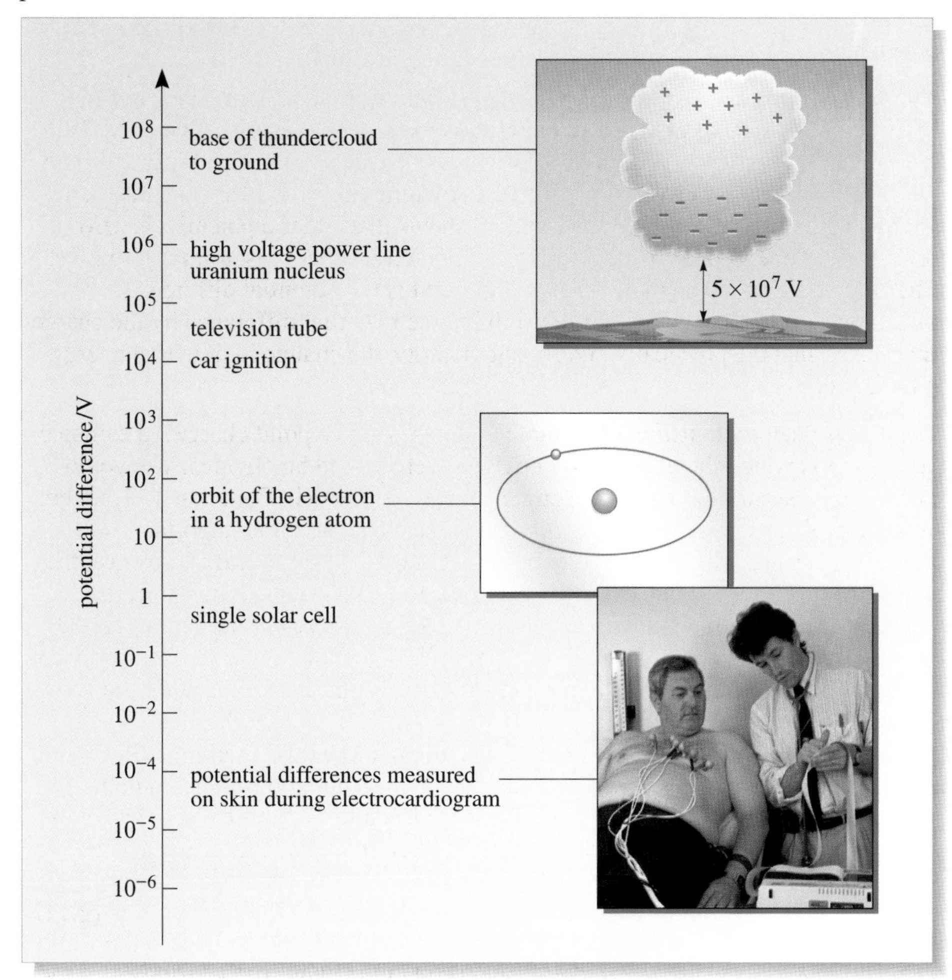

Figure 2.10 Some potentials and potential differences.

Example 2.2

Two point charges are separated by 6 m as shown in Figure 2.11. What is the electric potential at point A which is 4 m from the + 4 μC charge and 2 m from the – 6 μC charge? (1 μC = 1 × 10^{-6} C; take the zero of potential energy to be at infinity.)

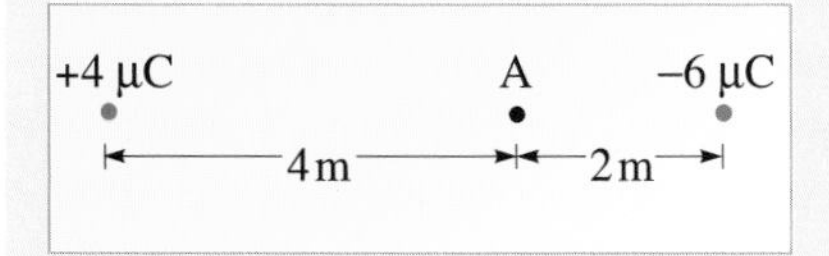

Figure 2.11 For Example 2.2.

Solution

Preparation The only equation we need is:

$$V(\boldsymbol{r}) = \left(\frac{1}{4\pi\varepsilon_0}\right)\frac{Q}{r}. \qquad \text{(Eqn 2.11)}$$

Let $Q_1 = +4\,\mu\text{C}$, $r_1 = 4\,\text{m}$, $Q_2 = -6\,\mu\text{C}$ and $r_2 = 2\,\text{m}$: $\dfrac{1}{4\pi\varepsilon_0} = 9 \times 10^9\,\text{N}\,\text{m}^2\,\text{C}^{-2}$.

Working Potential is a scalar so we can simply add the potential at A due to the $+4\,\mu\text{C}$ charge and the potential at A due to the $-6\,\mu\text{C}$ charge:

$$V(\boldsymbol{r}) = \frac{1}{4\pi\varepsilon_0}\frac{Q_1}{r_1} + \frac{1}{4\pi\varepsilon_0}\frac{Q_2}{r_2}$$

$$= \frac{1}{4\pi\varepsilon_0}\left(\frac{Q_1}{r_1} + \frac{Q_2}{r_2}\right).$$

Substituting the numbers and units gives

$$V(\boldsymbol{r}) = 9 \times 10^9\,\text{N}\,\text{m}^2\,\text{C}^{-2}\left(\frac{(4 \times 10^{-6}\,\text{C})}{4\,\text{m}} + \frac{(-6 \times 10^{-6}\,\text{C})}{2\,\text{m}}\right)$$

$$= -1.8 \times 10^4\,\text{V}$$

$$= -2 \times 10^4\,\text{V} \text{ to one significant figure.}$$

Checking Point A is closer to the negative charge. The negative charge also has a greater magnitude than the positive charge, so we would expect the potential at A to be negative.

Question 2.8 What is the potential energy change when 2 C of charge flow from the positive terminal of a 9 V battery to the negative terminal through a radio? ■

3.2 The relationship between field and potential

You should now be familiar with three relationships between the various quantities describing gravitational or electrostatic systems. These are summarized in the following diagram and are written out explicitly in the subsequent box.

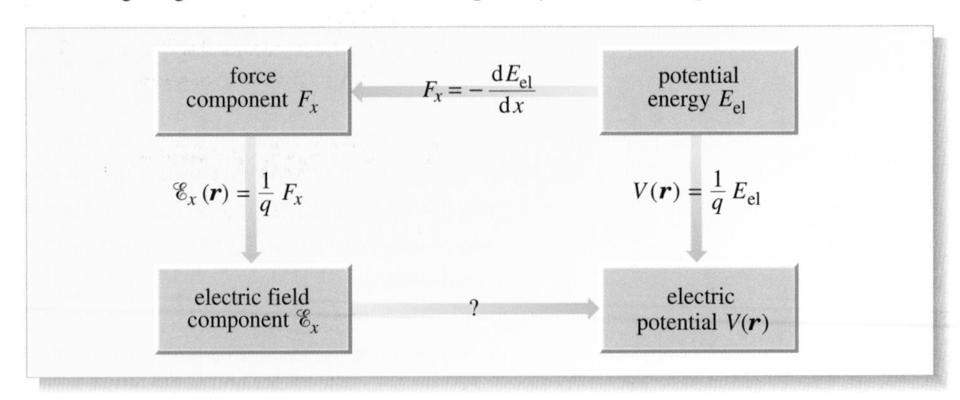

The force on a particle in any direction is minus the gradient of the potential energy curve in that direction, e.g. for the x-component of the force,

$$F_x = -\frac{dE_{el}}{dx}. \quad (2.14)$$

The electric field at a point defined by position vector $\boldsymbol{r}$ is the electrostatic force per unit test charge at point $\boldsymbol{r}$,

$$\text{e.g. for the } x\text{-component of the field, } \mathscr{E}_x = \frac{F_x}{q}. \quad (2.15)$$

The electric potential at a point $\boldsymbol{r}$ is given by the electrostatic potential energy per unit test charge placed at that point $\boldsymbol{r}$

$$\text{i.e.} \quad V(\boldsymbol{r}) = \left(\frac{1}{q}\right) E_{el} \quad (\text{with } q \text{ at } \boldsymbol{r}). \quad (\text{Eqn 2.9})$$

From these equations, we can complete the diagram by deriving a fourth relationship, namely that between the field and the potential. We shall write this explicitly for the electric field. From Equation 2.9, it follows that the change of the electric potential with distance in the x-direction is equal to the change of the potential energy per unit charge with position. That is,

$$\frac{dV(\boldsymbol{r})}{dx} = \left(\frac{1}{q}\right)\frac{dE_{el}}{dx}. \quad (2.16)$$

Now using Equations 2.14 and 2.15, we can rewrite Equation 2.16:

$$\frac{dV(\boldsymbol{r})}{dx} = \left(\frac{1}{q}\right)(-F_x) = -\mathscr{E}_x.$$

Thus, at any point, the electric field component in any direction is equal to minus the gradient of the electric potential in that same direction:

$$\mathscr{E}_x = -\frac{dV(\boldsymbol{r})}{dx}. \quad (2.17a)$$

Of course, we can write similar equations for the y- and z-components:

$$\mathscr{E}_y = -\frac{dV(\boldsymbol{r})}{dy} \quad (2.17b)$$

$$\mathscr{E}_z = -\frac{dV(\boldsymbol{r})}{dz}. \quad (2.17c)$$

In practice, the electric field component is usually determined by measuring the change in potential ΔV over a small displacement Δx, rather than measuring the gradient of the potential at a point. In such cases, we can use the approximation

$$\mathscr{E}_x \approx -\left(\frac{\Delta V}{\Delta x}\right) \quad (2.18)$$

and this approximation becomes more and more accurate as the displacement Δx becomes smaller and smaller. Finally, you should note that, from Equation 2.18, electric fields can be expressed in units of $\mathrm{V\,m^{-1}}$, as well as in units of $\mathrm{N\,C^{-1}}$ with which you are already familiar. (If you are unsure about this, use the conversions $1\,\mathrm{V} = 1\,\mathrm{J\,C^{-1}}$ and $1\,\mathrm{J} = 1\,\mathrm{N\,m}$ to check that the two units are equivalent.) In fact, in most situations, it is more conventional to quote the values of electric fields in units of $\mathrm{V\,m^{-1}}$ rather than $\mathrm{N\,C^{-1}}$, and from here on you will most probably find that this is the convention followed.

The following questions illustrate the application of Equation 2.18.

Question 2.9 Suppose that the electric potential difference between the terminals of a battery is 12 V and that their separation is 1.5 cm. What is the average magnitude of the electric field between them?

Question 2.10 The maximum electric field strength that can be sustained in air without breakdown occurring is $3 \times 10^6\,\mathrm{V\,m^{-1}}$. What is the magnitude of the maximum electric potential difference that can be sustained, without breakdown, between two points 1 cm apart, assuming that the electric field between them is uniform and has its maximum value at all points? ■

Uniform electric fields

The relationship (Equation 2.18) between electric field and rate of change of potential with distance is particularly easy to apply when the electric field is uniform, as in the region between two oppositely charged, parallel, conducting plates. In this case, if the potential difference between the plates is ΔV and the distance between the plates is d, then the magnitude of the electric field between the plates is just $\Delta V/d$.

Question 2.11 In Question 2.7, you were given that the electric field between two oppositely charged, parallel, conducting plates was $10^4\,\mathrm{N\,C^{-1}}$ and that the plates were 1 mm apart. (a) Calculate the potential difference between the plates. (b) Now recalculate, using Equation 2.13 this time, the change in potential energy of an electron when it moves from one plate to the other. ■

3.3 Equipotentials and field representations

In this section, we will see how electric fields and potentials, when represented diagrammatically, lead to the important idea of equipotential surfaces. We will develop this idea in more familiar territory — the gravitational field of the Earth. To do this, we need to complete our comparison between electric and gravitational fields by defining the gravitational equivalent of the electric potential.

Gravitational potential

Just as the electric potential is defined as the electric potential energy per unit charge, the **gravitational potential**, as you might expect, is defined as the gravitational potential energy per unit mass. For a test mass m at a distance r from another mass M, then

$$V_{\mathrm{grav}}(\boldsymbol{r}) = \left(\frac{1}{m}\right)E_{\mathrm{grav}} = \frac{1}{m}\left(\frac{-GmM}{r}\right) = \frac{-GM}{r} \qquad (2.19)$$

and this is the gravitational potential due to the mass M. It is clearly similar to the expression for the electric potential due to a point charge:

$$V(\boldsymbol{r}) = \frac{Q}{4\pi\varepsilon_0 r}. \qquad \text{(Eqn 2.11)}$$

Near the surface of the Earth, where the gravitational field is approximately uniform, it is convenient to express the gravitational potential energy, E_{grav}, of a mass m as:

$$E_{\text{grav}} = mgh \qquad \text{(Eqn 2.8a)}$$

with $E_{\text{grav}} = 0$ at height $h = 0$. In this case, the gravitational potential energy per unit mass is given by

$$V_{\text{grav}}(h) = \frac{E_{\text{grav}}}{m} = gh\,. \qquad (2.20)$$

Near the Earth's surface, g is approximately constant, so, on Earth, the gravitational potential is (approximately) directly proportional to the height.

We can also relate the component of the gravitational field in a certain direction to the gradient of the gravitational potential in that direction, i.e.

$$g_x = \frac{-\mathrm{d}V_{\text{grav}}}{\mathrm{d}x};\quad g_y = \frac{-\mathrm{d}V_{\text{grav}}}{\mathrm{d}y};\quad g_z = \frac{-\mathrm{d}V_{\text{grav}}}{\mathrm{d}z}. \qquad (2.21)$$

These relationships are directly analogous to those relating the electric field to the gradient of the electric potential (Equation 2.17).

Question 2.12 What are the units of gravitational potential? ■

On an Ordnance Survey map, you will see contour lines showing how the height above sea-level varies. Each contour line is labelled by the height h. Contour lines connect all points on the ground surface that have the same gravitational potential, i.e. all points at the same height, as shown in Figure 2.12. It could equally well have been labelled by the gravitational potential gh, but geographers and walkers are more accustomed to the idea of height. In Figure 2.12, the height (and therefore the gravitational potential) changes most rapidly along the direction AB; this is the direction in which the contours are closest together and the slope is steepest. It will therefore require greater muscular effort (i.e. greater force parallel to the slope) to climb in this direction than in direction DE, where the contours are more widely spaced. If we take the x-direction to be in some direction along the surface of the ground, then $g_x = -\mathrm{d}V_{\text{grav}}/\mathrm{d}x$ is the component of the gravitational field in that direction. The component of the gravitational field

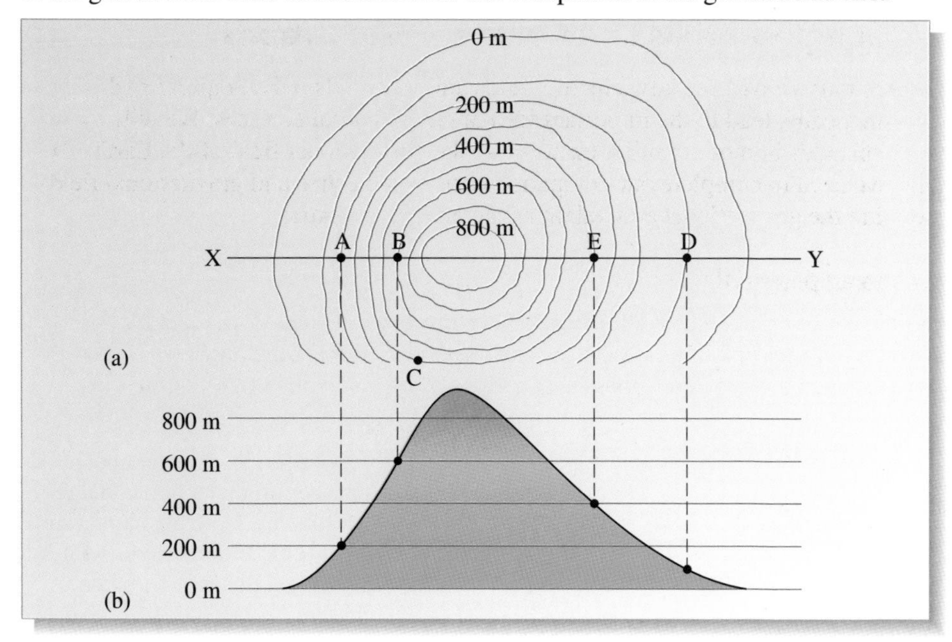

Figure 2.12 (a) A contour map of a mountain, and (b) a cross-section along line XY of the same mountain. The slope is steepest where the contour lines are closest together.

parallel to the surface is greatest where the contours are closest together and in the direction perpendicular to those contours, that is, where the magnitude of dV_{grav}/dx is greatest. In addition, it points 'downhill', from high potential to low. So you can see that, from a diagram showing lines of equal height, we have arrived at a conclusion about the way in which the effort involved with walking varies with position and direction: we have constructed a force field for the *component* of $\boldsymbol{g}$ parallel to the ground surface. (The gravitational field itself is, of course, uniform and pointing vertically downwards.) In addition, if a walker started at B and walked only along the 600 m contour, i.e. neither going uphill nor downhill, then no work would be needed against the gravitational field. At no time will the gravitational field oppose such a walker, for the path taken is horizontal and the gravitational field acts vertically. Thus the contour joins points for which the gravitational potential has the same value: it is a gravitational *equipotential.*

Equipotentials

These results are not restricted to the gravitational field. Imagine two point charges q_1 and q_2 separated by a distance R. Suppose q_1 is held fixed at the origin, but q_2 may be moved. We know from Section 3.2 that the potential due to q_1 is:

$$V(\boldsymbol{r}) = \frac{q_1}{4\pi\varepsilon_0 r}. \qquad \text{(Eqn 2.11)}$$

Since the electric field and potential due to a point charge are spherically symmetric, that is, the same in all directions, V depends only on the magnitude of $\boldsymbol{r}$, not its direction, so we can write simply

$$V(r) = \frac{q_1}{4\pi\varepsilon_0 r}.$$

At the location of q_2, where $r = R$, the potential due to q_1 is $V(R)$, so the potential energy of q_2 is $q_2V(R)$. Moreover, q_2 may be moved from its initial position to any other point on the surface of an imaginary sphere of radius R, centred on q_1, without any change in the energy of the system. All the points on this imaginary sphere are therefore at the same potential: the sphere describes an **equipotential** surface. Figure 2.13 shows spherical equipotential surfaces around a charge of 10^{-11} C at 1, 2, 3, 4 and 5 V (assuming $V = 0$ at infinity).

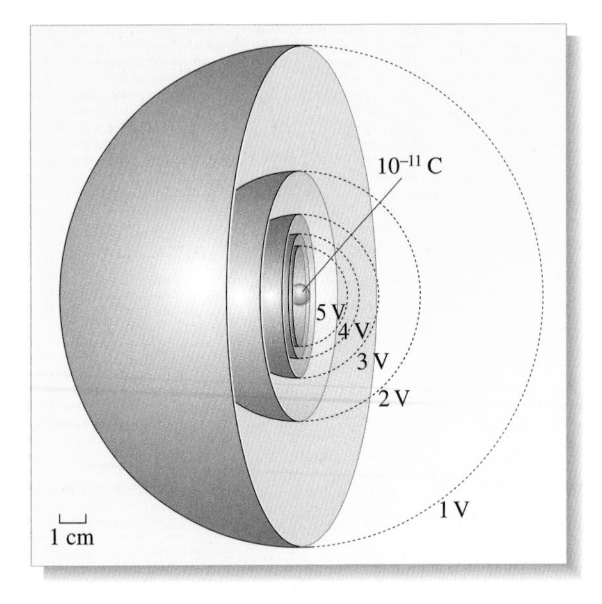

Figure 2.13 The equipotential surfaces around a point charge are spheres. Note the increasing separation of the spheres with increasing distance from the charge.

Question 2.13 What can you say about the relationship between the spacing of the equipotential surfaces in Figure 2.13 and the magnitude of the electric field? From what you know about the direction of the electric field associated with a point charge, what can you say about the relative orientations of the equipotential surfaces and the electric field lines for a point charge? ■

In this example, the radial electric field is perpendicular to the spherical equipotential surfaces. But is it generally true that field lines cut equipotentials at right angles? To answer this question, consider two adjacent points A and B on the same equipotential (Figure 2.14a) and suppose the direction from A to B determines the direction of the x-axis. By definition, A and B must be at the same potential. Since ΔV is therefore equal to zero, Equation 2.18 ($\mathscr{E}_x = -\Delta V/\Delta x$) tells us that there can be no component of $\mathscr{E}$ in the direction joining the two points. If the electric field exists at all in this region, then in order to have zero x-component, the field must be at right angles to the x-direction. Hence:

Electric field vectors always cut equipotential surfaces at right angles.

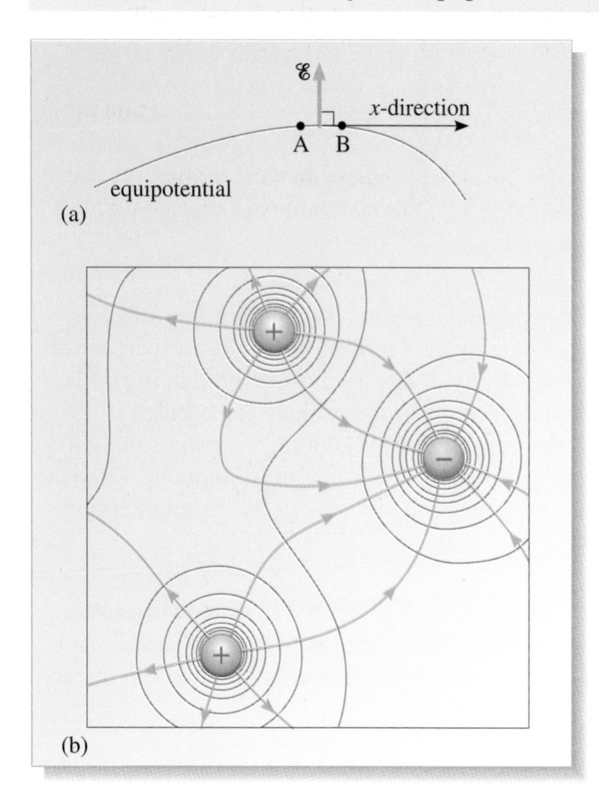

Figure 2.14 (a) Since A and B are at the same potential, there can be no force component (and therefore no electric field component) along the line between them. Hence the electric field vector $\mathscr{E}$ must always cut equipotentials at right angles. (b) The field lines and equipotentials around an array of three charges.

The relationship between equipotentials and the electric field is very important (Figure 2.14b): it is easier to measure potential than it is to measure electric fields. However, if we are given a map of the equipotential surfaces, we can deduce the direction and relative strength of the field at any point.

Returning to the subject of gravitation, the relationship between the gravitational field and gravitational equipotentials is exactly the same as in the electrostatic case. The gravitational field is always perpendicular to the equipotential surfaces. It is directed from high gravitational potential to low, and its magnitude is greatest where the equipotentials are closest together. Near the surface of the Earth, gravitational potential is given (approximately) by $g \times h$ and hence the equipotentials are equally spaced horizontal planes and the gravitational field is uniform and acts vertically downwards. Contours on a map are lines drawn where the equipotential planes and the land surface intersect. These horizontal planes are, of course, really small parts of huge spheres centred on the centre of the Earth, which, like electric equipotentials around a point charge, grow more widely spaced with distance from the Earth (see Figure 2.15). On this large scale, the gravitational field is directed radially towards the centre of the Earth, and its magnitude decreases with increasing r.

Figure 2.15 Gravitational equipotentials around the Earth are concentric spheres with increasing spacing. On the smaller scale of everyday life near the surface of the Earth, these concentric spheres approximate to equally spaced planes.

Question 2.14 Large electric fields can be dangerous. When lightning strikes a tree, the electric potential of the tree trunk may be 1 million volts with respect to the zero of potential which is taken to be at infinity. It has been observed that if a cow takes shelter under a tree during a thunderstorm, it is more likely to be killed if it is standing along the line of a radius to the tree than if it is standing sideways on to the tree. Why is this? (*Hint:* Imagine looking down on the situation from above. Draw in a set of equipotentials round the tree trunk; then put one cow in the radial direction, and one cow in a sideways direction.)

Question 2.15 Figure 2.16 shows the electric equipotential lines between two parallel charged plates. Sketch on the figure the field lines both in the central region and at the edges of the plates. What can be said about the strength of the field in each region? ■

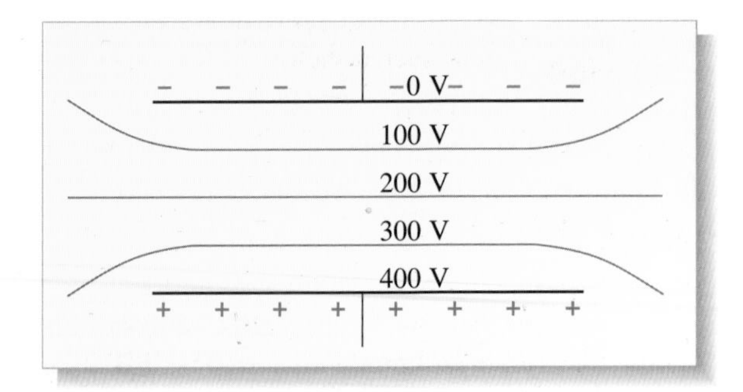

Figure 2.16 For use with Question 2.15.

3.4 Electrostatic screening

So far in this chapter, we have been concerned with the variation of the potential in the neighbourhood of charges situated in free space. We are now going to discuss how potentials may be manipulated by the presence of **conductors**.

A perfect conductor is one in which the charges are able to move so freely that it is impossible to sustain a potential difference within the conductor: if a potential difference is created in the conductor, the charges immediately move so as to cancel it out. Thus, at all points within a perfect conductor the potential is uniform. This is, of course, an idealization but is a good model for many real conductors in fields that are static or slowly varying.

Question 2.16 In what direction must electric field lines leave the surface of a perfect conductor? ■

A further consequence of the equipotential nature of conductors is that not only must the potential be constant within the body of a conductor, but it must also be constant inside a hollow conducting shell that contains no free charges. This means that there can be no electric field inside a closed, perfectly conducting shell that contains no free charges. We shall now prove that this is the case.

First, let us assume (incorrectly) that an electric field *does* exist within the interior of the closed conducting shell shown in cross-section in Figure 2.17. This implies that there must be some electric field lines in the enclosed volume. Now, as electric field lines cannot form closed loops, but must run from positive charges to negative charges, and as we are assuming that there are no free charges in the volume enclosed by the shell, the only possible lines are those that connect different parts of the inside surface of the shell. But, as you know, field lines point in the direction of the electric field which, in turn, points down the potential gradient. So, the existence of a field line running, for example, from point X to point Y in Figure 2.17, would imply that point X was at a higher potential than Y. However, since conductors are equipotentials, X and Y must be at the *same* potential.

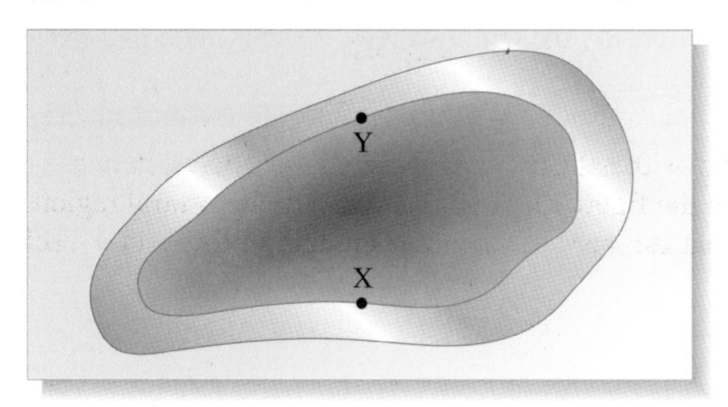

Figure 2.17 A cross-sectional 'slice' through a closed conducting shell containing no free charges. Points X and Y must be at the same potential because the shell is a perfect conductor. No electric field can exist inside such a shell.

Thus, our original assumption has led to a contradiction and must therefore be false, and so we conclude that:

> No electric field can exist inside a perfectly conducting shell that contains no free charges.

This is a result of great technological importance. Some sensitive electronic circuits must be kept isolated from externally produced electric potential differences, that is, they must be *screened* from external electric fields. As we have just shown, this is most easily achieved by enclosing the circuits inside conducting boxes (Figure 2.18).

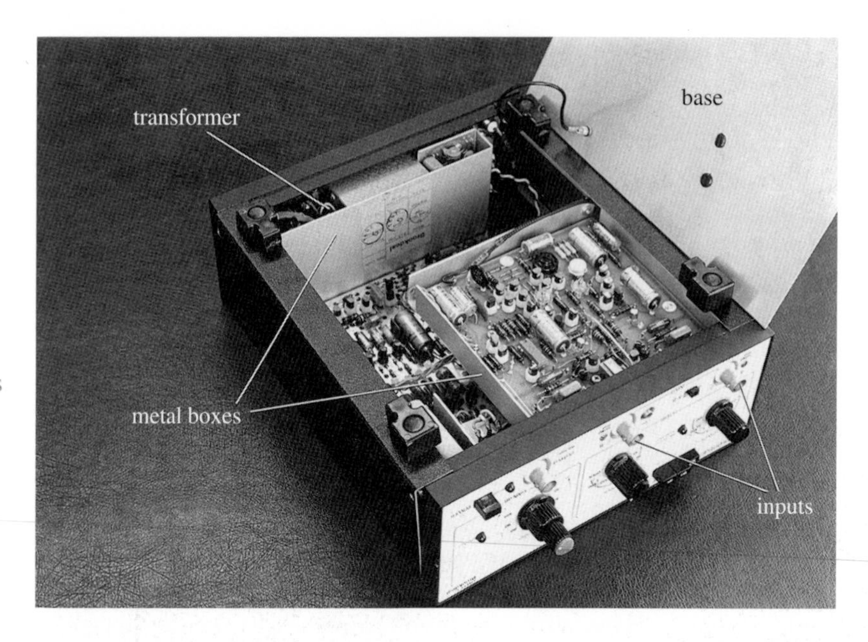

Figure 2.18 Electrostatic screening and earthing of an electrical appliance: when the lid is closed, the most sensitive parts of this high-gain amplifier are enclosed in an earthed metal box. The transformer is similarly encased. The weak signal is thus protected from corruption by electrical 'noise'.

Sometimes these screens fulfil an additional safety function. If the conducting box is **earthed**, that is, connected by a conductor to the Earth, the electric potential of the box will be equal to that of the Earth. Then, if the circuit should touch the box, charge will flow from the circuit into the box itself, but this will not materially change the potential of the box. This is because the good conducting path between the box and the Earth will maintain the box at Earth potential in spite of the weaker conduction path from the box to the circuit. Now, had the box been at a much higher potential than the Earth, a person touching it might have been electrocuted, because the human body provides a fairly good conducting path to Earth (through the feet, or some other part of the body touching water pipes etc.). Thus, most high voltage equipment, and some domestic appliances like washing machines, kettles, etc., are contained inside earthed conducting boxes, which protect the user from electric shocks, screen the equipment from externally produced fields, and also screen other devices from the electric fields produced by the equipment.

Question 2.17 When an aeroplane is struck by lightning, why are the passengers and crew not electrocuted? ■

4 Capacitance and energy storage in electric fields

In many practical applications, such as in electrical circuits, it is necessary to produce an electric potential difference by separating positive and negative charges and to store them (temporarily) on different objects, or on different parts of the same object.

Historically, the first method of producing a potential difference was by using mechanical energy to separate positive and negative charges, and for some time rubbing was the only known way of producing charge separation. The rubbing techniques did, however, achieve considerable sophistication, and electrical experimentation became quite a fashionable pursuit (Figure 2.19).

Figure 2.19 An engraving of an electrical machine constructed by Otto von Guericke, Bürgermeister of Magdeburg, Germany, in 1666. The sulfur ball at the right was rotated at high speed and rubbed with a cloth or dry hands. The ball, an insulator, acquired an electrostatic charge: when anyone approached the charged ball, they experienced a tingling sensation and noticed crackling noises and even sparks.

A significant advance towards developing a means of producing a potential difference was made when Professor Pieter van Musschenbroek of Leyden University demonstrated a method of separating charge. His device, which became known as a Leyden jar, is illustrated in Figure 2.20. This jar was the first version of what would now be called a *capacitor*. Capacitors are extremely important components of many electrical circuits.

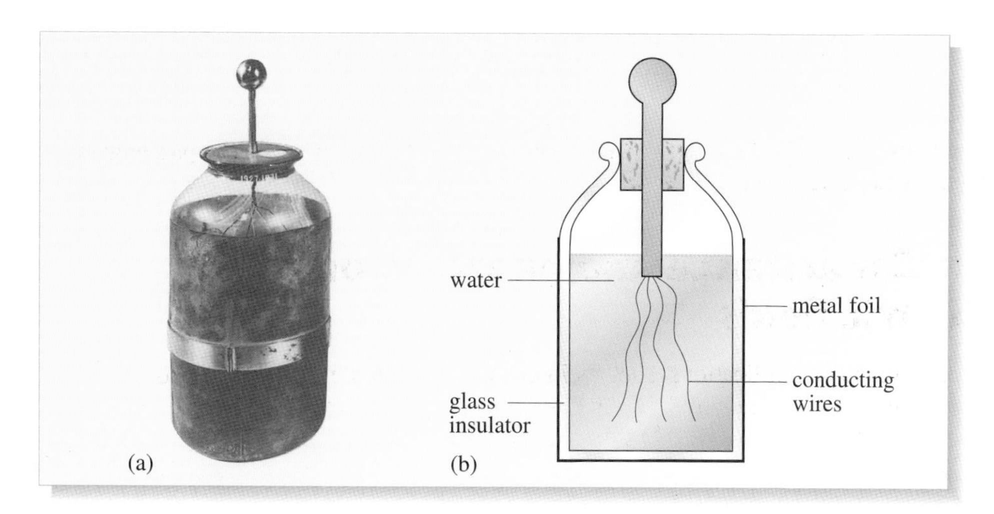

Figure 2.20 (a) A Leyden jar in the Science Museum. (b) A schematic diagram of a Leyden jar. Following earlier work by another scientist, E. G. von Kleist, Musschenbroek found that a bottle partly filled with water and having a metal rod projecting through the neck could be charged from an electrical machine. When he held the jar in one hand and touched the rod with his free hand, he received a severe shock. In his report, he referred to a 'new but terrible experiment which I advise you on no account to repeat yourself'. Of course, people did.

In Chapter 1, you saw that when charge is placed on a body, an electric field is produced in the region around the body. The electric field in turn makes it harder to place more charge of the same sign on the body. Thus, energy is required to establish any particular charge distribution, and this energy can be recovered by allowing charges of opposite sign to be attracted to the charge distribution and neutralize it. As the energy used in establishing a charge distribution is recoverable, it may be described as stored energy. In the rest of this section, you will see how capacitors store electrostatic potential energy by maintaining a separation of positive and negative charges.

4.1 Capacitors

The electric potential of a conductor is related to the amount of charge residing on it. The more positive charge there is on a body, the more energy is required to bring up extra positive charge and deposit it on the body.

Many electrical circuits contain devices whose primary function is to store energy by separating charge in the circuit. Such devices are called **capacitors**. A commonly used type of capacitor is the **parallel plate capacitor**. In its simplest form, this consists of two parallel, conducting plates, each of area A and separated by a distance d, in a vacuum (Figure 2.21). A charge q is transferred to one plate and a charge of the same magnitude and the opposite sign is transferred to the other, resulting in a potential difference V between the plates. The ratio of the charge on either of the plates to the potential of that plate relative to the other plate, is called the **capacitance** of the capacitor, and is denoted by the letter C:

The difference between two potentials is more logically written as ΔV, not V. However, by convention, the Δ symbol is usually omitted, particularly for electric circuits. You should remember that V may be positive or negative depending on whether the potential increases or decreases along the direction being considered.

$$C = \frac{q}{V}. \tag{2.22}$$

Note that if q is positive, V is also positive, whereas if q is negative, V is negative. The capacitance is, therefore, always positive. Capacitance is a constant that depends on the geometry and materials of the capacitor but is itself independent of the charge q and potential difference V, even though it is equal to their ratio.

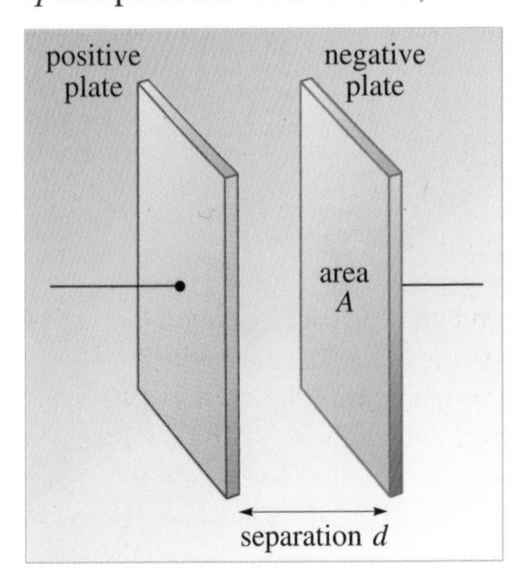

Figure 2.21 A schematic diagram of a parallel plate capacitor.

In SI notation, the unit of capacitance is the **farad** (which has the symbol F). The unit is named in honour of Michael Faraday, who devised the electric field concept. The farad is defined as

$$1\ \text{farad} = \left(\frac{1\ \text{coulomb}}{1\ \text{volt}}\right), \qquad \text{i.e. } 1\,\text{F} = 1\,\text{C}\,\text{V}^{-1}.$$

The factors that determine the capacitance of a given parallel plate capacitor, with a vacuum between the plates, are the area of the plates, A, and the distance between the plates, d. The derivation of an expression for the capacitance of a parallel plate capacitor in terms of A and d is beyond the scope of this course, so the result will be simply quoted: the capacitance of a parallel plate capacitor in a vacuum is given by

$$C = \varepsilon_0 \frac{A}{d}. \tag{2.23}$$

where ε_0 is the permittivity of free space.

It follows that, for a parallel plate capacitor with a vacuum between the plates,

$$\frac{q}{V} = \varepsilon_0 \frac{A}{D}.$$

But what happens when we fill the intervening space between the capacitor plates with an insulator (sometimes called a dielectric)? In dielectrics, the electrons are not free to move throughout the body of the material but rather are bound to particular atoms. Normally, the positive and negative charges are distributed in such a way that they cancel each other out. However, a different situation obtains if we put such a material between the parallel plates of a capacitor in which positive and negative charges have been separated on to the two plates.

When a dielectric is placed between the plates of a charged capacitor, its constituent atoms are in an electric field so the charge distribution associated with each atom is distorted slightly, with negative charges moving slightly towards the positive plate and positive charges moving towards the negative plate. This is illustrated schematically in Figure 2.22. At the surface of the dielectric, therefore, charges appear. These charges are of opposite sign to the charges on the adjacent plates, and the result is that there appears to be a partial cancellation of the charges on the plates. The potential difference between the plates is therefore reduced. Since the capacitance is inversely proportional to the potential difference (Equation 2.22), this must mean that the capacitance is increased when the dielectric is introduced between the plates.

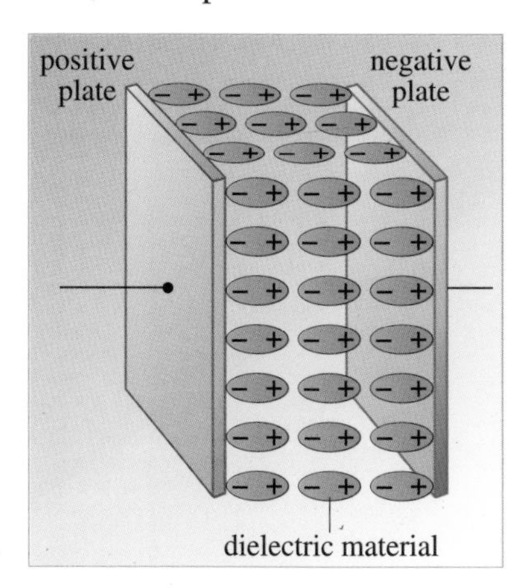

Figure 2.22 When the space between the plates of a parallel plate capacitor is filled with a dielectric, the charges within the atoms of the dielectric are displaced slightly, such that a net positive charge appears on the face closest to the negative plate of the capacitor, and a net negative charge appears on the face closest to the positive plate of the capacitor.

Even if the potential difference between the plates is held constant by, for example, connecting them to the terminals of a battery, the capacitance is still increased when the dielectric is introduced. This is because the surface charges induced on the dielectric offset, to some extent, the mutual repulsion of the charges on the plates and so enable a greater amount of charge to accumulate there before the magnitude of the potential difference between the plates reaches the battery voltage.

The amount by which the potential difference is reduced and the capacitance is increased when a dielectric replaces a vacuum between the plates of a capacitor is known as the **relative permittivity**, ε_r, of the particular medium. We can define the relative permittivity as the capacitance of a capacitor, with the space between the plates filled with the dielectric, divided by the capacitance of the same capacitor with the plates separated by a vacuum.

Table 2.1 shows the value of ε_r for a few common materials. Since the capacitance is increased by a factor ε_r when a dielectric fills the space between the plates, Equation 2.23 becomes

$$C = \varepsilon_r \varepsilon_0 \frac{A}{d} . \qquad (2.24)$$

Table 2.1 Values of ε_r for some materials.

Material	Relative permittivity, ε_r
Air	1.005
Glass	6.7
Nylon	3.7
Wood	2.1

Thus, the requirements for large capacitance are large plates (A), a small gap (d) and a material with a high relative permittivity (ε_r). To fulfil the first criterion and still keep the device compact, the plates are often made in the form of thin strips that are rolled up into cylinders (Figure 2.23). The characteristics of the dielectrics used to separate the plates are high values of relative permittivity, together with high breakdown electric fields; this second criterion means that the capacitor plates can be much closer together than would be possible if they were separated by air.

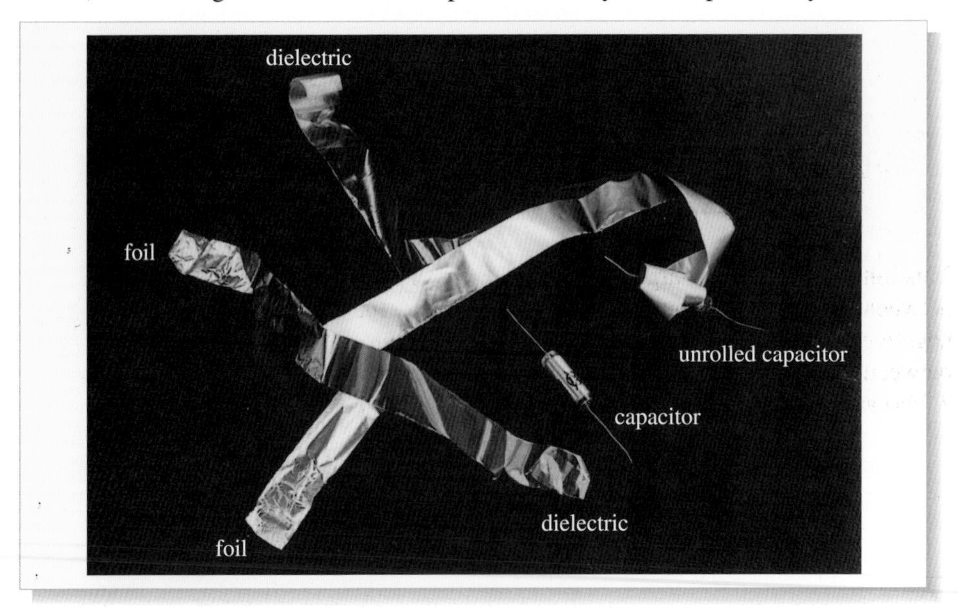

Figure 2.23 A partially unrolled capacitor and an intact capacitor of the same capacitance. The two strips of foil and two strips of dielectric were arranged in a foil–dielectric–foil–dielectric sandwich before being rolled together. Each strip is about 70 cm long and 2 cm wide.

Question 2.18 An aeroplane flies through a thundercloud and acquires a charge of 10^{-6} C. When the plane lands, its wings and the ground beneath them can be modelled as a parallel plate capacitor. If the total area of the wings is $300\,m^2$ and they are 3 m above the ground, what is the capacitance of the wings of the plane? By calculating the potential difference between the wings and the ground, can you guess why great care has to be taken to earth the plane before it is refuelled? (Assume that $\varepsilon_r = 1$ for air.) ■

4.2 Energy storage in capacitors

We have seen that a capacitor can be used to separate positive and negative charges, and this is equivalent to saying that a capacitor will store energy. Energy is required to push electrons onto an already negatively charged body and this energy is stored in a capacitor until it is recovered by letting some of the excess negative charge flow away, driven by the electrostatic force. The aim of this section is to calculate the amount of energy stored in a capacitor.

Consider a parallel plate capacitor as shown in Figure 2.24. We shall imagine the charge separation process (often inaccurately termed the 'charging' process) as a sequence of steps, each of which transfers a positive charge Δq from one plate of the capacitor to the other. Figure 2.24a is a schematic diagram of the capacitor as the first small bundle of charge is transferred.

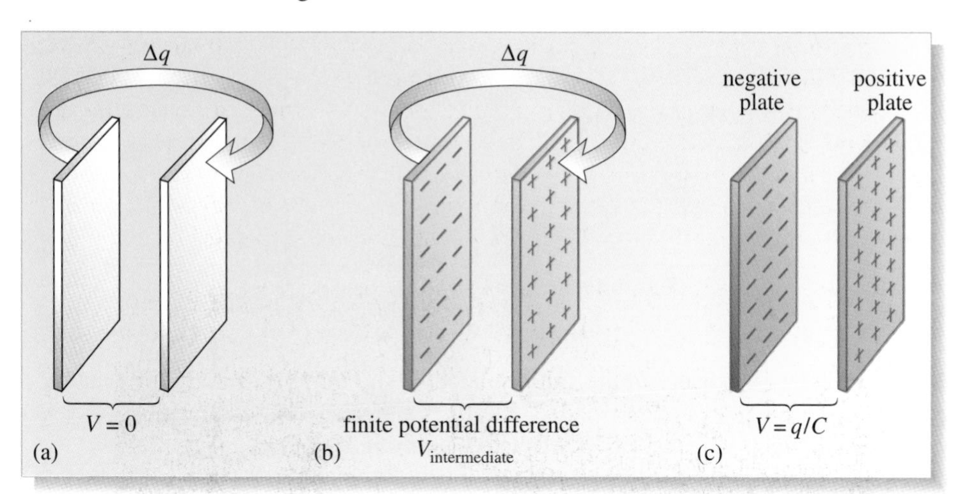

Figure 2.24 The process of charge separation in a capacitor viewed as a sequence of transfers of a small amount of charge of the same sign Δq. This leaves the capacitor with a negative charge on one plate, a positive charge on the other, and a potential difference V between the plates. The charged capacitor stores potential energy.

This initial transfer of charge requires no energy since there is no electric field between the plates to oppose the movement of the charge (i.e. the plates are not yet negative or positive). Figure 2.24b shows the capacitor at some intermediate stage between the state when the potential difference between the plates is zero (Figure 2.24a) and the state when the capacitor is 'fully charged' to a final potential difference V (Figure 2.24c). In this intermediate case, energy is required to transfer the small additional charge Δq. The amount of energy required is given by

$$\Delta E_{\text{el}} = (\Delta q) \times \begin{pmatrix} \text{electric potential difference of the positive} \\ \text{plate relative to the negative plate.} \end{pmatrix}$$

This is equal to the area of the shaded strip in Figure 2.25. As each packet Δq of charge is transferred, the energy needed is equal to the area of the appropriate strip and, therefore, the energy required to achieve maximum separation of charges is equal to the sum of the areas of all the strips in Figure 2.25. This may remind you of a result that was derived in Chapter 2 of *Predicting motion*, where you saw that the energy required to stretch a spring was equal to the area under the force versus extension graph. In the same way, the energy stored in the capacitor is equal to the area under the potential difference versus charge graph. This triangular area is given by $\frac{1}{2}$ × base × height, which is

$$E_{el} = \frac{qV}{2}$$

Substituting for q from Equation 2.22 gives

$$E_{el} = \frac{CV^2}{2}.$$

We can also substitute for V in the expression for E_{el}, which gives

$$E_{el} = \frac{q^2}{2C}.$$

Thus, drawing all of these together

$$E_{el} = \frac{qV}{2} = \frac{CV^2}{2} = \frac{q^2}{2C}. \tag{2.25}$$

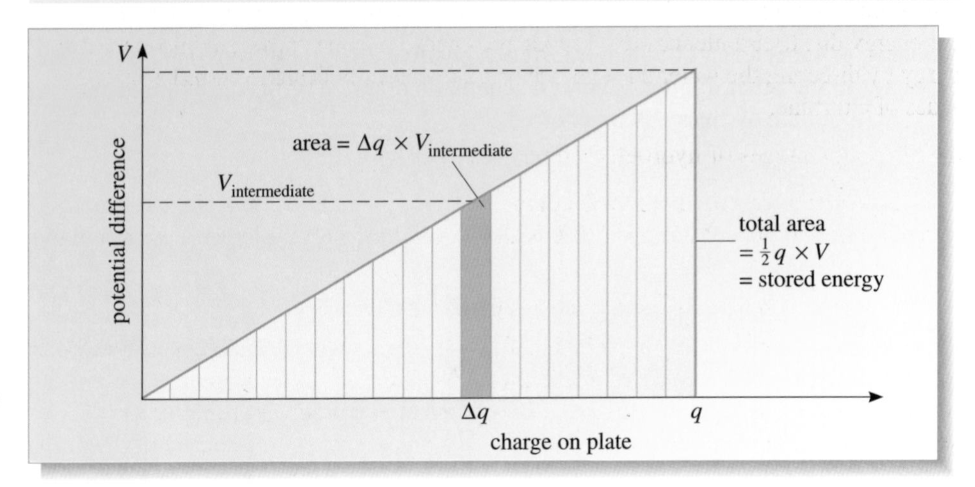

Figure 2.25 The potential difference of the positive plate of a capacitor with respect to the negative plate plotted against the charge on the positive plate. The total potential energy is equal to the area under the line. That is, stored energy = $qV/2$.

Question 2.19 A sheet of mica (an insulator) is slid into the, initially evacuated, space between the plates of an isolated, parallel plate capacitor, 'charged' to a potential difference V. It just fits. The relative permittivity of mica is 5. By how much does the potential difference between the plates change? By how much does the recoverable energy stored in the capacitor change? ■

Open University students should leave the text at this point and use the multimedia package *Forces, fields and potentials: part 2*, which is about electrical and gravitational potential and its representation. The activity will take about two hours.

5 Applications of potential and potential energy

Examples of electric potentials and potential differences are all around and within us. Changes in the potential differences between the inside and outside of nerve fibres are the means by which electric signals are carried along the nerves. In the next chapter, you will see how potential differences between the terminals of batteries cause currents to flow through electric circuits. The build-up of electric charge in thunderclouds leads to potential differences between clouds and the Earth of hundreds of millions of volts.

The cathode ray tube, which forms the basis of your television set, is a device in which electrons are emitted from a source and accelerated by a series of positive electrodes through a potential difference of about 20 000 V towards the screen. This makes the television set potentially by far the most lethal piece of electrical equipment in the home.

We shall now consider some specific examples of the application of potential and potential energy.

5.1 Hydroelectric power

The gravitational potential energy that water in mountain lakes possesses relative to lower levels is exploited in hydroelectric schemes. From early times, this energy was used in waterwheels. However, this source is not dependable. The modern waterwheel is the water-driven turbine of a hydroelectric generation plant. The basic requirement for a usable source of hydroelectric power is a river with sufficient flow to constantly replenish the lake. Compared with a site at the foot of the mountain, the water has a large amount of gravitational potential energy. If, therefore, the water is allowed to flow under gravity in a controlled manner to a lower level, it will lose potential energy and gain a comparable amount of kinetic energy. (There will be some loss of energy in the form of heat energy due to turbulence etc.) The kinetic energy may be converted into electrical energy by directing the water through one or more carefully designed nozzles on to the blades of a turbine.

The main advantages of hydroelectric power *in general* are:

1. A high efficiency of about 90% compared to the much lower efficiencies of conventional power stations (30% to 40%) in which much energy is wasted in the form of heat.
2. The source is renewable and there are no harmful emissions.

Hydroelectric power is widely exploited in those countries with the necessary mountainous terrain, for example Norway, Sweden and Switzerland. In the United Kingdom, the utilization is necessarily limited (to about 2% of total power production). However hydroelectric power plants do provide a readily available source of electrical energy to meet sudden surges in demand. The hydroelectric supply may be switched into the grid when needed. In off-peak times, the plant can be used to pump water back to the higher level from a lower level reservoir. Figure 2.26 shows a photograph of the Cruachan Dam station in Scotland.

Figure 2.26 The Cruachan Dam and intake–outfall at Loch Awe, Scotland. The station itself is buried deep underground in the heart of the mountain.

Box 2.1

Cragside (Figure 2.27) near Rothbury in Northumberland, was, in 1878, the first house in the world to be lit by hydroelectricity. It was the seat of William George Armstrong (1810–1900, the first Lord Armstrong, Figure 2.28) who was involved with many industrial and technological innovations in the nineteenth century. In 1880, forty-five of Joseph Swan's newly developed incandescent lamps were installed at Cragside — 'the first proper installation' of such lamps, according to Swan. Electricity for the lamps was generated by a dynamo driven by a turbine operated by water flowing from the nearby Debdon Lake. This was probably the first hydroelectric power plant in Britain. In 1886, hydroelectric power at Cragside was expanded by the development of Burnfoot Powerhouse, using water from lakes 104 m above the powerhouse.

Figure 2.27 Cragside, Northumberland, the first house in the world to be lit by hydroelectricity.

Figure 2.28 William George Armstrong (1810–1900).

Question 2.20 A hydroelectric station is designed to deliver 10 MW of power using a dammed lake 1000 m above the location of the plant. Assuming 90% efficiency of energy conversion, what must be the rate of flow of water (in kg s^{-1}) from the lake? ■

5.2 White dwarfs, red giants, neutron stars and black holes

Gravitational potential energy plays a part on a gigantic scale in the formation and evolution of stars. The process starts with a cloud of gas mostly consisting of hydrogen but also containing a small amount (about 1%) of dust (recall Figure 2.1). Because of the gravitational attraction of each particle of matter for every other, the cloud contracts. During the collapse, the gravitational potential energy of the particles is converted to kinetic energy. In other words, the material heats up. This proceeds until the matter is hot enough and dense enough for nuclear fusion processes to be initiated. Fusion can occur only at very high temperatures because the positively charged nuclei require very large kinetic energies in order to overcome their mutual electrostatic repulsion.

We may think of the onset of fusion as the birth of a star. Stars convert nuclear potential energy into electromagnetic radiation, which shines out across space. A star may remain in equilibrium at about the same size and brightness for a very long time — as in the case of our Sun. This is due to the balance that is achieved between the inward force of gravity, and the outward pressure forces. The two sources of pressure are kinetic pressure and radiation pressure. In the former, the very rapidly moving matter in the interior of the star collides with matter in the outer layers and exerts a pressure similar to the pressure exerted by the molecules of a gas confined by a piston. In the latter, radiation that is produced in the very hot interior of the star escapes outwards and, as it does so, it collides with the matter it meets, giving an outward pressure effect. Can this equilibrium be maintained forever? The answer to this question is 'no', because the available nuclear potential energy is limited. When the star runs out of nuclear fuel, it must collapse. This collapse can occur in a number of different ways, largely depending on the mass of the star.

Average stars, including our Sun, are powered by the fusion of hydrogen to form helium, and can persist in equilibrium for billions of years. A stage will eventually be reached, however, when all the hydrogen has been used up and the core of the star will consist largely of helium. At this point, there is no source of energy to maintain the interior pressure, and the core of the star will collapse under gravity. Consequently, the core temperature will rise as the gravitational potential energy released by this collapse is converted to heat, and the resulting radiation pressure may be sufficient to expand the outer layers of the star. This will cause the star to become a **red giant**. If the star has enough gravitational potential energy, that is, if it is massive enough, the core temperature becomes greater than about 10^8 K, and helium fusion can be initiated to produce heavier elements. Increasingly massive stars are thus able to initiate fusion of increasingly heavy nuclei. This process is the Universe's main source of chemical elements other than hydrogen. Eventually, however, there will be no further available sources of nuclear energy and the core of the star will collapse under gravity whilst the outer layers (and thus a large proportion of the star's mass) are blown off in the process.

For stars of mass up to just a few times the mass of the Sun (M_{Sun}), the collapsing core shrinks under self-gravity to about the same size as the Earth. At this density, a further reduction in gravitational potential energy is prohibited by a quantum-mechanical effect in the atomic electrons known as **Pauli pressure**. The star comes to a state of equilibrium in which it is known as a **white dwarf**.

For more massive stars, in which the collapsing core has a mass greater than 1.4 M_{Sun}, the Pauli pressure is insufficient to prevent *sudden* collapse in which electrons and protons combine to form neutrons. This process can occur only in extreme conditions such as those pertaining during the collapse of a massive star. During this violent event, a great deal of gravitational potential energy, and some nuclear potential energy, are released. This powers an explosion so dramatic that the single collapsing star can produce enough electromagnetic energy to outshine the billions of other stars in the galaxy. Such an explosion can be observed in distant galaxies as a **supernova**. A nearby supernova was observed in 1987, as shown in Figure 2.29.

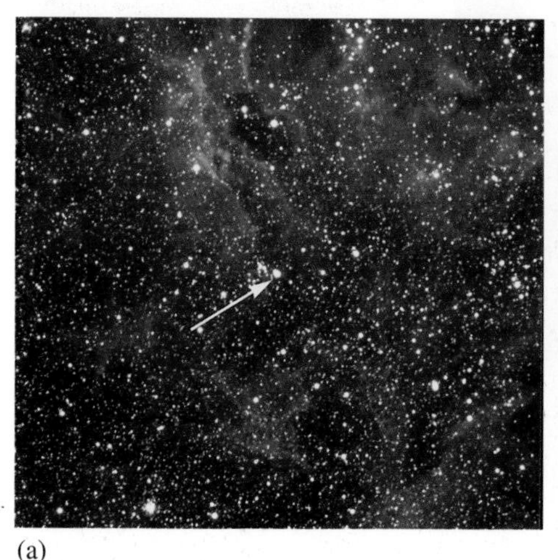
(a)

(b)

Figure 2.29 (a) The precursor star of supernova SN 1987A as it appeared until 1987. (b) In February 1987, the supernova's explosion was detected. The brightness of the star was then about a hundred million times that of the Sun and, even though it was 163 000 light-years away, it was visible by naked eye from the Earth.

The supernova core undergoes extreme gravitational compression, to form a **neutron star** with a radius of the order of 10 km. A neutron star has an overall density comparable to that of an atomic nucleus: one teaspoonful would weigh about 100 million tons! Like white dwarf stars, neutron stars are prevented from further collapse under self-gravity by Pauli pressure, but this time the effect is operating on the neutrons rather than the electrons. Table 2.2 shows some properties for comparison.

Table 2.2 A comparison of neutron stars, white dwarfs and the Sun. Mass of the Sun is 1.99×10^{30} kg

Characteristics	Neutron stars (predicted values)	White dwarfs (observed values)	Sun
Typical mean density/kg m^{-3}	10^{14} to 10^{18}	10^8	1400
Typical mass	$> 1.4\,M_{Sun}$	$< 1.4\,M_{Sun}$	M_{Sun}
Typical radius/km	10	6000	7×10^5
Typical rotation period	0.1 s	0.02 to 100 days	25 days
Magnetic field strength/T	10^5	up to 1000	10^{-4}
Surface temperature/K	10^6	up to 50 000	6000

Do we have any evidence that neutron stars exist? Figure 2.30 shows the supernova remnant Cas A as observed in X-ray radiation by NASA's *Chandra* satellite. The glowing wisps of gas are traces of the outer layers of the star blown away in the supernova explosion. The bright point near the centre is the glowing neutron star — the hot relic of the core, which powered the star before its violent death. The energy these X-rays carry away from the neutron star is supplied by the internal heat, just as coals glow in the dark after a fire goes out. By this cooling process, the neutron star will become dimmer until finally it is undetectable.

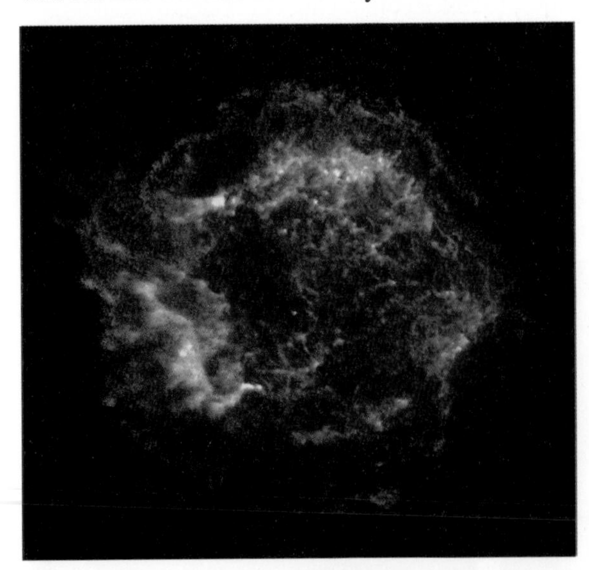

Figure 2.30 An image of the supernova remnant Cas A showing the high energy X-ray radiation emitted by the glowing gaseous wisps which are the remains of the disrupted outer layers of the star that exploded. The bright point source near the centre is the hot neutron star formed by the collapsed stellar core.

It is generally believed that stars we observe as **pulsars** (*puls*ating *r*adio *s*tars) are rapidly rotating neutron stars.

The final possibility for the fate of a star is the case where the mass of the collapsing core is so great that no equilibrium can exist and the gravitational collapse does not come to an end. This is the case for stellar cores of mass greater than about $3M_{\text{Sun}}$. If such an object were to form, it would be so dense that its gravitational field would prevent even light escaping from it. In classical terms, its escape velocity exceeds the speed of light — the highest speed permissible according to relativity theory. Such a star is called a **black hole** because of this property, since, if light cannot escape, then it would be undetectable by means of radiation emitted by the star itself. The existence of black holes has been strongly inferred in several binary systems in our Galaxy. One particular case is the binary V404 Cygni. In a binary system, two stars orbit about their common centre of mass. If one of the stars is a black hole, it can attract matter from its partner. The matter will be compressed and heated, and under the conditions in the neighbourhood of a black hole, the radiation given off will be in the X-ray region of the electromagnetic spectrum. Such X-rays have been detected from V404 Cygni, and the orbital motion of the partner means the black hole mass must be greater than $6M_{\text{Sun}}$, that is more than twice the maximum possible neutron star mass.

To sum up, the evolution of stars is governed by the ubiquitous effect of gravitation. The tendency of a star to collapse under its own gravitational attraction is halted only by other physical effects: the nuclear fusion that powers stars for the majority of their luminous lifetimes; the Pauli pressure of the electrons that offers final stability to low mass stars as white dwarfs; and the Pauli pressure of the neutron medium in neutron stars. The stellar matter in black holes suffers the ultimate unknown (or unknowable) gravitational fate — complete gravitational collapse.

5.3 Ion thrusters

Geosynchronous satellites (Figure 2.31) used for communication purposes must orbit the Earth in such a way that their position relative to the Earth's surface does not change. To do this, they must travel in a circular orbit with an orbital period of one day. A satellite in such an orbit is perturbed by the gravitational attraction of the Moon, and so regular corrections must be made to its position to ensure the continued usefulness of the satellite. Geosynchronous satellites are therefore equipped with thrusters, which can exert the small forces necessary to maintain the prescribed orbit. One of the devices used for this purpose is the ion thruster. This ejects **ions** with a certain momentum in one direction, and imparts an equal and opposite momentum to the satellite. (Remember that the total momentum of an isolated system is a conserved quantity.) We shall now calculate the means of doing this that will consume the minimum amount of energy.

Figure 2.31 A geosynchronous satellite in orbit above the Earth.

The arrangement used is shown schematically in Figure 2.32. Positive ions (i.e. atoms that have had one or more electrons removed) are emitted by a source S and accelerate in the x-direction towards the negative electrode N, the electric potential

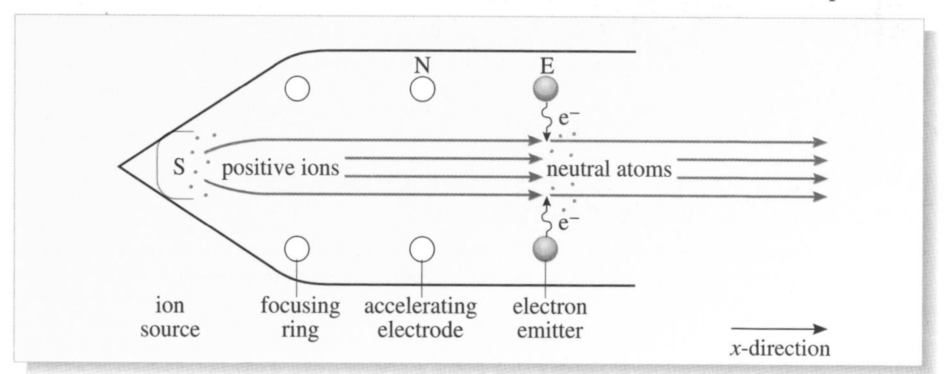

Figure 2.32 Schematic diagram of an ion thruster.

of which is $|\Delta V|$ volts below that of the source (ΔV is negative). The beam of ions then passes through the ring E, which is an electron source. The electrons neutralize the ions. If the ring E were omitted, the loss of positive ions would leave an excess negative charge on the satellite. The positive ions would then be drawn back to the satellite and that would prevent any permanent transfer of momentum from the ions to the satellite. What are the optimum values for the charge and mass of the ions in order to achieve the maximum momentum transfer for any given power input?

Let us consider one single ion of mass m and positive charge q_{ion} moving through the ion thruster. As it is accelerated towards the electrode N, the ion acquires translational kinetic energy $\frac{1}{2}mv^2$ at the expense of a loss in electrostatic potential energy $|q_{\text{ion}}| \times |\Delta V|$. Since, in this example q_{ion} is positive, we can drop the modulus sign.

Thus, using the principle of conservation of energy,

$$\tfrac{1}{2}mv^2 = q_{\text{ion}}|\Delta V|$$

i.e.

$$mv = \sqrt{2mq_{\text{ion}}\,|\Delta V|}$$

This ion carries with it momentum of magnitude $mv = (2mq_{\text{ion}}|\Delta V|)^{1/2}$ and momentum of the same magnitude, but in the opposite direction, is imparted to the satellite. During this process the electrical system has given up the electrostatic potential energy $q_{\text{ion}}|V|$. The performance of the thruster may be defined as the ratio:

$$\text{performance} = \frac{\text{(magnitude of momentum given to the satellite)}}{\text{(electrical energy used)}}$$

$$= \frac{(2mq_{\text{ion}}\,|\Delta V|)^{1/2}}{q_{\text{ion}}\,|\Delta V|} = \left(\frac{2m}{q_{\text{ion}}\,|\Delta V|}\right)^{1/2}$$

Thus, the best performance will be obtained using the heaviest ions available with the lowest charge q_{ion}.

In this example, we have been calculating energy and momentum changes when particles move through a given potential difference. Often, in such problems, it is convenient to use, instead of the joule, a different unit of energy, the **electronvolt** (eV). This is simply the magnitude of the energy change of an electron moving through a potential difference of 1 volt. The definition involves taking the modulus of each side of Equation 2.13:

$$|\Delta E_{\text{el}}| = |q\Delta V| \qquad \text{(Eqn 2.13)}$$

thus

$$1\,\text{eV} = |-e \times 1\,\text{V}|$$

i.e.

$$1\,\text{eV} = 1.602 \times 10^{-19}\,\text{J}.$$

Example 2.3

Suppose that an ion thruster uses argon ions (Ar^+) (of mass 6.6×10^{-26} kg and charge $+e$) and that it fires 10^{19} ions per second through an electric potential difference of 1 kV. Calculate (a) the momentum given to the ions each second and hence the force on the satellite, (b) the kinetic energy (in electronvolts) given to each argon ion, and (c) the power consumed by the ion thruster.

Solution

Preparation Assume the ions, each with charge q_{ion}, are fired in the x-direction (see Figure 2.33). Useful equations are:

force = rate of change of momentum

$$\Delta E_{\text{el}} = q\Delta V;\ E_{\text{trans}} = \tfrac{1}{2}mv^2$$

power = rate of transfer of energy.

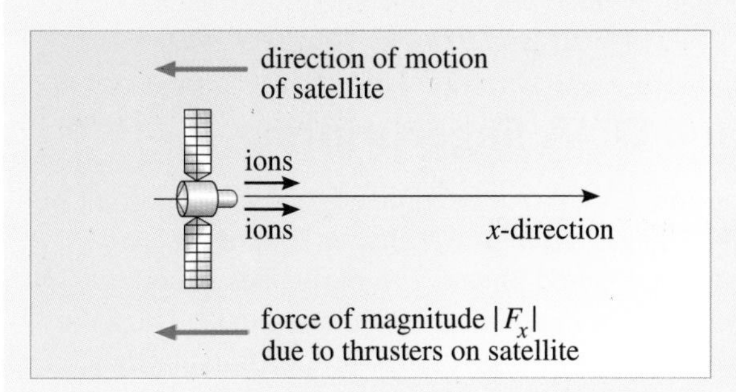

Figure 2.33 For use with Example 2.3.

Working (a) From conservation of energy:

The magnitude of the change in the translational kinetic energy of each ion = the magnitude of the change in the electrostatic potential energy of each ion:

$$\tfrac{1}{2}mv_x{}^2 = |q_{\text{ion}}||\Delta V|.$$

Since q_{ion} is positive, we can drop the modulus sign. So

$$v_x = \left(\frac{2q_{\text{ion}}|\Delta V|}{m}\right)^{1/2}.$$

The momentum acquired by each ion is

$$p_x = (2mq_{\text{ion}}|\Delta V|)^{1/2}.$$

Force on satellite = momentum change of satellite per second

= – momentum change of ions per second

$$= -\,(\text{number of ions fired per second}) \times \begin{pmatrix}\text{momentum change}\\ \text{of each ion}\end{pmatrix}.$$

F_x is negative because the force on the satellite is in the $-x$-direction. Thus,

$$F_x = -10^{19}\,\text{s}^{-1} \times (2mq_{\text{ion}}|\Delta V|)^{1/2}$$

$$= -\,10^{19}\,\text{s}^{-1} \times (2 \times 6.6 \times 10^{-26}\,\text{kg} \times 1.602 \times 10^{-19}\,\text{C} \times 10^3\,\text{V})^{1/2}$$

$$= -\,0.046\,\text{N}.$$

(b) Each ion has charge $+e$ and moves through a potential difference of 1 kV and therefore the change in electrical potential energy is $+e \times 1$ kV (see definition of electronvolt), i.e. 1 keV and so the kinetic energy given to each argon ion is 1 keV.

(c) Power = number of ions fired per second × energy per ion
$= 10^{19}\,\text{s}^{-1} \times 1\,\text{keV} \times 1.6 \times 10^{-19}\,\text{J}\,\text{eV}^{-1} = 1.6\,\text{kW}.$

(This power can be obtained from batteries, which are continuously recharged from solar cells.)

Checking (a) If more ions were fired per second (i.e. more than 10^{19}), the force would be greater, as expected. If the ions were replaced by ones with greater charge or mass, the force would be greater, as also expected.

(b) Check units: [kilovolts] × [units of electronic charge *e*] = [kiloelectronvolts] which are units of energy.

(c) If more ions are fired per second or the energy per ion is increased, the power would be increased, as expected.

5.4 Electrophoresis and DNA fingerprinting

One application of the electrostatic force, that may become even more important in the future, is **electrophoresis**. This is a technique in which a stream of mixed materials in solution is passed between charged plates. The molecules become charged by adjusting the pH of the medium in which they are contained. Thus, they will be subject to a force towards one or other of the plates, proportional to their charge, so the materials will tend to separate because they move at different rates.

Electrophoresis works well for organic materials, notably proteins. A mixture of proteins can be placed in a layer of stiff jelly or starch, between a pair of electrodes. Different proteins have different total charges and hence move at different rates. On a larger scale, this technique is already proving useful for preparing rare and expensive drugs in pure form. Unfortunately, the electrostatic force has such a weak effect in this situation, especially in a medium such as water, that the separation effects may be spoilt by convection, so electrophoresis is often carried out in very narrow tubes to aid the dissipation of heat and avoid the problems caused by convection.

A particularly well-known application of electrophoresis is its use in DNA fingerprinting. The DNA molecule is the basis of human genetic material. Genes consist of long strings of four different chemical bases (abbreviated A, C, G, T) along the length of a double-stranded helix. The order of the bases (e.g. ATGACG … etc.) constitutes a code for the development of proteins in cells. This results in the passing on of inherited characteristics.

The use of electrophoresis to analyse DNA arose from an observation by Professor Alec Jeffreys who noticed that there are certain sequences in DNA (called mini-satellites) that do not contribute to the function of a gene but are repeated within the gene and in other genes of a DNA sample from a particular person. He discovered that each person has a unique pattern of these mini-satellites, the only exceptions being multiple individuals from a single zygote (e.g. identical twins). The strands can be separated by electrophoresis using the following procedure:

1. A sample of DNA-containing cells is obtained. This may consist of skin, blood, hair etc.
2. The DNA is extracted and purified.
3. The pure DNA is treated with specific enzymes whose action is to cut the DNA chain at characteristic places (e.g. between T and G). This produces fragments of varying length depending on where these points occur in the particular genetic material treated.
4. The fragments are put in a gel and subjected to electrophoresis. The shorter fragments move more quickly towards the positive anode and so are separated according to length.

5 The sorted double strands are treated to split them into single strands. These are transferred to a nylon sheet.
6 These strands are now subjected to autoradiography in which they are exposed to synthesized, radioactive DNA, which binds to the mini-satellites.

X-ray film is exposed to the fragments and a dark mark is produced at any point at which a radioactive probe has become attached. It is this developed X-ray film that is analysed. A typical analysed film comparison is shown in Figure 2.34.

If a sample found at the scene of a crime, for example, has a pattern identical to that of a sample taken from a known individual, then, with a *very* high degree of probability, that individual left the sample at the scene of the crime. This is the reason for the popular name *DNA fingerprint*. In addition, since the genes are responsible for inherited characteristics, relatives will have many features in common on a DNA fingerprint. The fingerprint may therefore be used to establish whether individuals are related or not (i.e. have a common ancestor). This test established that remains found buried in a Russian forest were those of the murdered Romanov family by comparing their DNA fingerprints with those of members of the British royal family, who had a common ancestor with the Romanovs in Queen Victoria.

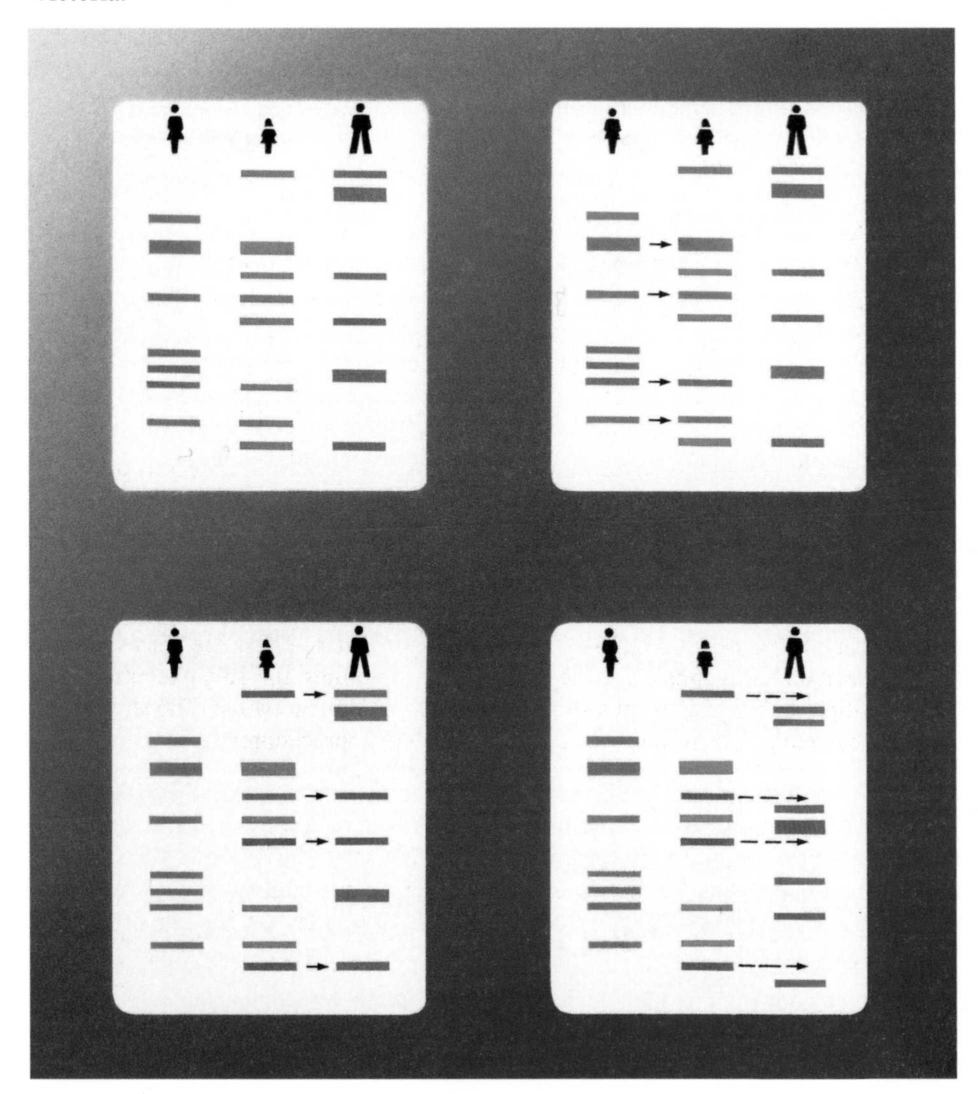

Figure 2.34 Paternity testing by analysis of DNA fingerprints. *Top left*: DNA fingerprints of mother (green) child (red) & putative father (blue). *Top right*: the mother's DNA bands are identified in the child's DNA fingerprint; the remaining bands in the child's fingerprint must come from the father. *Bottom left*: these remaining bands in the child's fingerprint match up with the bands in the DNA fingerprint of the real father; paternity is proved. *Bottom right*: the child's remaining bands do not match up with bands in the alleged father's fingerprint; paternity is disproved.

5.5 Cathode ray and television tubes

The basis of the modern television tube has its origin in the experiments of J. J. Thomson who discovered the electron in 1897. He worked with highly evacuated vessels across which a very large potential difference was maintained between a (negative) cathode and a (positive) anode. He discovered that, if the vessel was very highly evacuated, so that the amount of residual gas was very low indeed, particles were drawn from the cathode and attracted toward the anode. Because of this attraction he was able to deduce that the particles were negative and he also measured their charge to mass ratio. As noted in the previous chapter, a separate measurement by Millikan was needed to determine the charge carried by each electron. The electron beam was observed to cause the emission of light when it struck a specially treated fluorescent screen. It is this property that is the basis for the production of a television image on the screen.

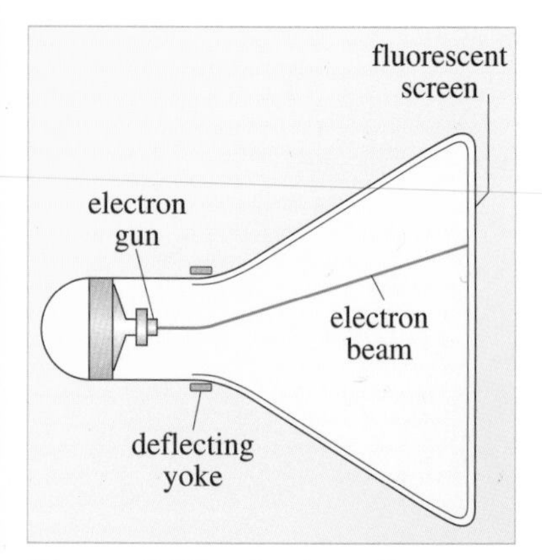

Figure 2.35 A schematic diagram of a television tube.

Figure 2.35 shows a schematic diagram of a television tube. At the left end of the tube is the so-called **electron gun**. This is the source of an electron beam. It contains both the cathode and the anode, but the anode allows the electrons to escape through it into the space beyond. If not further influenced, the electron beam would strike the fluorescent screen at the right-hand end of the tube and a spot of light would be observed at its centre. In order to produce a recognizable image on the screen, the electron beam must:

1. be caused to impact on every point of the (rectangular) screen rather just at its centre, and
2. the density of electrons must vary according to whether the light intensity at each point on the image is bright or dark.

The first objective is achieved by means of the deflecting yoke shown in Figure 2.35. In principle, this consists of two pairs of deflection plates between each of which a varying potential difference may be applied. One pair of plates deflects the electron beam in a horizontal direction (i.e., out of the plane of Figure 2.35) and the other in the vertical direction. A steadily increasing potential difference across the horizontal plates causes the electron beam to deflect, striking the fluorescent screen at a continuously moving point, so that a line of light is observed as shown in Figure 2.36a.

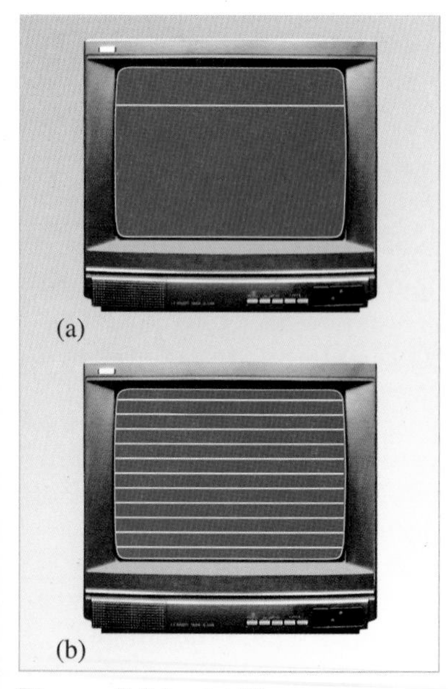

Figure 2.36 (a) The fluorescence produced on the television screen by one horizontal scan of a steady electron beam. (b) An example of a complete set of scans of the screen (in the real case, there are many more).

The scan is very rapid so that the eye retains the image of all the individual electron strikes, leaving the impression of a continuous line (Figure 2.36a). When the beam reaches the right-hand side of the screen, the voltage is *very rapidly* returned to its starting value (during which time the electron gun is switched off) and then the steady increase is resumed. On its own, this would have the effect of producing the same horizontal line over and over again. However, *at the same time*, the voltage applied to the vertical plates is, relatively slowly, but steadily, increased, so that the lines on the screen will resemble those shown in Figure 2.36b. Nearly horizontal lines will eventually cover the entire screen. With a fast enough rate of scan, and sufficient lines, the screen would, with a steady electron beam, look uniformly illuminated. The quality of the illumination is improved by a second scan, with the lines (the B lines) produced between those of the first scan (the A lines). The actual rate of scanning depends on the frequency of the electrical supply available. With a 50 Hz supply, the coverage of the screen is made up of 625 lines at the rate of 25 frames per second.

In order to create a picture, the television signal from the transmitter must be carried to the electron gun via the television antenna (or cable). This causes the density of electrons emitted to vary according to the intensity of the light recorded at each point of the picture.

To create a 'black and white' picture, the phosphor on the screen is chosen to be one that emits white light when struck by the electron beam. For a colour picture, the basic principle is the same but the screen is coated with a regular array of phosphors of three different kinds which emit light of the three primary colours — red, blue and green — when struck by an electron beam. Each element of the array consists of one of each 'colour' phosphor at the corners of a minute equilateral triangle. The electron gun now has three electron-producing elements — one for each colour. By a precise arrangement of holes in a mask (about 200 000 of them!) behind the fluorescent screen, electrons from each beam can strike only the appropriate phosphor. As with the black and white set, the speed of operation is so fast that the eye is deceived into perceiving a continuous realistic picture from what is in fact a myriad of rapidly flickering red-, blue- and green-coloured dots.

6 Closing items

6.1 Chapter summary

1 We first recall from *Predicting motion* that the force on a body is equal to minus the gradient of the potential energy versus position graph. This expression is quite general for conservative fields, and may therefore be applied to both gravitational and electrostatic cases,

$$F_x = -\frac{dE_{pot}}{dx}. \qquad (2.1a)$$

Potential energy changes due to movement in a conservative force field are equal to minus the work done by the conservative force.

2 For a uniform gravitational field, such as the gravitational field near the Earth, the gravitational potential energy E_{grav} is equal to mgh for a mass m at a height h above the surface of the Earth, where g is the magnitude of the acceleration due to gravity. The zero of gravitational potential energy in this case is taken to be at the surface of the Earth.

3 For a 'point' mass, the gravitational force is radial, so the force–potential energy relationship can be written

$$F_r = -\frac{dE_{pot}}{dr}. \qquad (2.3)$$

The gravitational force law, which may be written in the form

$$F_r = -\frac{Gm_1m_2}{r^2},$$

then implies that the equation for the gravitational potential energy of two point masses in free space is

$$E_{grav} = -\frac{Gm_1m_2}{r} \qquad (2.6)$$

where E_{grav} is defined to be zero when $r = \infty$.

4 For a point charge, the electrostatic force is radial and the force–potential energy relationship can again be written

$$F_r = -\frac{dE_{el}}{dr}.$$

The Coulomb force law, which may be written in the form

$$F_r = \frac{q_1q_2}{4\pi\varepsilon_0 r^2}$$

then implies that the equation for the electrostatic potential energy of two point charges in free space is

$$E_{\text{el}} = \frac{q_1 q_2}{4\pi\varepsilon_0 r} \qquad (2.7)$$

where E_{el} is defined to be zero when $r = \infty$.

5 For a charge q in a uniform electric field in, say, the y-direction, the change in potential energy on moving the charge through a displacement Δy is given by $\Delta E_{\text{el}} = -q\mathscr{E}_y \Delta y$.

6 The gravitational potential at a point is defined as the gravitational potential energy per unit test mass placed at that point, i.e.

$$V_{\text{grav}}(\boldsymbol{r}) = \left(\frac{1}{m}\right) E_{\text{grav}} \text{ (with } m \text{ at position defined by } \boldsymbol{r}) \qquad (2.19)$$

and thus the gravitational potential due to a point mass M is given by

$$V_{\text{grav}}(\boldsymbol{r}) = -\frac{GM}{r}. \qquad (2.19)$$

7 The electric potential at a point is defined as the electrostatic potential energy per unit test charge placed at that point, i.e.

$$V(\boldsymbol{r}) = \left(\frac{1}{q}\right) E_{\text{el}} \text{ (with } q \text{ at position defined by } \boldsymbol{r}) \qquad (2.9)$$

and thus the electric potential due to a point charge Q is given by

$$V(\boldsymbol{r}) = \frac{Q}{4\pi\varepsilon_0 r}. \qquad (2.11)$$

8 The electrical potential energy of a charge q at a position $\boldsymbol{r}$ where the electric potential is V, is given by

$$E_{\text{el}} \text{ (with } q \text{ at } \boldsymbol{r}) = qV(\boldsymbol{r}) \qquad (2.10)$$

and the change in electrical potential energy of a charge q when it moves through a potential difference ΔV is given by

$$\Delta E_{\text{el}} = q\Delta V. \qquad (2.13)$$

9 The component of the electric field in the x-direction is related to the electric potential by

$$\mathscr{E}_x = -\frac{\mathrm{d}V(\boldsymbol{r})}{\mathrm{d}x} \qquad (2.17)$$

and similar equations apply for the other components.

10 The following scheme sums up the various equations that relate force, field, potential energy and potential for small Δx for the electrostatic case. Similar equations apply in the case of gravitation

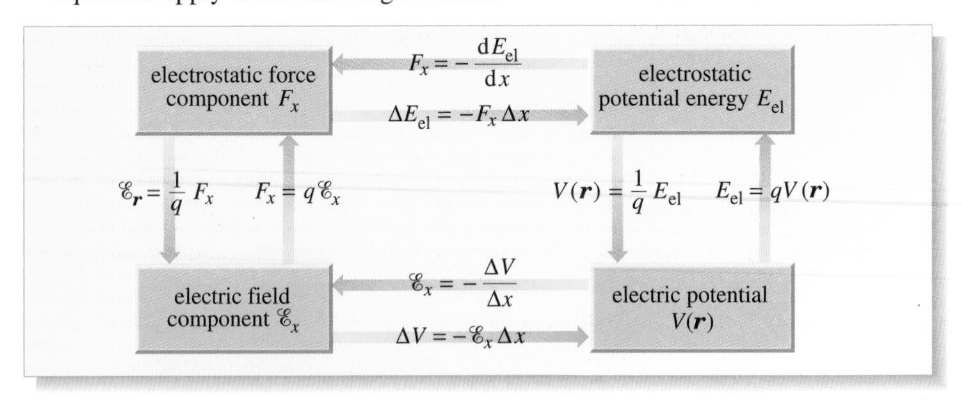

11 A particularly useful way of representing electric or gravitational potential is by plotting equipotentials. Field lines always cut equipotentials at right angles.

12 A perfect conductor is an equipotential. As a result, there can be no electric field inside a perfect conductor, and no electric field can exist inside a perfectly conducting shell that contains no free charges. This is the basis for the technique of electrostatic screening of electrical devices.

13 Capacitors are devices for storing charge and electrostatic potential energy.

14 The capacitance of a parallel plate capacitor is the ratio of the charge on one of the plates to the potential of that plate relative to the other:

$$C = \frac{q}{V} \tag{2.22}$$

where C represents capacitance, q is the charge and V is the electric potential difference. The unit of capacitance is the farad, F. Thus, we have $1\,\text{F} = 1\,\text{C}\,\text{V}^{-1}$.

15 For a parallel plate capacitor with plates of area A separated by a distance d, the capacitance is given by

$$C = \varepsilon_r \varepsilon_0 \frac{A}{d} \tag{2.24}$$

where ε_r is the relative permittivity of the dielectric material between the two plates.

16 The electrostatic potential energy in a capacitor is given by

$$E_{el} = \frac{qV}{2} = \frac{CV^2}{2} = \frac{q^2}{2C}\,. \tag{2.25}$$

17 The diversity of the application of the concepts of potential and potential energy is illustrated in five examples from a wide range of possibilities: (a) the generation of hydroelectric power; (b) the evolution of stars; (c) the principles of operation of the ion thruster; (d) the technique of DNA identification; (e) the principles of the formation of television pictures.

6.2 Achievements

Now that you have completed this chapter, you should be able to:

A1 Understand the meaning of all the newly defined (emboldened) terms introduced in this chapter.

A2 Recall the relationship between force and potential energy at a point in a conservative force field.

A3 Recall the equation for the gravitational potential energy of two point masses and use this equation in simple calculations.

A4 Recall the equation for the electrostatic potential energy of two point charges and use this equation in simple calculations.

A5 Define gravitational potential and relate it to gravitational potential energy and gravitational field, and use these relationships in simple calculations.

A6 Define electric potential, relate it to electrostatic potential energy and electric field, and use these relationships in simple calculations.

A7 Define the volt and electronvolt.

A8 Represent the gravitational or electric potential using equipotentials, and relate these equipotentials quantitatively to the corresponding field.

A9 Draw comparisons between the electrical and gravitational interactions, in particular the form of the potential, and potential energy both for point particles and in the case of uniform fields.

A10 Define the concept of capacitance and use it to solve simple problems relating charge, potential and stored electrostatic energy to one another.

A11 Outline some applications of potential and potential energy.

6.3 End-of-chapter questions

Question 2.21 The gravitational potential energy of a boulder at the top of a 30 m cliff is 3.5×10^5 J relative to the beach below. What is the mass of the boulder? Should a person on the beach below be worried that the boulder might fall?

Question 2.22 An electron has a charge of 1.6×10^{-19} C and a mass of 9.1×10^{-31} kg. If it falls from rest through a potential difference of 1 kV, what will be its final kinetic energy and speed? Give the value of the kinetic energy in units of joules *and* electronvolts.

Question 2.23 In Question 2.2, you calculated that the mean potential energy of the Earth–Moon system is -7.8×10^{28} J. How does this compare to the kinetic energy of the Moon in its orbit? (The mean distance of the Moon from the Earth is 3.8×10^8 m and its mass is 7.4×10^{22} kg. You may assume that the Earth is stationary and the lunar month is exactly 28 days.)

Question 2.24 A capacitor of capacitance 10 μF receives a charge of 2.5×10^{-5} C. How much stored energy does this represent?

Question 2.25 *Outline* the sequence of changes of potential energy and the sources of energy that come into play in the evolution of a star with the same mass as the Sun. What will be the final state of such a star? ■

Chapter 3 Electric currents

1 An electric world

The matter that makes up our Universe is intrinsically electrical in its nature: the electrons and protons within all atoms possess the property of electric charge. Atoms themselves are electrically neutral, but it is not uncommon for atoms either to lose or to gain electrons. Motion, either of a detached electron, or of a charged atom (which is known as an *ion*), constitutes a flow of electric charge, which is known as an *electric current.*

Such motion occurs in a very wide variety of situations (Figure 3.1). For example:

- in and above the atmosphere of the Earth, the motion of charged particles produces such phenomena as lightning and the aurora;
- within our bodies, the flow of ions across the walls of cells allows us to activate our muscles, and even to think;
- in a bewildering variety of technological devices, electric currents are the agent through which we interact with the world. Consider electric lights and motors, televisions or radios, CD players and computers.

Figure 3.1 Electric currents in the everyday world: the current during a lightning flash is very large; the motion of charged particles in the upper atmosphere is responsible for the aurora; complex circuitry inside a computer; pylons carry cables that distribute electricity from power stations to our homes and factories.

Thus, one can fairly say that the flow of electrically charged particles represents an important topic for study.

For most of this chapter we will look at a phenomenological description of electric currents. In the following chapter we will build on the description developed here in order to understand the relationship between electric currents and the generation of magnetic fields.

2 What is an electric current?

An electric current is a 'flow of electric charge'. However, electric charge does not exist by itself: it is a property of the fundamental particles that constitute matter. Thus, the nature of an electric current differs from one circumstance to another depending on *which* charged particles are contributing to the flow, and the situation of these charged particles. For example, in metals, an electric current consists of a flow of negatively charged electrons through a background of positive ions, but in a solution (such as the liquid in a car battery) an electric current consists of a flow of *both* negatively charged ions and positively charged ions, each with different masses and possibly different concentrations.

In most of this chapter we will not need to consider the microscopic nature of the charge flow: we will find that we can understand a good deal of the phenomena associated with current flow by knowing just the net rate at which charge is transported. But first we will try to get a flavour for the very different processes that constitute an electric current in different circumstances. In particular, in this section, we will consider in turn the nature of current flow in a metal, an insulator, a solution, and a plasma.

2.1 Currents in metals

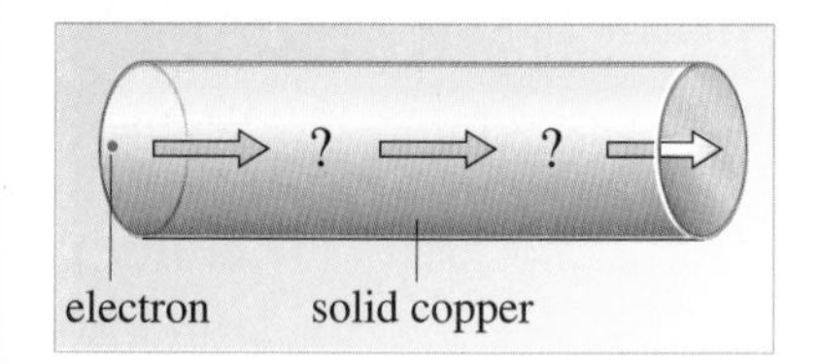

Figure 3.2 How can a material particle, an electron, 'flow' through solid metal?

The idea that electric currents 'flow' through metals is now commonplace. Yet, consider yourself in the position of an electron about to leave the electrical outlet in your home to travel to, say, your kettle. To the electron, the wire ahead looks like a block of solid metal many centimetres thick. Can electrons *really* 'flow' through solid metal? The answer is "Yes", electrons can flow through a metal, but not much else would be able to. And the mere fact that they *can* tells us a great deal about the nature of current flow in a metal.

The situation of electrons within a metal is described more fully in Chapters 1 and 2 of *Quantum physics of matter*, but Figure 3.3 illustrates a simple model that allows us to understand many of the properties of metals. In this model, one or more electrons become detached from each atom, and move randomly around the interior of the metal, leaving a positive ion behind them. The electrons can be considered to

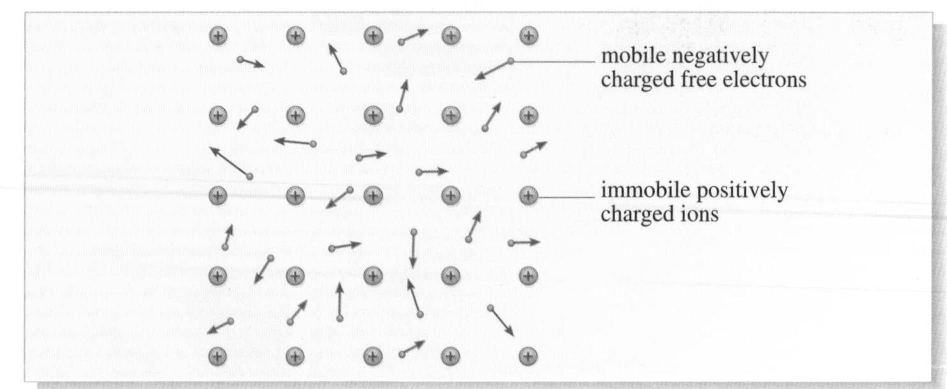

Figure 3.3 A simple model of a metal. Each atom loses one or more electrons, which are then able to move relatively freely through the lattice of positive ions. The arrows indicate the random directions of the motions of individual electrons.

form a kind of 'gas' trapped within the boundaries of the metal. The number density of electrons in this 'gas' is the same as the number density of ions, which is of the order of $10^{28}\,\mathrm{m}^{-3}$, or around one electron in a cubic box of side 0.5 nm. The charge density due to the electron 'gas' is very large, around $10^{9}\,\mathrm{C\,m}^{-3}$, but is compensated by the exactly equal and opposite background charge of positive ions. The ions can vibrate about fixed positions, but are unable to move through the metal. If the electron 'gas' is set in motion at even a small speed, the large charge density means that a very large amount of charge moves down the wire.

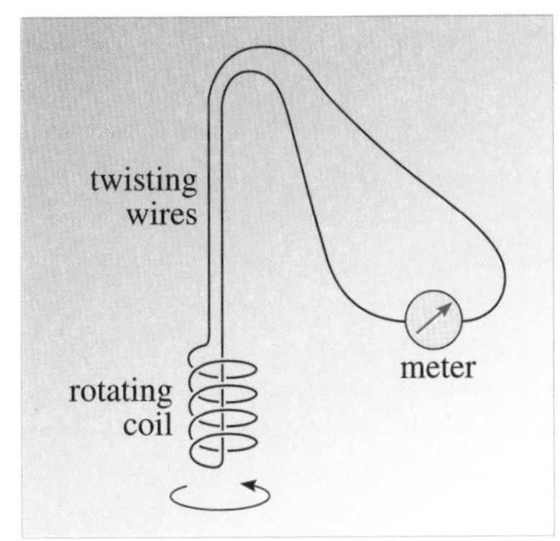

Figure 3.4 The Tolman–Stewart experiment. A rotating metal coil was connected to a meter by long wires that were able to twist up as the coil spun round. The coil was rotated at 4000 revolutions per minute and then brought to rest in a fraction of a second. The conduction electrons carried on moving after the bulk of the metal (the ions) had been stopped. These moving electrons produced a tiny pulse of current through the coil which Tolman and Stewart were just able to measure. They analysed their data to find the charge to mass ratio of the current carriers in various metals and found close agreement with the known charge to mass ratio of the electron in free space.

The question of whether or not electrons can really flow through a metal was answered definitively by an experiment devised by Tolman and Stewart in 1915. This experiment directly confirmed the validity of the idea of an electron gas, able to move through a background of positive ions. Figure 3.4 shows a diagram of the Tolman–Stewart apparatus and describes the experiment in more detail.

Example 3.1 in Section 3.1 will give you some feel for the speeds involved in electron transport through metals. They turn out to be only about $1\,\mathrm{mm\,s}^{-1}$.

2.2 Currents in insulators or semiconductors

In insulators (e.g. glasses or plastics), electrons remain electrically bound to their individual atoms. They are not free to move through the material in response to applied electric fields, and a current cannot readily be passed through such materials. A simple model of an insulator is shown in Figure 3.5a, and Figures 3.5b and c present an analogy to the distinction between what happens to an insulator and to a metal when an electric field is applied. In a semiconductor the situation is basically the same as for an insulator, except that there is a very small number of mobile charge carriers that are able to respond to an applied electric field.

It is worth noting that even in insulators, currents do flow in response to applied electric fields, albeit very much smaller currents than in metals. Values are given in Table 3.1 in Section 3.3, but typically, in a given electric field, the current through an insulator might be a factor 10^{16} smaller than through a piece of metal of similar shape and size. It is also interesting to note that large currents can be made to flow through an insulator if sufficiently large electric fields are applied, typically greater than $10^{7}\,\mathrm{V\,m}^{-1}$. However, the insulator is often destroyed in a process called electrical breakdown.

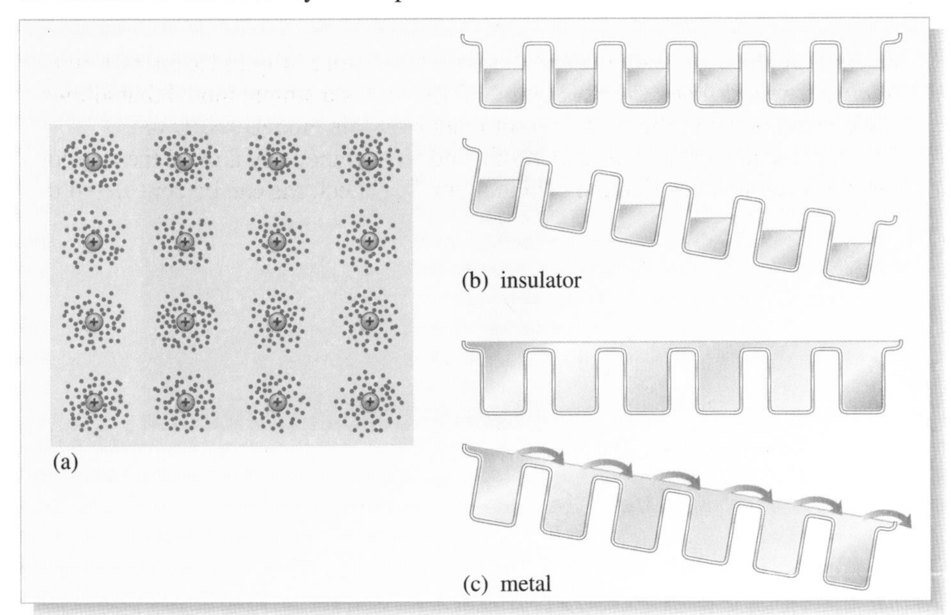

Figure 3.5 (a) A simple model of an insulator. All electrons are bound to, and localized around a particular atom. None is free to move as in a metal. (b), (c) A simple analogy of the difference between the ways an insulator and a metal respond to an external electric field. If we consider water in a container similar to that used for making ice cubes, then, if we tilt the container to try to make water flow from one end to the other, it will do so only if the water level is sufficiently high. Similarly, in an electric field that tries to force electrons from one end of a piece of material to another, the electrons will move only if they are sufficiently weakly bound to the atoms.

2.3 Currents in solutions

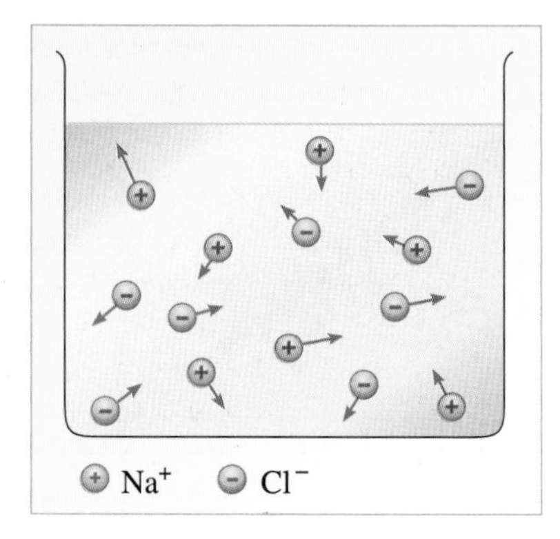

Figure 3.6 When salt, NaCl, is dissolved in water it splits up into sodium (Na^+) ions and chloride (Cl^-) ions. The sodium ion is a sodium atom minus one electron, and the chloride ion is a chlorine atom with an extra electron. These ions can move through the liquid in response to an electric field.

The situation within an ionic solution, such as a solution of salt in water, is illustrated in Figure 3.6. Ions (atoms with either an excess or a deficiency of electrons) are relatively free to move through the liquid, and so such solutions conduct electricity quite well. The electric charge in a solution is carried by ions, of which there are always both positive and negative types. Comparing the situation with that in a metal, there are three important differences:

1. The number density of ions in a solution is generally much less than the concentration of free electrons in a metal;
2. Ions move much less easily through a solvent (e.g. water) than electrons through a metal. This is because they are typically at least 50 000 times more massive than electrons, and because they must physically displace solvent molecules in order to move;
3. *Both* types of ion move in response to an applied electric field, but in opposite directions: positive ions moving in the direction of the electric field and negative ions moving in the opposite direction.

Points 1 and 2 above mean that, compared with a metal, ionic solutions are generally poorer conductors, perhaps by a factor of a few hundred or a few thousand, but they are still far better conductors than insulators.

Ionic conduction also exists in so-called soft solids, such as gels or pastes. More important still, ionic conduction occurs in many of the liquid/solid mixtures from which our bodies are composed. Indeed, it is now in vogue to consider that our very thoughts are nothing more than ionic currents flowing between living cells within our brains.

2.4 Currents in plasmas

Most of you will be reading this on Earth, where most of the matter that we encounter can be fairly classified as solid, liquid or gas. However, in the Universe as a whole, solids, liquids and gases are not at all common: around 99.9% of visible matter in the Universe is in the **plasma** state, which is essentially an ionized gas. Stars are mainly composed of plasma, as are many of the glowing clouds (Figure 3.7b) sometimes seen between them. In this state, atoms are broken apart, and electrons and positive ions move separately, though the mixture of ions and electrons remains electrically neutral overall (Figure 3.7a).

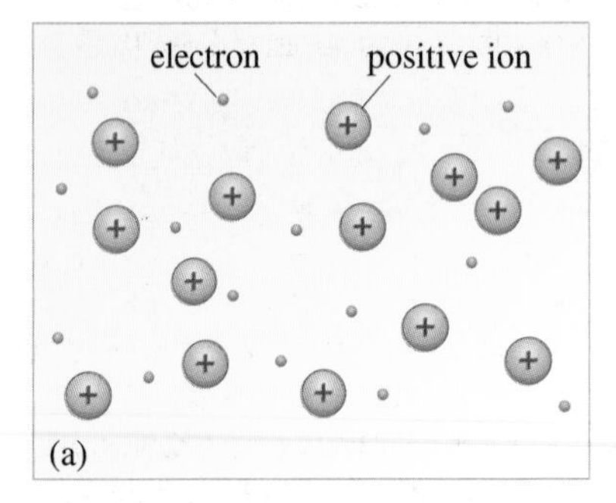

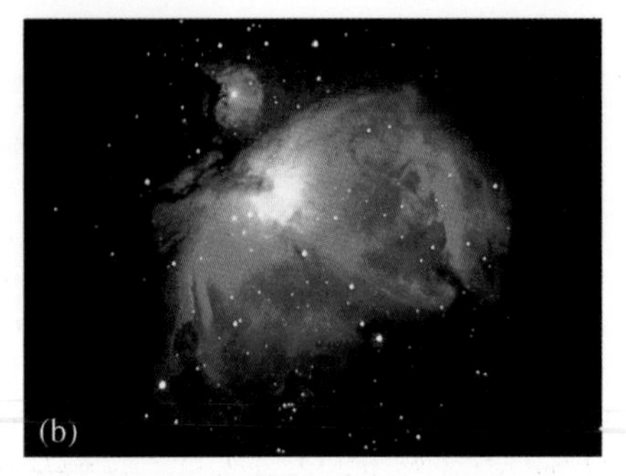

Figure 3.7 (a) Schematic illustration of a plasma and (b) the Orion nebula.

Comparing a plasma with an ionic solution, we see that both states contain positive and negative conducting particles, but there are two key differences:

- The negatively charged particle in a plasma is an electron, which has a much lower mass than its positively charged ionic partner;

- The charged particles in a plasma do not have to displace any solvent molecules in order to move.

Thus, we would expect that a plasma with a similar number density of charge carriers to an ionic solution would be a rather better conductor. We will consider the properties of charged particles moving through a plasma in a magnetic field in Section 5 of the next chapter.

In summary, we have briefly described several, quite different, physical systems in which charged particles are able to move in response to an applied electric field. However, in what follows we will only rarely need to concern ourselves with these microscopic details. We will find that, for many purposes, it is sufficient merely to characterize the amount of electric charge that is transported. This lends a generality but also a certain abstraction to what follows.

3 Currents in simple circuits

3.1 Basic definitions

The **electric current**, i, flowing along a wire at any instant is the rate at which electric charge passes through a plane perpendicular to the axis of the wire. If an amount of charge Δq, moving from left to right, crosses the shaded area in Figure 3.8 in a time Δt, the average current is $\Delta q/\Delta t$. As Δt gets smaller and smaller, we reach the limiting case of an instantaneous current, which we can write as

$$i = \frac{\mathrm{d}q}{\mathrm{d}t}. \tag{3.1}$$

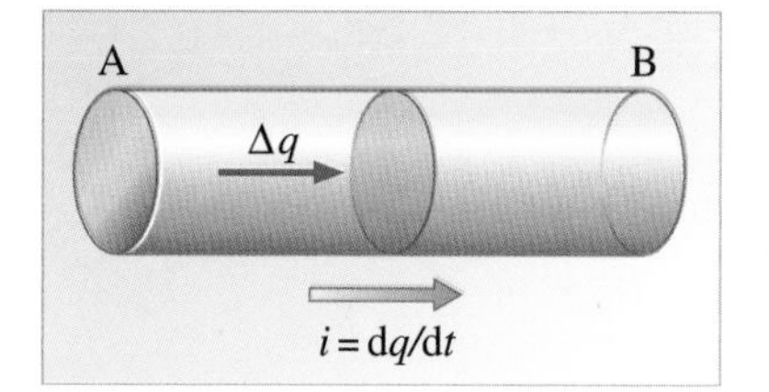

Figure 3.8 The instantaneous current at a point along a wire is defined as the net rate at which charge passes through an area perpendicular to the axis of the wire at that point.

Note that there is a sign convention incorporated into the expression $\Delta q/\Delta t$ (and hence into Equation 3.1) in that we must first set up a positive sense for the direction of flow. The current can then be positive or negative depending on the sign of Δq and the direction of flow. A *positive* value for i, in the direction shown in Figure 3.8, means that during the interval Δt either:

- a certain amount of *positive* charge moves to the *right* through the shaded area in Figure 3.8

or, equivalently,

- a certain amount of *negative* charge moves to the *left* through the shaded area in Figure 3.8.

Corresponding statements apply for negative values of i, that is, either positive charge moves to the left or negative charge moves to the right.

In solid wires, the only particles with positive charges are protons in the nuclei of the atoms of the wire. These atoms are only able to vibrate about their fixed equilibrium positions. So in any solid material, an electric current can be carried only by electrons, which have a *negative* charge. Consequently, when we say that a positive current is flowing in a certain direction in a wire, we are actually describing a flow of electrons in *the opposite direction*. However, when indicating the direction of the current flow on a circuit diagram, all that is necessary is to show the direction in which (hypothetical) positive charge *would* flow. This direction is known as the direction of **conventional current** flow. Only occasionally will we need to bear in mind that this current is actually produced by electrons moving in the opposite direction.

The unit of current, the **ampere**, takes its name from André-Marie Ampère, a French scientist of the early nineteenth century. In the SI system of units, the ampere (often abbreviated to amp, but more properly represented by the symbol A) is a fundamental unit like the kilogram and the metre. In Section 5.4 of the next chapter, you will see how the amp is defined in terms of an effect discovered by Ampère. For the moment, it is quite correct to say that

1 ampere represents a charge flow of 1 coulomb per second, so $1\,A = 1\,C\,s^{-1}$.

The orders of magnitude of the currents involved in some common phenomena and devices are shown in Figure 3.9.

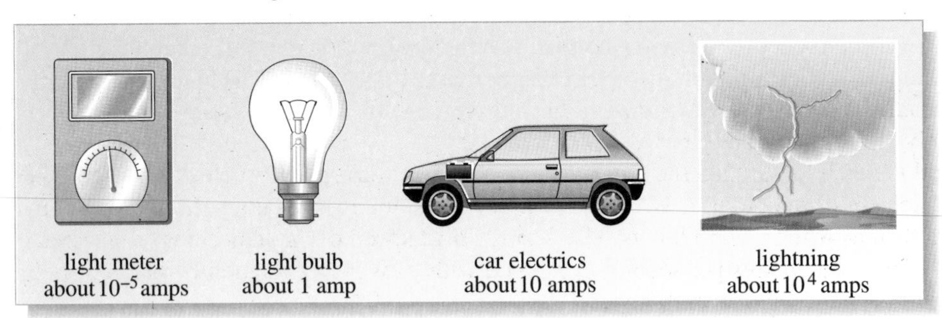

Figure 3.9 The orders of magnitude of currents involved in a variety of processes.

Example 3.1

If the number density of electrons in a wire is $n = 10^{28}\,m^{-3}$, at what speed must they travel through a wire of diameter $d = 1$ mm to deliver a current of 1 A?

Solution

Consider the flow of electrons past a particular cross-section of a wire with uniform area A (Figure 3.10). If the electrons flow from right to left with average speed v, then in a time Δt all the electrons in a volume $Av\,\Delta t$ will flow past the cross-section.

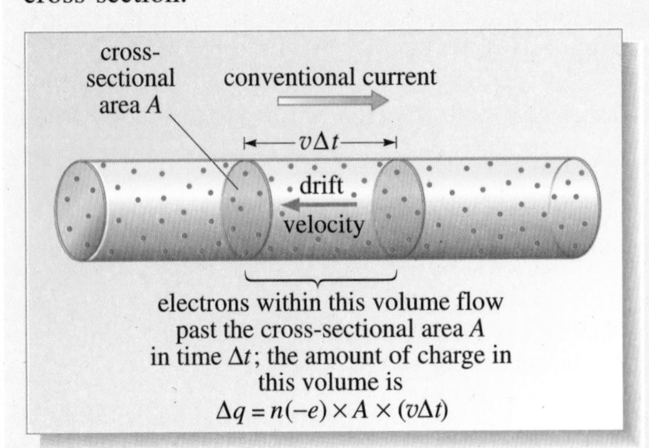

Figure 3.10 For use in Example 3.1.

The number of electrons in this volume will be $nAv\,\Delta t$ and the charge Δq carried past the cross-section by these electrons is $n(-e)Av\,\Delta t$. Because we have been asked only for a speed, not a velocity, we need only be concerned with magnitudes, thus:

$$|\Delta q| = neAv\,\Delta t.$$

This expression can be rearranged to yield $|\Delta q|/\Delta t$, the amount of charge per second carried by electrons past a cross-section in a wire, that is, the magnitude of the current, i:

$$i = \frac{|\Delta q|}{\Delta t} = neAv. \qquad (3.2)$$

Substituting $\pi d^2/4$ for the cross-sectional area A, we find

$$i = nev\left(\frac{\pi d^2}{4}\right).$$

Solving for v, we find

$$v = \frac{4i}{ne\pi d^2} = \frac{4 \times 1\,\text{A}}{10^{28}\,\text{m}^{-3} \times 1.6 \times 10^{-19}\,\text{C} \times \pi \times 10^{-6}\,\text{m}^2}$$

$$v = 7.96 \times 10^{-4}\,\text{m s}^{-1}$$

$$v \approx 1\,\text{mm s}^{-1}.$$

In other words, the electrons flow down the wire with an average speed of around only 1 millimetre per second. This speed is known as the **drift speed** of the electrons and is much lower than the typical (randomly directed) speeds of electrons in metals, which are of the order of $10^6\,\text{m s}^{-1}$. One can consider as an analogy the speeds of air molecules when a wind blows. The average drift speed of the molecules (wind speed) is only a few metres per second, compared with their typical randomly directed speed of a few hundred metres per second.

3.2 Why currents flow

In an isolated wire there is no net charge flow. (It is interesting to note that 'isolated', in this context, means beyond *all* electrical and magnetic influences, and not just out of contact with other matter. This is because, as you will see in Chapter 1 of *Dynamic fields and waves*, it is possible to induce a current to flow in an otherwise isolated wire by changing the magnetic field around the wire.)

We can understand why there is no charge flow in an isolated wire. The ions vibrate about fixed positions, and the 'free' electrons move as commonly in one direction as in another. Yet we know that it is possible to arrange for there to be a net movement of electrons towards one end of the wire. How is this done?

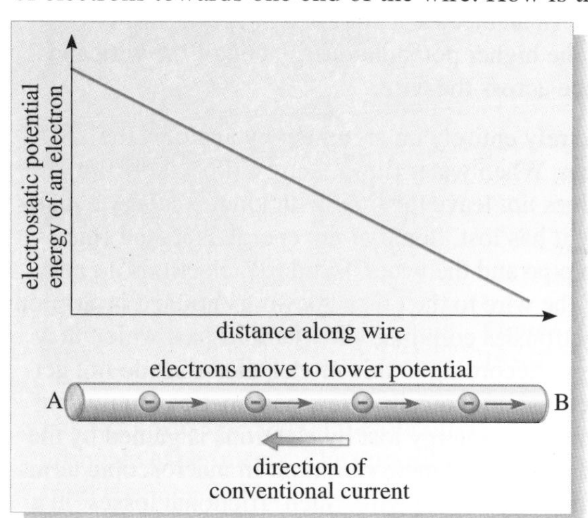

Figure 3.11 Electrons will move along a wire from A to B if there is a consequent reduction in electrostatic potential energy.

The simplest answer follows from energy arguments. Electrons will flow from A to B in a wire if, by so doing, their potential energy is reduced. This is analogous to water flowing down a pipe. However, because electrons have such a small mass, the amount of *gravitational* energy involved is too small for gravitational forces to cause electrons to settle at the lowest point of a wire. It is the *electrostatic* potential energy that is important here (Figure 3.11). If some means is found of arranging for electrons at A to have a higher electrostatic potential energy than at B, electrons will tend to flow from A to B.

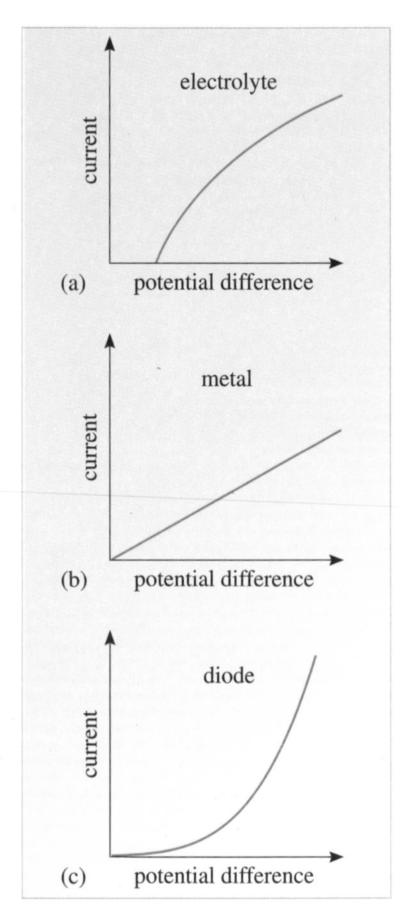

Figure 3.14 Ohm's law is not a fundamental law of nature. Current versus potential graphs are shown for: (a) an **electrolyte**, such as salt water, through which a current is passed through two dissimilar electrodes, (b) a metal, and (c) a diode, a device that passes current much more easily in one direction than the other. Only for the metal does Ohm's law hold, i.e. the current is directly proportional to the potential difference.

Ohm's law is *not* a fundamental law of physics. It is merely an empirically tested and useful relationship which is obeyed by most metals in normal circumstances and at constant temperature. It is not obeyed by all materials or by all components in electrical circuits (Figure 3.14).

Question 3.1 A torch bulb is connected to a battery that maintains a potential difference of 1.5 V across the bulb's filament. If the current through the bulb is 0.3 A, what is the resistance of the filament? What is the magnitude of the current that will be drawn from the battery if a bulb of half the resistance replaces the first one? (Assume that the only resistance in the circuit is that provided by the bulb filament.) ■

Resistivity

Short fat pipes carry water more easily than long, thin pipes. Similarly, short fat wires carry current more easily than long, thin wires. In general, therefore, the resistance of a wire at a given temperature is determined by its geometry and by the material from which it is made. This dependence can be expressed by the relationship

$$R = \frac{\rho L}{A} \tag{3.4}$$

where L is the length of the wire, A is its cross-sectional area and ρ is called the **resistivity**. Resistivity is numerically equal to the resistance when $L = 1$ m and $A = 1\ \text{m}^2$. Thus:

> The resistivity of a conducting material is numerically equal to the resistance of a slab of that material of length 1 m and cross-sectional area 1 m^2.

Because area is measured in m^2 and length is measured in m, the unit of ρ is Ω m.

The resistivity of a material is characteristic of that material. It is constant as long as the temperature and the strain in the material remain constant. An increase in temperature increases the amplitude of ionic vibration, and in a metal this results in increased 'frictional losses' and an increased resistivity. Similarly, straining (i.e. stretching) a conductor may introduce microscopic defects that also affect the resistivity.

When we know the value of ρ, for any material, we can use Equation 3.4 to calculate the resistance of a wire, of any length and cross-sectional area, that is made of that material. Table 3.1 lists the resistivities of some substances at around room temperature.

Table 3.1 The resistivity of some substances at around room temperature unless stated otherwise. Superconductors have a resistivity $<10^{-25}\,\Omega\,m$, (experimentally indistinguishable from zero) but only at much lower temperatures.

Substance	Resistivity/Ω m	Notes
silver	1.6×10^{-8}	the best conductor
copper	1.7×10^{-8}	
gold	2.4×10^{-8}	
aluminium	2.7×10^{-8}	
brass	6.3×10^{-8}	an alloy of copper and zinc
platinum	9.8×10^{-8}	at 0 °C
platinum	4.8×10^{-7}	at 1200 °C
tungsten	4.9×10^{-8}	at 0 °C
tungsten	3.9×10^{-7}	at 1200 °C
lead	2.1×10^{-7}	
carbon (graphite)	3.5×10^{-5}	
salt solution	9.6×10^{-3}	solution containing 5.85 g litre^{-1} NaCl'
germanium	4.7×10^{-1}	semiconductor
'pure' water	10^{2}–10^{5}	depends critically on impurities
silicon	2.3×10^{3}	semiconductor
glass	10^{9}–10^{12}	
diamond	10^{10}–10^{11}	

Question 3.2 A connection between two circuit elements in an integrated circuit computer chip is made using a metal connector 10 μm long, 1 μm wide and 0.1 μm thick. What would be the resistance of such a connector if it were made of (a) copper or (b) aluminium? (1 μm = 10^{-6} m). ■

3.4 Kirchhoff's laws

When your computer stops working, one glance inside the cover will probably convince you that expert help is called for. Many electronic systems are dauntingly complex, and specialized knowledge and equipment are needed to design or test them. However, quite a number of circuits, such as the wiring in a house or in older cars, contain only a few circuit elements and can be analysed on the basis of fairly rudimentary knowledge. In this subsection we look at one of these important circuit elements, the **resistor** — a circuit element that has been designed to obey Ohm's law — and examine the two ways in which resistors may be combined in a circuit.

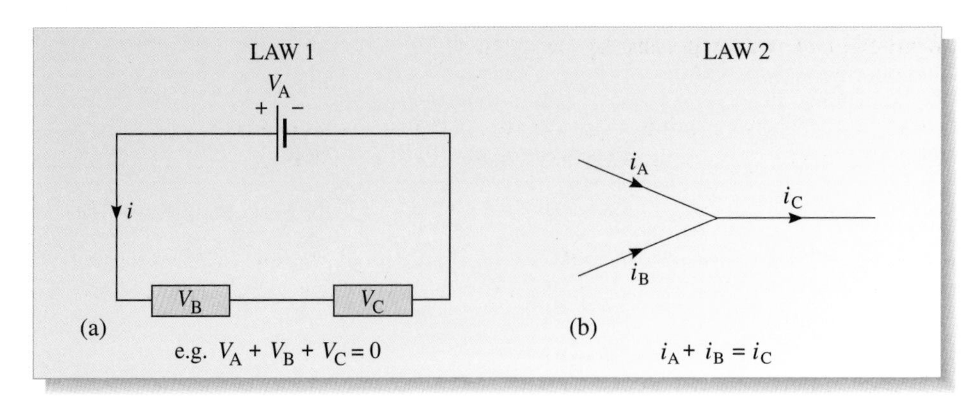

Figure 3.15 Kirchhoff's laws. Note the symbol used for a battery in a circuit diagram: the longer vertical line represents the positive terminal and the shorter one the negative terminal.

All circuits, irrespective of their complexity, adhere to the two principles summarized in Figure 3.15. These principles — which are essentially those of conservation of energy and of charge — are known as **Kirchhoff's laws**.

Kirchhoff's first law

The sum of the potential changes in a given sense around a circuit is zero.

If Kirchhoff's first law were not true, it would be possible to start from some point in a circuit where the electrical potential was known, and, having gone round the circuit adding up the potential differences, come back to the same point and find it had a different potential. Clearly, this is impossible.

Kirchhoff's second law

At a junction of two or more wires, the current flowing into the junction is equal to the current leaving the junction.

If Kirchhoff's second law were not true, then electric charge would have to be created or destroyed at the junction which, again, cannot happen.

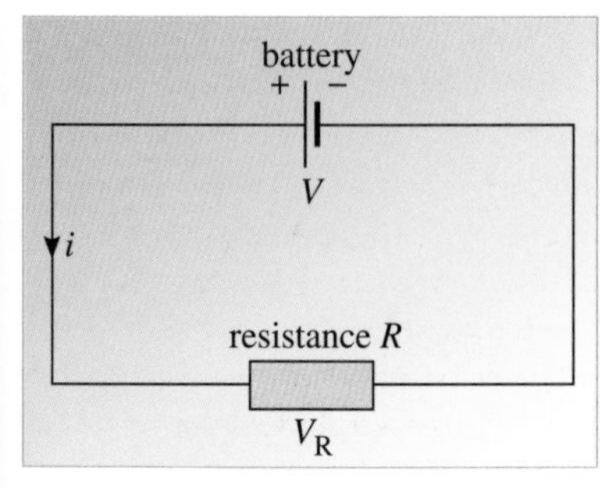

Figure 3.16 A simple circuit consisting of only a battery and a resistor with resistance R. The current i is maintained by the potential difference V across the terminals of the battery. A potential difference V_R develops across the ends of the resistor.

The simplest resistor circuit consists of a source of electrical energy, such as a battery, and a single resistor (Figure 3.16). By convention, these are represented by the symbols shown in the figure. The longer line in the battery symbol represents the positive terminal, i.e. the one at which the electrical potential is higher. In analysing the battery–resistor circuit, we shall assume that the battery has no resistance and maintains a potential difference V between its terminals, irrespective of the current it supplies. This is often not true in practice, owing to an effect known as *internal resistance*. In Section 4.2, we will discuss the validity of this assumption in more detail. For the present, we shall also assume that the connecting wires have zero resistance. Neither of these assumptions will ever be completely true, but usually the error involved will be small.

For the circuit in Figure 3.16, the conventional current flow is anticlockwise. In order to apply Kirchhoff's laws, consider a 'sense' round the circuit. Generally, it is best to take the 'sense' as the direction of current flow, if this is known for certain. Then, the potential difference across a component is taken as negative if the potential drops in the direction chosen, or positive if it rises. If the current direction is not known, the direction is assigned arbitrarily. Then if, when evaluated, a current turns out to be negative, we know that our assumption about the direction of current flow was incorrect. We will now apply Kirchhoff's laws to the circuit in Figure 3.16. Applying the first law we obtain:

$$V + V_R = 0.$$

But from Ohm's law as expressed by Equation 3.3,

$$V_R = -iR$$

where we have included a minus sign to indicate that the voltage *drops* in the direction of the current flow through the resistor. Thus $V - iR = 0$, or

$$V = iR \tag{3.3a}$$

which is an alternative and more commonly used statement of Ohm's law that allows the current i in the circuit to be calculated if the battery voltage V and the resistance R are known.

Resistors in series

In analysing more complicated circuits, it is often useful to be able to construct a simple **equivalent circuit** such as that of Figure 3.16. As an example of this approach, consider a circuit in which there are two resistors rather than one. The two resistors may be placed in a line so that the same current goes through each (Figure 3.17a). In this arrangement, the resistors are said to be **in series**. From Kirchhoff's first law, the sum of the potential differences around the circuit going in an anticlockwise direction must be zero, i.e.

$$V + V_1 + V_2 = 0.$$

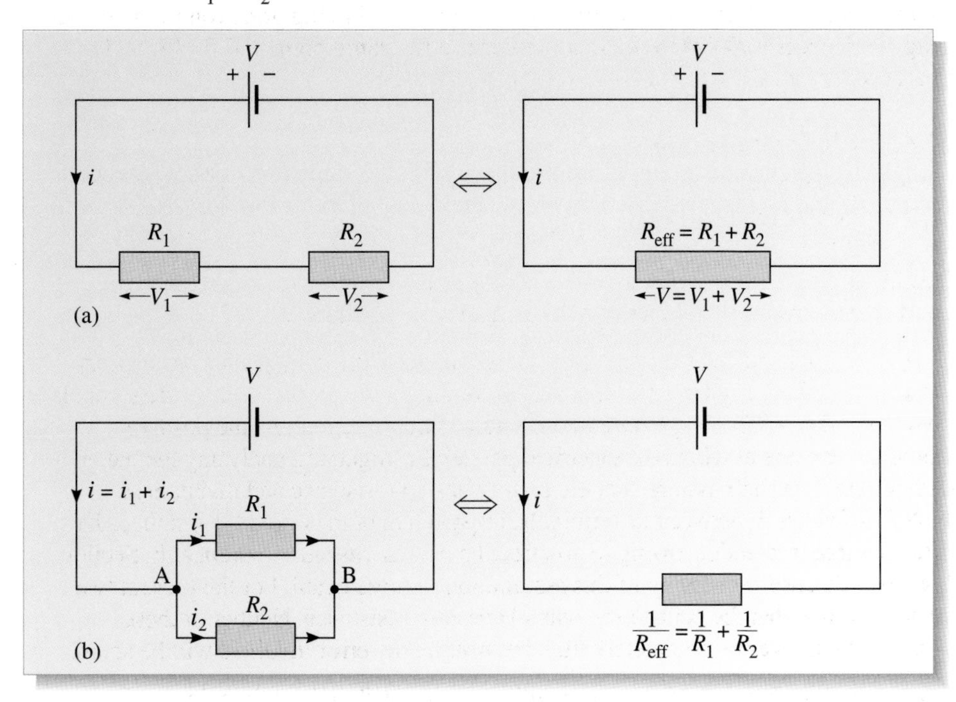

Figure 3.17 (a) Series resistors: the battery would supply the same current if the two resistors with resistances R_1 and R_2 were replaced by a single resistor with effective resistance $R_{\text{eff}} = R_1 + R_2$. (b) Two parallel resistors with resistances R_1 and R_2 are equivalent to a single resistor with an effective resistance R_{eff} that is given by $1/R_{\text{eff}} = 1/R_1 + 1/R_2$.

Using Ohm's law for the two resistors, the potential change across R_1 is $-iR_1$ and the potential change across R_2 is $-iR_2$, and so this equation can be rewritten as:

$$V - iR_1 - iR_2 = 0$$

or

$$V = iR_1 + iR_2$$

$$= i(R_1 + R_2).$$

A comparison with the equation for a single resistor, $V = iR$, shows that the effective resistance, R_{eff}, of the pair of resistors in series is equal to the sum of their individual resistances:

$$R_{\text{eff}} = R_1 + R_2. \tag{3.5}$$

With this equation, the current through the circuit may be calculated by using the effective resistance in the equation $V = iR_{eff}$.

Resistors in parallel

Instead of being in series, the resistors may be placed side by side (Figure 3.17b). When their ends are connected to the same points in the circuit like this, the resistors are said to be **in parallel**. Again, we can calculate the value of resistance that a single resistor would have to have in order to have the same effect as these two resistors. We shall do this in the following example.

Example 3.2

Using Kirchhoff's second law and Ohm's law, find an expression for the effective resistance R_{eff} in terms of the individual resistances R_1 and R_2 in Figure 3.17b.

Solution

Applying Kirchhoff's second law at either point A or point B in Figure 3.17b:

$$i = i_1 + i_2. \tag{3.6}$$

When the battery potential difference V is placed across the effective resistance R_{eff}, the current is given by

$$i = \frac{V}{R_{eff}}.$$

Similarly, the currents i_1 and i_2 are produced with the same potential difference V across R_1 and R_2 respectively. Therefore, again using Ohm's law,

$$i_1 = \frac{V}{R_1} \quad \text{and} \quad i_2 = \frac{V}{R_2}.$$

With these expressions for the currents, Equation 3.6 can be rewritten as

$$\frac{V}{R_{eff}} = \frac{V}{R_1} + \frac{V}{R_2}.$$

Dividing by V gives

$$\frac{1}{R_{eff}} = \frac{1}{R_1} + \frac{1}{R_2}. \tag{3.7}$$

This may be written in the form

$$R_{eff} = \frac{R_1 R_2}{R_1 + R_2}$$

but Equation 3.7 is usually easier to remember, to use, and to generalize for more than two resistors in parallel.

Note that, according to Equation 3.7, R_{eff} for parallel resistors is always *less* than the smaller of the two resistances R_1 and R_2. For example if R_1 is 10 Ω and R_2 is 100 Ω then R_{eff} = 9.09 Ω. This may seem counterintuitive, since adding an extra resistance to a circuit might seem to imply an increase in the resistance. However, in this case, the name 'resistor' is somewhat misleading. Adding a resistor in parallel provides an additional path for the current to follow. If the resistor is one of high resistance then not much current will flow along that path and the overall resistance will be reduced by only a relatively small amount, but it will still be reduced. Returning to the flowing water analogy, this is like adding an extra pipe through which the water can flow out of the tank. Even if the extra pipe is long and narrow, the water will empty out a bit more quickly than with only one pipe.

These derivations may be generalized for any number of resistors in series or parallel to give the following expressions:

series resistances

$$R_{\text{eff}} = R_1 + R_2 + R_3 + \ldots = \sum_i R_i \tag{3.5a}$$

parallel resistances

$$\frac{1}{R_{\text{eff}}} = \frac{1}{R_1} + \frac{1}{R_2} + \frac{1}{R_3} + \ldots = \sum_i \frac{1}{R_i} \tag{3.7a}$$

Question 3.3 If you were given three 10 Ω resistors, what effective resistances could you build into a circuit?

Question 3.4 Figure 3.18 shows a collection of identical resistors, each with resistance R. What is the effective resistance between points A and B? ■

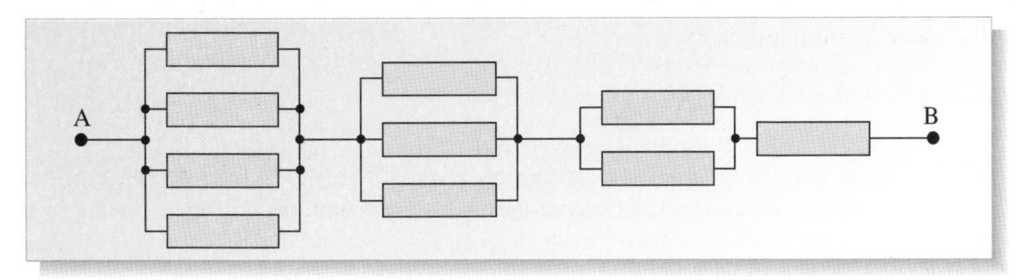

Figure 3.18 For use with Question 3.4.

Kirchhoff's laws can be used in more complicated circuits where all the directions of the currents are not necessarily known, as the following worked example demonstrates.

Example 3.3

Figure 3.19 shows a circuit containing four resistors with resistances R_1, R_2, R_3 and R_4, a voltmeter and a battery. This network is particularly important and is known as a Wheatstone bridge circuit. By applying Kirchhoff's first law to the closed paths ABDA and BCDB, show that, in general, if the voltmeter reads zero, then $R_1/R_3 = R_2/R_4$. (Assume that all connecting wires have negligible resistance.)

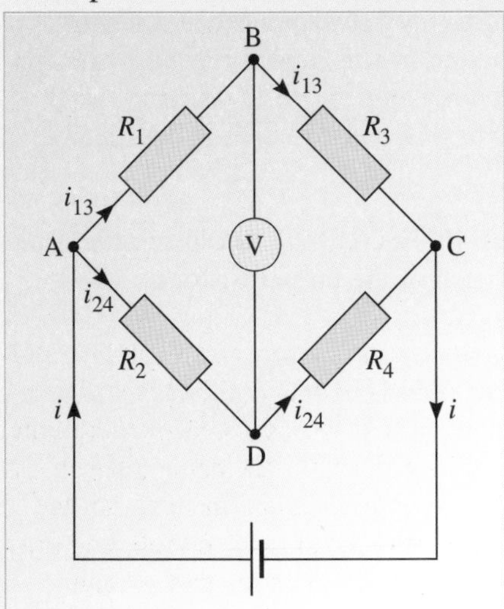

Figure 3.19 A Wheatstone bridge circuit.

Solution

The assumed current directions are shown on the figure. In numerical problems, the choice of directions of currents and potential drops is arbitrary. If we choose the wrong direction, the current will come out negative, which means that a current (or potential drop) is in the opposite direction to the one we chose. Note that when a circuit loop is specified, such as BCDB, the sense round it is determined by the ordering of the letters.

The total current flowing through the battery is labelled i, and i_{13} flows along ABC and i_{24} along ADC. No current flows along BD since there is zero potential difference between B and D (the voltmeter reads zero).

Applying Kirchhoff's first law to the loop ABDA

$$-i_{13}R_1 + i_{24}R_2 = 0$$

so $$i_{13}R_1 = i_{24}R_2. \qquad \text{(i)}$$

Around the loop BCDB

$$-i_{13}R_3 + i_{24}R_4 = 0$$

so $$i_{13}R_3 = i_{24}R_4. \qquad \text{(ii)}$$

The negative signs are required because, with the currents in the directions chosen and the chosen senses round the loops, there would be a potential drop across R_1 and across R_3.

Since, in both of the equations (i) and (ii), the left-hand side is equal to the right-hand side (by definition), the ratio of the left-hand side of (i) to the left-hand side of (ii) must be equal to the ratio of the right-hand side of (i) to the right-hand side of (ii).

Thus $$\frac{R_1}{R_3} = \frac{R_2}{R_4}$$

as required.

Other than checking back over the algebra and the substitutions, we do not need to make any other checks in this problem because we have derived the required result.

Comment: *The Wheatstone bridge circuit is important because it allows the accurate determination of an unknown resistance (say R_4) with only one standard resistor (R_2). The absolute values of R_1 and R_3 need not be known, only their ratio. If the connection at B is, for example, a sliding contact on a length of resistance wire, then this ratio is simply the ratio of the lengths of wire on either side of the contact.*

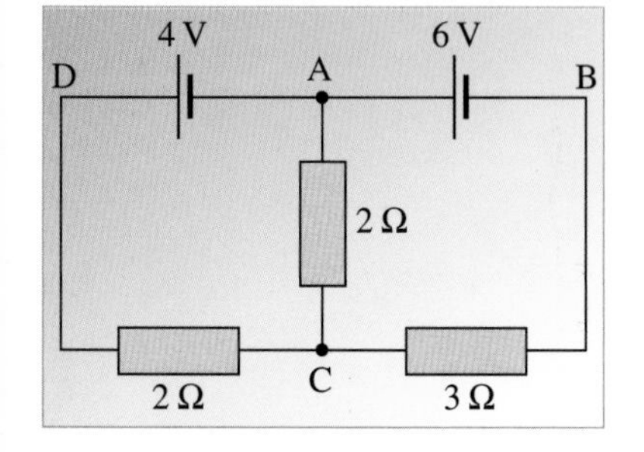

Figure 3.20 For use with Question 3.5.

Question 3.5 For the circuit network shown in Figure 3.20, find the value, and the direction, of the currents flowing through each of the three resistors. ■

3.5 Circuits with a capacitor

Study note: *This might be a good time to revise the discussion of capacitors from Section 4 of the previous chapter.*

Currents from capacitors

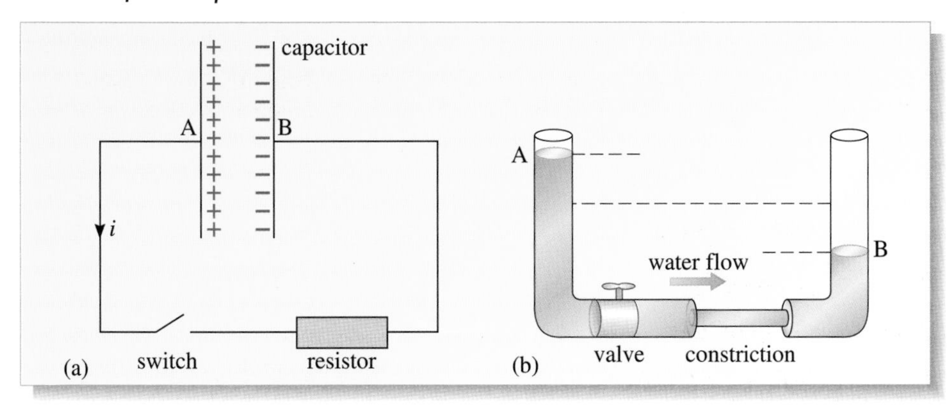

Figure 3.21 (a) When the two plates of a capacitor are charged, there is a potential difference between them. When the plates are connected together, this potential difference will cause a current to flow as shown until they are at the same potential. (b) Two containers of water connected by a tube, a valve and a constriction. When the valve is opened, the difference in gravitational potential between A and B causes the liquid to move in the direction of the arrow. The water will flow until the gravitational potential in both arms is the same. The length of time this process takes depends on the amount of water in the reservoir and the dimensions of the constriction.

As we discussed in Section 4 of Chapter 2, charge can be stored on the plates of a parallel plate capacitor. When the plates are connected together by closing a switch in the connecting wire (Figure 3.21a), electrons will flow from the negatively charged plate, through the wire and any resistances present, to the positively charged plate. Following convention, the current is said to flow from the positive to the negative plate. To understand the reason for this flow, it is useful to return to our analogy of water in a pipe. Consider two containers connected by a tube, a constriction, and a valve (Figure 3.21b). The containers are filled with water such that the water level in the left-hand container is higher than that in the right-hand container. The water is prevented from flowing by the closed valve. The difference in the gravitational potential between the water at A and the water at B corresponds to the difference of electric potential between the capacitor plates, and the valve is equivalent to the open switch. When the valve is opened, a 'current' of water flows from the left- to the right-hand side until the level is equal in both arms, i.e. until levels A and B are at the same gravitational potential. Similarly, an electric current (conventional) will flow from A to B when the capacitor plates are connected by a wire until the electric potential at plate A is equal to that at plate B.

Qualitatively, it is fairly easy to see what happens when the switch is closed. Each plate of the charged capacitor acts as a 'reservoir of charge'. The current is driven by the potential difference between the plates of the capacitor, which is proportional to the stored plate charge, since $q = CV$. As the current flows, the 'reservoir' empties and the potential difference falls. But it is the potential difference that is driving the current and so, as the originally separated charges redistribute themselves evenly between the two plates, the current also drops. The result is shown in Figure 3.22.

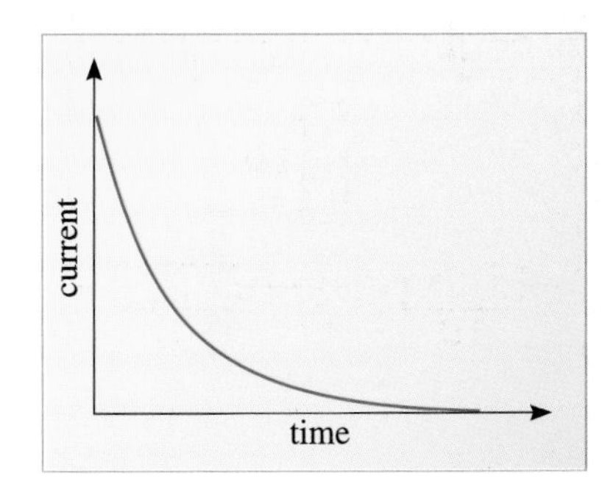

Figure 3.22 The decay of the current obtained by discharging the capacitor.

What factors determine how long the process of discharging the capacitor will take? Using our water analogy, we would expect the process to take a long time if (a) the reservoir is large and (b) the constriction through which the water flows is narrow. Electrically, we would therefore expect the process to take a long time if both the capacitor and the resistor were large. (Remember that the greater the resistance of a resistor, the smaller the current that flows through it, for a given potential difference across its terminals.) These expectations are indeed borne out by the following quantitative treatment of the process of discharging a capacitor.

We know that the charge stored on each plate of the capacitor is related to the potential difference, V, between the plates by the usual equation:

This equation comes from rearranging Equation 2.22: $C = q/V$.

$$V_{\text{capacitor}} = \frac{q}{C}. \tag{3.8}$$

By Kirchhoff's first law, the potential differences around the circuit in Figure 3.21a must add to zero, i.e.

$$V_{\text{capacitor}} + V_{\text{resistor}} = 0.$$

Substituting for the potential differences in the last equation using Ohm's law and Equation 3.8 gives

$$\frac{q}{C} - iR = 0$$

which can be rearranged to give

$$i = \frac{q}{RC}.$$

Now, the current i through the resistor consists of charge that has left the capacitor, and so i is determined by the rate of change of the plate charge. Since a positive value for the current, in the direction shown, means that the charge on the positive plate of the capacitor is *decreasing*, it follows that

$$i = -\frac{\mathrm{d}q}{\mathrm{d}t}.$$

Combining the previous two equations for i we get

$$\frac{\mathrm{d}q}{\mathrm{d}t} = \left(-\frac{1}{RC}\right) \times q. \tag{3.9}$$

So, the rate of change of the plate's charge is proportional to the amount of charge left stored on it.

Processes in which the rate of change of some quantity is proportional to the quantity itself are relatively common, not only in physics but also in other disciplines, and are called **exponential processes**. In Chapter 2 of *Predicting motion* we discussed exponential decay caused by damping of simple harmonic oscillations. In Chapter 1 of *Quantum physics of matter* we will discuss radioactive decay, in which the number of nuclei disintegrating each second is proportional to the number of radioactive nuclei left in the sample. Because of the general importance of such processes, it is worth spending a little more time discussing the example of the exponential decay of the charge on the plates of a capacitor.

Exponential processes

Equations such as Equation 3.9 have the general form

$$\frac{\mathrm{d}y}{\mathrm{d}t} = By \tag{3.10}$$

where B is a constant. The solution to such an equation is a function and it generally includes one or more arbitrary constants that do not appear in the equation itself. In the case of Equation 3.10, the general solution is a function of the form

$$y = A\mathrm{e}^{Bt} \tag{3.11}$$

where A is an arbitrary constant. One can readily check that this is indeed the solution to an equation of the form of Equation 3.10. The rule for differentiating the exponential $y = A\mathrm{e}^{Bt}$ has been worked out to be

$$\frac{\mathrm{d}y}{\mathrm{d}t} = \frac{\mathrm{d}}{\mathrm{d}t}(A\mathrm{e}^{Bt}) = BA\mathrm{e}^{Bt}.$$

Since we know $y = Ae^{Bt}$, we can substitute for dy/dt and y into Equation 3.10 to see that:

$$(BAe^{Bt}) = B \times (Ae^{Bt})$$

which is true. So, an expression for y of the form of Equation 3.11 does indeed satisfy an equation of the form of Equation 3.10.

In other words, if the rate of change of the quantity y is directly proportional to the quantity y itself, then the quantity y will either increase or decrease exponentially with time, i.e. $y = Ae^{Bt}$. The **time constant** τ for the change is the time taken for y to fall to 1/e of its initial value and is equal to $1/B$. The value of A will be equal to the initial value of y at $t = 0$.

Example 3.4

Show by substitution that:

$$q = q_0 e^{-t/RC} \tag{3.12}$$

is a solution to Equation 3.9

$$\frac{dq}{dt} = \left(-\frac{1}{RC}\right) \times q. \qquad \text{(Eqn 3.9)}$$

Hence show that the initial charge on the capacitor is q_0 and the time constant is RC.

Solution

If we differentiate Equation 3.12 we find that

$$\frac{d}{dt}(q) = \frac{d}{dt}\left(q_0 e^{-t/RC}\right) = -\frac{1}{RC} q_0 e^{-t/RC}.$$

Substituting this result, and the expression for q given by Equation 3.12, into Equation 3.9, we arrive at

$$-\frac{1}{RC} q_0 e^{-t/RC} = \left(-\frac{1}{RC}\right) \times q_0 e^{-t/RC}$$

which is clearly true, and so $q = q_0 e^{-t/RC}$ is indeed a solution to Equation 3.9.

We can find the initial charge on the capacitor by putting $t = 0$ in Equation 3.12:

$$q = q_0 e^{-0/RC}$$

$$q = q_0 e^0$$

$$q = q_0 \times 1$$

$$q = q_0$$

We can find the time constant of the circuit by finding the time at which the charge has fallen to 1/e of its initial value, q_0, that is, when

$$\frac{q}{q_0} = \frac{1}{e} = e^{-1}.$$

Then, substituting this expression for q/q_0 into Equation 3.12 and comparing the exponents of e on either side, we see that

$$-1 = -\frac{t}{RC}$$

which implies that the time constant, τ, is given by

$$\tau = RC.$$

Question 3.6 Prove that the SI unit of the product RC is the second. (Hint: Use the equations $V = iR$ and $q = CV$, which define R and C.) ■

So we have seen that, when a capacitor is discharged, the charge on the plates decays exponentially, according to Equation 3.12. Recalling that the current through the resistor is given by:

$$i = -\frac{dq}{dt}$$

and noticing that we have differentiated q already, we can write this expression as:

$$i = -\frac{dq}{dt} = \frac{1}{RC}q = \frac{1}{RC}q_0 e^{-t/RC}$$

i.e. $$i = i_0 e^{-t/RC} \quad (3.13)$$

where $i_0 = q_0/RC$. Here, i_0 is the current that flows when the connection is first made (at $t = 0$). So the current through the resistor obeys a similar exponential law to that obeyed by the stored charge.

4 Power, internal resistance and electromotive force

4.1 Power dissipated in a resistor

Having spent some time considering circuits with both resistors and capacitors, let's return to a simple circuit containing a single resistor, such as that shown in Figure 3.16. The resistor often takes the form of a piece of carbon or metal, with dimensions tailored to give the required resistance (Figure 3.23).

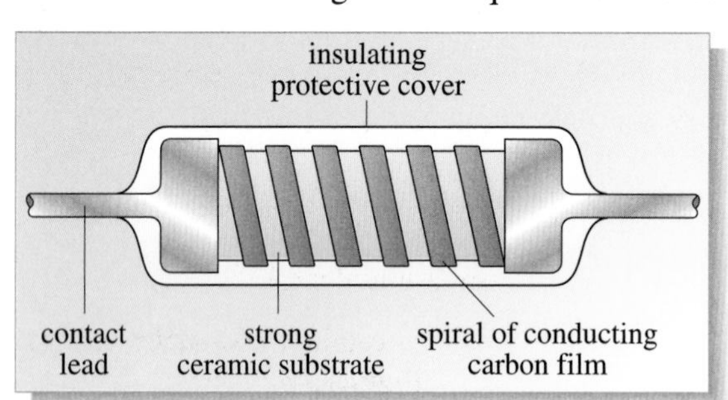

Figure 3.23 Diagram of a resistor with part of the insulating protective cover removed.

When a current is flowing in a resistor such as this, the charge is moving to positions where its electrostatic potential energy is lower. There is a *loss* of electrostatic potential energy. Of course, 'lost' energy does not just disappear. That would contravene the law of conservation of energy. Rather, it is transformed into heat: the electrons are accelerated by the electric field, gaining kinetic energy; this energy is subsequently lost in collisions with the ions, causing them to vibrate more. The rate of loss of electrostatic potential energy is easily found. If a charge Δq flows through a resistor across which there is an applied potential difference V (as in Figure 3.16), then there is a loss of potential energy given by

$$\text{loss of potential energy} = \Delta q \times V. \quad (3.14)$$

This follows directly from the definition of potential in Section 2 of the previous chapter. You should recall from the definition of the volt that if Δq is measured in coulombs and V in volts, then to satisfy the above equation the unit of potential energy must be the joule. Power can be defined as the rate at which work is done, or energy is transferred, and the unit of power, the watt, is equal to one joule per second.

In the electrical case, the electrostatic potential energy is dissipated as heat and the rate at which this occurs is given by

$$\text{power} = \frac{\text{loss of potential energy}}{\text{time interval}}.$$

Substituting from Equation 3.14 and letting the time interval be Δt, we get

$$\text{power} = \frac{\Delta q \times V}{\Delta t}.$$

But, the rate of transfer of charge, $\Delta q/\Delta t$, is simply the average current, and in the limit of small Δt this becomes dq/dt, the instantaneous current, i. So, in electric circuits,

$$\text{power} = iV. \tag{3.15}$$

If i is measured in amps and V in volts, then the power calculated from Equation 3.15 will be in watts.

If a current i is passed through a resistor with resistance R, we can use Ohm's law to find a way of expressing Equation 3.15 for the electric power dissipated in the resistor in terms of i and R. A simple substitution is all that is needed. Equation 3.15 tells us that power = iV. But $V = iR$ and so, substituting for V,

$$\text{power} = iV = i \times iR = i^2R. \tag{3.16}$$

We can also substitute for i in Equation 3.15, to obtain the power in terms of V and R:

$$\text{power} = iV = \frac{V}{R} \times V = \frac{V^2}{R}. \tag{3.17}$$

The cost of energy

Electricity companies in the United Kingdom charge domestic customers for the amount of electrical energy that they use. This is just the product of the power and the time. They do not charge 'per joule': instead, their bills are written in terms of **kilowatt hours** (kW h):

$$1\,\text{kW h} = 1\ \text{kilowatt} \times 1\ \text{hour}$$

$$= (10^3\ \text{watts}) \times (3600\ \text{seconds})$$

$$= 3.6 \times 10^6\ \text{J}.$$

Although electricity bills may always seem too large, in fact, when compared with the cost of human labour, electricity is still extremely cheap. An adult human requires of the order 10 MJ of energy per day to 'tick over'. This energy is expended keeping one's metabolism in order and performing light manual work. A worker who shifts, say, 1000 shovel loads of soil in a day's work might need as much energy again. But we see from above that 10 MJ of electrical energy corresponds to roughly 3 kW h and would cost only about 21 pence.

Question 3.7 If an electric light bulb is rated at 100 W, 220 V, what current does it draw? What is the resistance of the filament? If 1 kW h of electricity costs 7 pence, how much does it cost to leave this light bulb on all day? ■

Current pulses from capacitors

Consider again the *RC* circuit described in Figure 3.21a in Section 3. Could such a simple circuit ever be used as a current source? Obviously, it could not provide *steady* currents, but when a *pulse* of current is required, it can indeed be useful. The length and power of the pulse can be tailored by varying both the resistance through which the current passes (the load resistor) and the value of the capacitor.

For example, if a 100 μF capacitor was discharged through a 0.1 Ω load, the time constant would be 10 μs. If the capacitor was 'charged' to a potential difference of 100 V, it would have an initial stored plate charge ($q_0 = CV$) of 10^{-2} coulombs. From Equation 3.13, the initial current would have a magnitude of q_0/RC = 1000 A. So, the initial power would be an astonishing Vi = 100 kW. Both V and i fall as the amount of plate charge falls. After 10 μs (a single time constant), both V and i would be reduced by a factor of e (~2.7); the power (given by the product of V and i) would therefore be reduced by a factor of e^2 (~7.4), to about 13 kW.

Question 3.8 A primitive spot-welding machine uses a 10^{-3} F capacitor, which is charged to a potential difference of 100 V and then discharged through the two pieces of metal to be joined. The brief but large current melts the metal in the region of contact and the pieces fuse. If the resistance of the two pieces of metal is 0.1 Ω throughout the operation, calculate (a) the initial stored plate charge, (b) the initial stored energy, (c) the initial current, (d) the time constant, (e) the initial power and (f) the power after 200 μs. ■

4.2 Batteries and electromotive force

The potential difference in an electrical circuit is often maintained by a battery. The physics and chemistry of how a battery works is rather complicated, but a summary of the key points is given below.

A **battery** (Figure 3.24a), consists of a number of **electrical cells** connected in series, each of which contains two conducting plates called electrodes. The two electrodes at either end of the series of cells are the battery terminals to which connections are made. Within each cell the electrodes are immersed in a conductor (called an **electrolyte**) in which ions are mobile. In order to allow ionic mobility, electrolytes usually consist of a liquid, such as dilute sulfuric acid, or some kind of gel or paste. The electrodes are made of two different materials, for example one may be copper and the other zinc (Figure 3.24b).

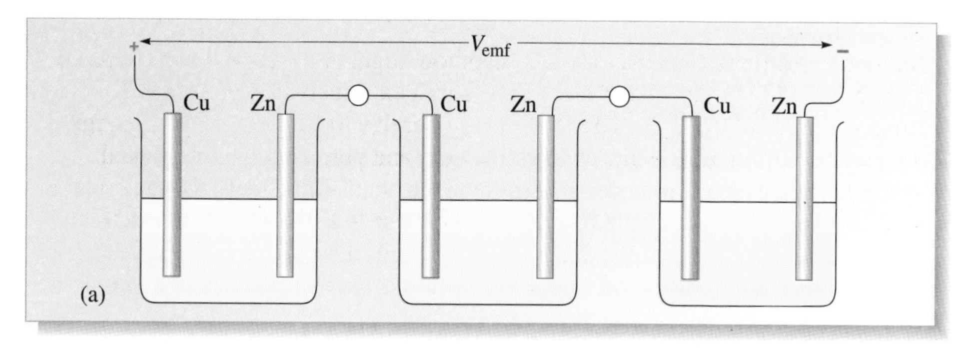

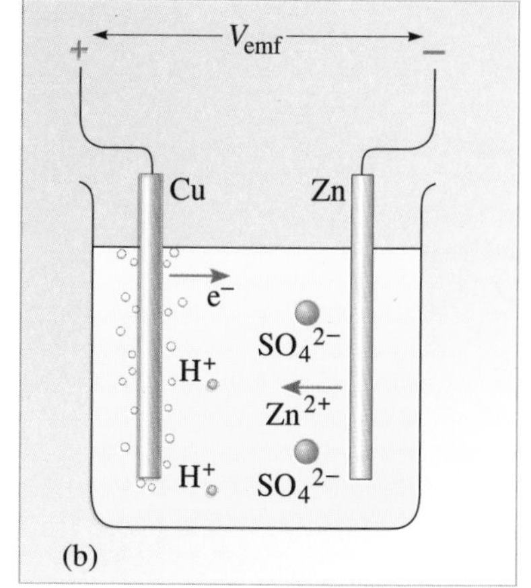

Figure 3.24 (a) A battery consists of several electrical cells connected in series. (b) A cell on open circuit (i.e. a circuit that does not form a closed loop). The copper and zinc electrodes are immersed in the electrolyte, which is sulfuric acid (H_2SO_4). At the copper electrode, hydrogen ions (H^+) from the electrolyte combine with free electrons from the metal and bubbles of hydrogen gas are liberated. At the zinc electrode, zinc ions (Zn^{2+})combine with sulfate ions (SO_4^{2-}) from the electrolyte to form zinc sulfate. As a result of these two reactions, electrostatic charge separation occurs. On open circuit, the potential difference between the terminals is V_{EMF}.

To be specific, let us consider what happens within a cell containing copper and zinc electrodes and a sulfuric acid electrolyte. As a result of a chemical interaction with the electrolyte, positively charged zinc ions flow into the electrolyte from the zinc electrode and leave it negatively charged. Similarly, electrons flow into the electrolyte from the copper electrode and leave it positively charged. A potential difference is thus set up between the electrodes.

In the electrolyte, the sulfuric acid partly dissociates into hydrogen ions and sulfate ions, leading to the formation of zinc sulfate at the zinc electrode and hydrogen gas at the copper electrode. The movement of ions within the electrolyte constitutes an electric current across the cell. The potential difference between the electrodes will be maintained even when a current is drawn by connecting the battery across a resistance, because further chemical reaction can take place between the electrodes and the electrolyte to continue the charging process. In principle at least, this can continue until the chemicals in the battery are consumed. Of course the *rate* at which charge can be supplied is limited by the rate at which ions can move through the electrolyte. Thus each battery has a maximum current that it can supply, with the limit set by the properties of the electrolyte and the geometry of the electrodes.

A device, such as a cell or battery, that is capable of producing an electrical potential difference from some other form of energy, is said to possess an **electromotive force** (V_{EMF}). The magnitude of the electromotive force may be defined in a rather non-intuitive way as follows:

> The electromotive force, V_{EMF}, is the ratio of the power supplied to the current delivered.

A more useful definition of V_{EMF} is given later. From the above definition, the unit of V_{EMF} is watts/amps, which is volts. Electromotive *force* is, therefore, a historical misnomer: V_{EMF} is *not* measured in newtons and is *not* a force. When we talk about a 12 V battery, strictly speaking we mean a battery whose electromotive force is 12 V (i.e. the battery generates 12 W of power for every 1 A it delivers). So, how is V_{EMF} related to the effective potential difference between the terminals of a battery when the battery is connected to a circuit and a current is flowing? To answer this question we need to understand what is meant by the *internal resistance* of a battery.

Internal resistance

As mentioned above, each cell has a maximum current that it can supply. Let's consider two extreme cases:

- When a small current is drawn from the cell — i.e. when a relatively large external resistance is connected across the terminals of the cell, then the rate at which charge is drawn from the cell is much less than the maximum. In this case, the chemical and diffusive processes within the cell are easily able to cope with the demand for current, and the current that flows is determined mainly by the external resistance.
- When a large current is drawn from the cell — i.e. when a relatively small resistance is connected across the terminals of the cell, then the rate at which charge is drawn from the cell can easily reach the maximum of which the cell is capable. In this case, the current that flows is determined only weakly by the external resistance, and strongly by the chemical and diffusive processes within the cell.

These processes effectively constitute an extra resistance in the circuit. We refer to this as the **internal resistance**, r, of the cell.

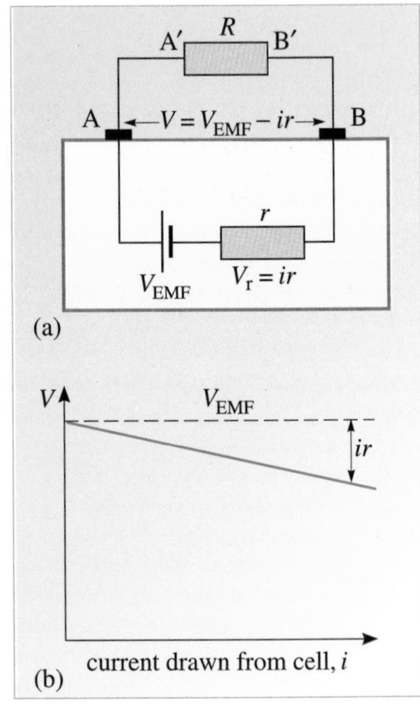

Figure 3.25 (a) Our model of a real cell including an internal resistance. A and B are the terminals of the cell. The connecting wires AA′ and BB′ have negligible resistance compared with R and r, and therefore A′ and B′ can be considered to be at the same potential as A and B, respectively. (b) When a current flows, the potential difference, V, at the terminals of a cell is reduced from its maximum value V_{EMF} because of its internal resistance r.

When a cell provides a circuit with current, the magnitude of the current is the same through the circuit and the electrolyte and between the electrolyte and the electrodes within the cell. We model the cell as an ideal source providing a potential difference of V_{EMF}, in series with an internal resistance, r (Figure 3.25). When the cell is connected to an external resistance R, a current i flows. We can now apply the principle of conservation of energy to this circuit:

energy supplied by cell per second = energy dissipated in resistors per second,

i.e. $$V_{EMF} \times i = i^2R + i^2r$$

$$V_{EMF} = i(R + r)$$

$$i = \frac{V_{EMF}}{(R + r)} \tag{3.18}$$

From Equation 3.18 we can deduce several important properties of any source of electromotive force:

- The maximum current that can be supplied by the cell is V_{EMF}/r, which would be supplied into a *short circuit* ($R = 0\,\Omega$) across the cell electrodes.
- The current delivered in normal use is always is *less* than the current V_{EMF}/R that would flow if the internal resistance, r, were zero.
- Let V be the potential difference across the terminals of the cell when the current is flowing. We can rewrite Equation 3.18 as

$$iR = V_{EMF} - ir$$

and hence

$$V = V_{EMF} - ir. \tag{3.19}$$

This equation indicates that the internal resistance causes the potential difference across the terminals of the cell to *decrease* as the current increases (Figure 3.25b). Conversely, as the current drawn from the cell is reduced, the potential difference between the terminals of the cell approaches V_{EMF}. In the limit, as i approaches zero, $V = V_{EMF}$.

From this last point, we can arrive at a second and more practical definition of V_{EMF}:

> The electromotive force, V_{EMF}, is equal to the limiting maximum potential difference between the terminals of a cell as the current drawn from the cell is reduced towards zero.

As the cell is used, electrolyte is chemically consumed and so there are fewer ions to carry the current and the internal resistance of the cell increases. Eventually, the potential difference between the terminals falls to a value that is too low to sustain the function of the circuit that the cell is powering.

● Voltmeters are designed so as to draw very little current. Why is testing the potential difference across a cell with a voltmeter a poor way of testing whether the cell is run down, but an effective way of determining V_{EMF} of the cell?

❍ The voltmeter draws a negligible current from the cell and therefore the effect of the internal resistance on the potential differences at the terminals is small and hence V_{EMF} is accurately determined by this procedure. However, even though a cell may have an acceptable potential difference at its terminals, when placed in a circuit requiring a larger current, it may be unable to supply that current because of a large internal resistance. ■

Question 3.9 A voltmeter has a resistance of 10 MΩ. It is used to measure the potential difference across a 100 kΩ resistor through which, in the absence of the meter, a current of 100 μA is passing. When the meter is connected across the resistor, the total current remains at 100 μA but a fraction of the current flows through the meter rather than the resistor (Figure 3.26). (a) Calculate the voltage across the meter. (b) Why should voltmeters have very high resistances? (c) Why should ammeters, which are connected in series with a circuit to measure the current, have low resistances?

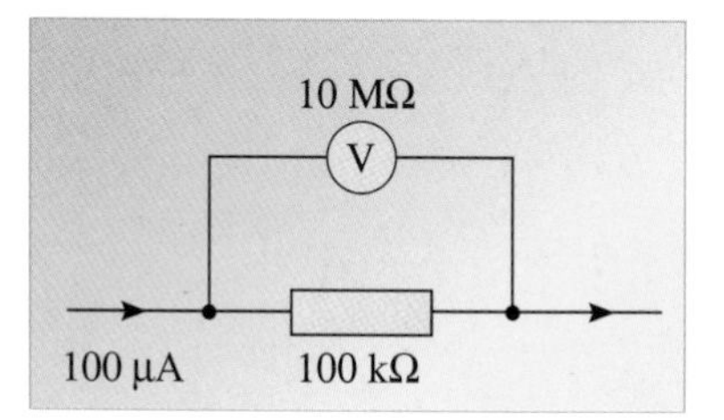

Figure 3.26 For use with Question 3.9.

Question 3.10 A run-down 2 V torch battery has an internal resistance of 10 Ω. Calculate the power dissipated in a 4 Ω light-bulb powered by this battery. How does it compare with the power available from a new battery with virtually zero internal resistance? (You should assume that the resistance of the bulb is independent of the current.)

Question 3.11 A voltmeter connected across a battery reads 11 V. When a 10 Ω load is placed across the same battery, a current of 1 A flows. Calculate the internal resistance of the battery. ■

Other sources of electromotive force

We have discussed internal resistance in the context of chemical cells and batteries, but in fact, the idea is far more general. *Any* source of electromotive force — e.g. electrical power supplies or dynamos — will have a maximum current that it can supply. Although the physics that dictates this maximum may differ from device to device, there is always some limit to the current that can be drawn, and conventionally this is characterized as an internal resistance.

5 Currents in the everyday world

In Sections 2 to 4 of this chapter, we have developed a number of ideas about current flow through the simplest of circuit elements — the resistor. In this section we will look at a number of examples of the way in which electric currents affect our world, both in technological applications and in nature.

5.1 Resistance sensors

In Section 3.3 we pointed out that the resistivity of a material was constant '...so long as the temperature and the strain in the material remain constant'. A resistance sensor is a resistor, or a collection of resistors, which responds to changes in its environment by changing its resistance. Resistance sensors exist that respond to changes in temperature, strain, pressure, wind speed, and light level, amongst other factors. We will take a closer look at the platinum resistance thermometer.

The platinum resistance thermometer

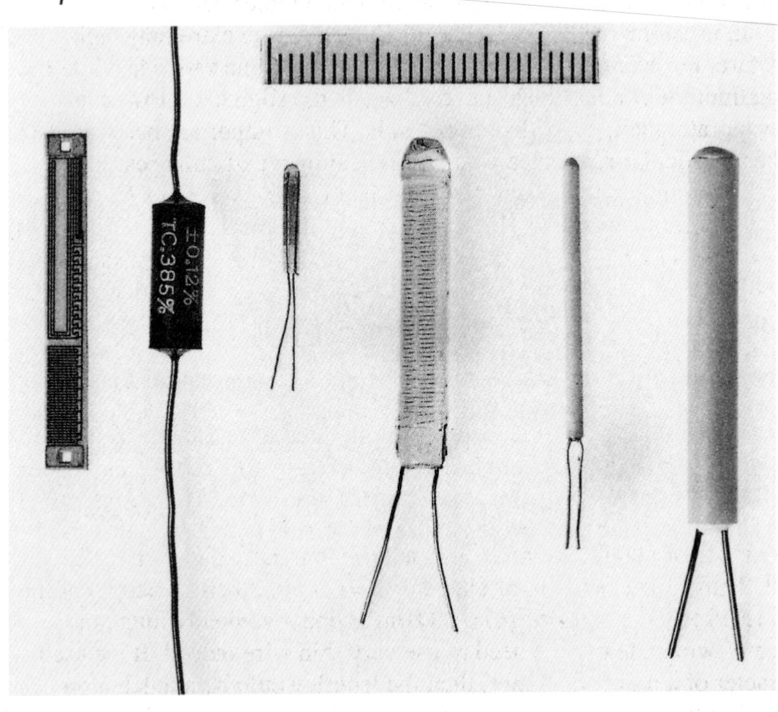

Figure 3.27 Some platinum resistance thermometers.

The notation $\rho(T)$ reminds us that ρ is a *function* of T.

A platinum resistance thermometer (PRT) is simply a piece of strain-free platinum with some electrical contacts (Figure 3.27). If the temperature-dependent resistivity of platinum is $\rho(T)$, then the (temperature-dependent) resistance of a piece of platinum wire of length L and cross-sectional area A is (from Equation 3.4)

$$R(T) = \frac{\rho(T)L}{A}. \qquad (3.4a)$$

Thus if we measure $R(T)$ and know L, A and $\rho(T)$ for our sensor, then we can determine the temperature of our sensor material. The resistivity of pure, strain-free platinum is a unique function of temperature, and all pieces of pure, strain-free platinum have a resistivity that varies in the same way. A great deal of work has been done to establish accurately how the resistivity of platinum changes with temperature from around 13 K (−260 °C) to greater than 1300 K (> 1000 °C) (Figure 3.28).

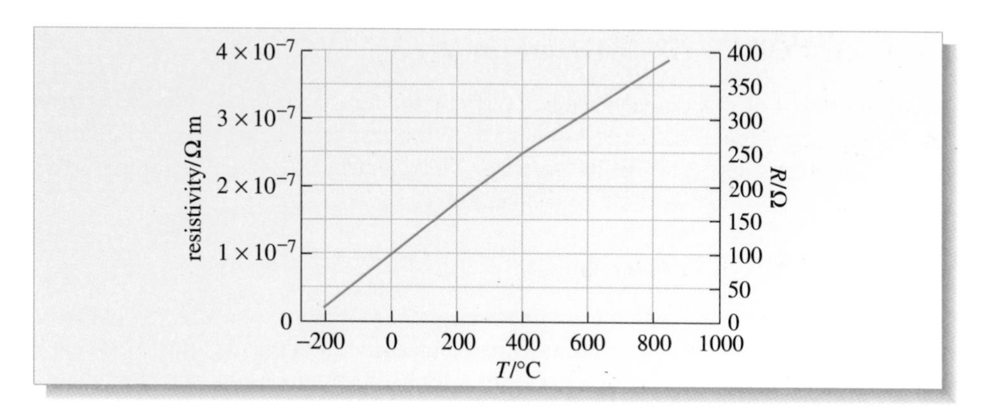

Figure 3.28 A graph of $\rho\,(T)$ against temperature for a PRT. The right-hand axis of the graph is scaled to show the resistance of a PRT, which has a resistance of 100 Ω at 0 °C.

Why platinum?

The resistivity of most substances changes with temperature, and so there is a wide choice of possible materials to use in a sensor. Platinum is one of the rarest and most expensive materials on Earth and so it might be thought a rather odd choice. However, platinum is astonishingly inert chemically and has an extremely high melting temperature. For example, it is possible to heat a platinum wire to white heat in air, and the platinum will not tarnish (i.e. oxidize) in the slightest. Most other metals would evaporate, melt, or oxidize in seconds. This is important because, if we are going to take a particular resistance value as indicating a particular resistivity, and hence a particular temperature, then it is important that the values of L and A do not change with time. If, for example, the cross-section of the wire were to gradually get smaller as it corroded, then its resistance would gradually increase. We might then erroneously interpret these changes as reflecting real changes in temperature, rather than changes in A. The choice of platinum means that we can use the thermometer up to 1000 °C without worrying about this effect.

Choosing L and A

The values of L and A are generally chosen so that at 0 °C the resistance of the device is some convenient value: 100 Ω and 273 Ω are common choices. From Table 3.1 (Section 3.3) we find that the resistivity of platinum at 0 °C is 9.81×10^{-8} Ω m, and so for a resistance of 100 Ω, Equation 3.4a tells us that the ratio L/A must be $1.019 \times 10^9\,\mathrm{m}^{-1}$. Thus, if we use wire of diameter $d = 0.1$ mm, then the length of the sensor must be $1.019 \times 10^9\,\mathrm{m}^{-1} \times (\pi d^2/4) = 8.00$ m, an inconveniently long (and expensive) piece of wire. Clearly we need to use very thin wire indeed. If we used wire of the diameter of a hair (~0.01 mm) then the length would be a much more convenient 8 cm, but the wire would be more delicate. In general, the wire would be wound on a thermally conducting ceramic bobbin and baked in thermally conducting cement. In some types of device, a film of platinum just a few micrometres thick is laid down on a ceramic substrate, and then patterned to yield the correct ratio L/A.

Terminal measurement

The resistance of a PRT with a value of 100 Ω at 0 °C changes by just 0.386 Ω for each degree Celsius change in temperature. So to determine the temperature with an accuracy of, say, 0.1 °C, we need to measure a resistance of around 100 Ω with an accuracy of better than 0.04 Ω. This requires special care, because the wires leading from the measuring instruments to the PRT typically have a resistance of several ohms. If we measured the resistance of these connecting wires in series with the PRT (Figure 3.29a) we would overestimate the resistance, and hence the temperature, possibly by several degrees. We could measure the resistance of the connecting

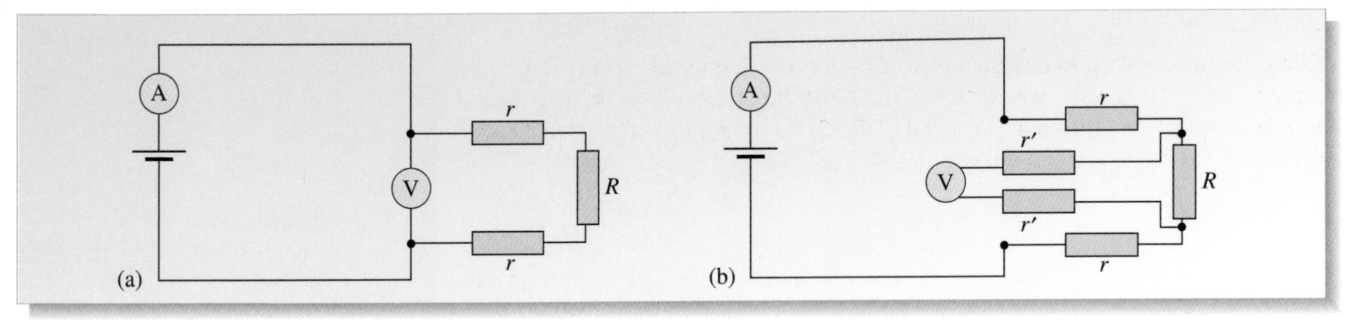

Figure 3.29 Circuits for measuring the resistance of a PRT. (a) A two-terminal measurement circuit that determines the resistance, R, of the PRT plus the resistance, r, of the wires leading to the device. (b) A four-terminal measurement circuit that determines the resistance of the PRT and is insensitive to the resistance of the wires leading to the device.

wires separately and subtract it from our result, but if the resistance of the connecting wires changed with temperature, this would quickly become an extremely time-consuming correction.

The problem is usually overcome by the use of four-terminal measurement (Figure 3.29b). In this circuit a current flows through the PRT, but the potential difference across the PRT is measured by a separate set of wires. Because the internal resistance of modern voltmeters is usually greater than $10\,\mathrm{M\Omega}$, only a tiny current (usually of the order of a nanoamp) flows through these wires. This makes the potential difference along these wires very small, usually of the order of a microvolt, and hence the error in the inferred resistance is very small.

5.2 Electric lights

The light-bulb is one of the most commonplace objects in the modern world. Without them, there would be a heavy restriction on activity during the night, which, in the polar regions of the Earth, extends to 24 hours at times! There would be no bright headlights for cars, and gas would have to be used for street lighting.

The incandescent light-bulb

An incandescent light-bulb (Figure 3.30) is essentially a heater: a small section of wire — the filament — is heated until its temperature is sufficient to cause it to glow. You might be surprised at how hot the filament becomes: in most light bulbs the temperature is around 1700 °C! Such extreme temperatures are required because the hotter the filament, the 'whiter' the light. However, this requirement severely restricts the materials that can be used for a filament. Most commonly, the element tungsten, which has a melting temperature of about 3400 °C, is used.

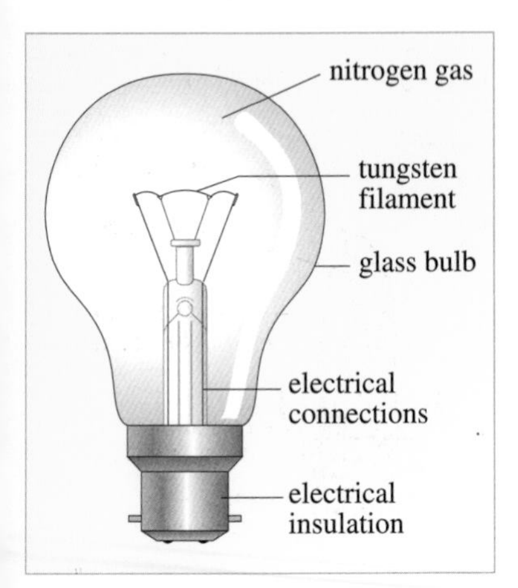

Figure 3.30 A diagram of a light-bulb.

Let's try to work out how to design a filament for a 60 W light-bulb to operate from a 220 V supply. This is the kind of light-bulb you might be using in a desk lamp as you read this.

The key to designing a light-bulb is to realize that it is essentially an energy *transducer*, i.e. it changes electrical energy into radiant energy in the form of electromagnetic waves. Thus, we would like to put in 60 W of electrical energy and get out 60 W of radiant energy.

Electrical energy

It is fairly straightforward to design a piece of wire that dissipates a specified amount of power. To do this, we consider first what the resistance of the wire needs to be. In Example 5 in Section 4.1, we stated that if the potential difference V across a resistance R is specified, then the power dissipated is:

$$P = \frac{V^2}{R}. \qquad \text{(Eqn 3.17)}$$

However, in Section 3.3 we stated that the resistance of wire was given in terms of its length, L, cross-sectional area, A, and its resistivity, ρ, by:

$$R = \frac{\rho L}{A}. \qquad \text{(Eqn 3.4)}$$

Thus, for a given filament material, a short thin wire would dissipate the same amount of energy as a long fat wire with the same value of L/A. However, in order to be useful, the filament must be heated by this electrical energy to a sufficiently high temperature that it will then radiate a significant proportion of it as visible light. The short thin wire has the advantage in this respect both of having a small heat capacity and a smaller surface area from which to lose the heat. There are several other factors that must be taken into account when designing the dimensions of a light bulb filament but, for a 60 W bulb operating at 220 V, the filament would need to be about two or three hundredths of a millimetre in diameter and a few centimetres long.

The effect of temperature

Given the ubiquity of light-bulbs, and the vast amount of energy we spend in lighting them, it is clearly important that they convert electrical energy into light efficiently. For that reason, it might surprise you to know just how inefficient incandescent light-bulbs are. At a typical filament temperature of 2000 K, only 1.4% of the radiation emitted by the filament is visible light: the remaining 98.6% is emitted as infrared radiation, which is invisible. Clearly, incandescent light-bulbs make excellent heaters, but are really rather inefficient at producing light!

However, if we could heat the filament to 2400 K, then around 3.9% of the energy would be emitted as light, i.e. nearly three times as much light. If all the normal incandescent light-bulbs could be replaced with lower power bulbs operating at 2400 K (but producing the same amount of light) then the cost of lighting would drop by around 65%.

The tungsten–halogen light-bulb

In fact, such 'high temperature' light-bulbs do exist (Figure 3.31), and are used commonly. However, such bulbs would not work at all if they were built in the same way as conventional light-bulbs. The reason that normal light-bulbs cannot be heated to higher temperatures is that, at these high temperatures, the tungsten slowly evaporates — you may have seen a blackening on the inside of an old bulb caused by this process. Heating the bulb through those extra few hundred degrees dramatically increases the rate of evaporation, and the tungsten evaporates so quickly as to make the bulbs useless.

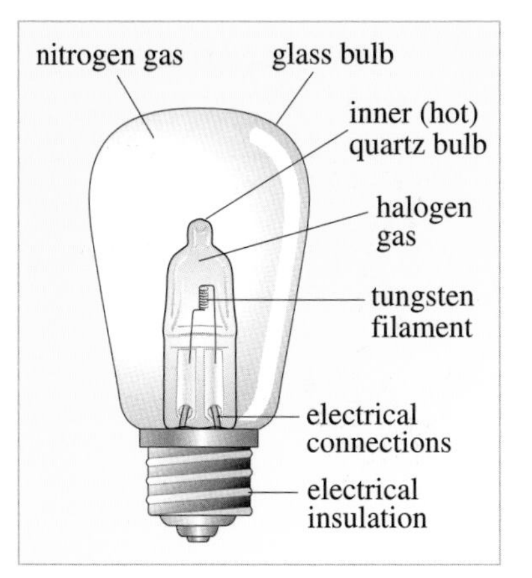

Figure 3.31 A diagram of a tungsten–halogen light-bulb.

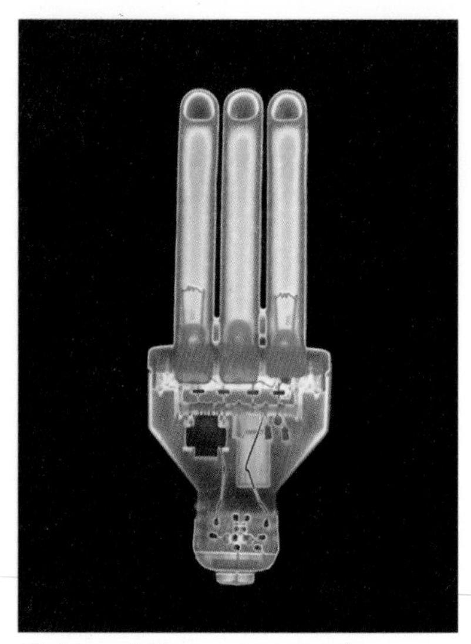
Coloured X-ray picture of a gooseneck tungsten–halogen light-bulb.

However, it was discovered that, with care, an extraordinary chemical process could be initiated. This process continually cleans the tungsten from the inside surface of the bulb, and redeposits it on the filament! In order to make this process work the light-bulb must satisfy the following criteria:

- The bulb cannot be made of ordinary glass, but must be made of quartz;
- The quartz must be heated to around 300 °C;
- The gas inside the bulb must be a halogen, for example chlorine or iodine.

When these conditions are satisfied, molecules of the halogen gas react with the tungsten at the quartz surface to make tungsten halide molecules. However, when these molecules eventually strike the filament, they decompose, leaving the tungsten on the filament, and releasing the halogen to clean some more of the quartz.

In order to satisfy the three conditions above, halogen bulbs must be made in a quite different way from normal bulbs. They are made much smaller, and they use a fraction of the heat dissipated in the bulb to keep the quartz at the correct temperature. There is generally an outer glass cover that prevents users being burned by the extremely hot surface of the bulb.

The fluorescent tube

Another method of converting electrical energy into light is the fluorescent tube, illustrated in Figure 3.32. In this device, electrical breakdown is initiated in a gas, and an electric current is then passed through the gas, driving it into the plasma state. In this state, both electrons and ions are accelerated through the plasma (in opposite directions) by the electric field applied along the tube. The conditions inside the tube are designed so that ions and electrons can be accelerated by the field over a sufficiently large distance that, in some of the collisions that they undergo (inelastic collisions), the energy of collision is sufficient (a few electronvolts) to excite the ions to emit light. This process is much more efficient than the thermal process used in an incandescent light bulb. However, the radiation emitted is mainly ultraviolet, which is harmful to most living things.

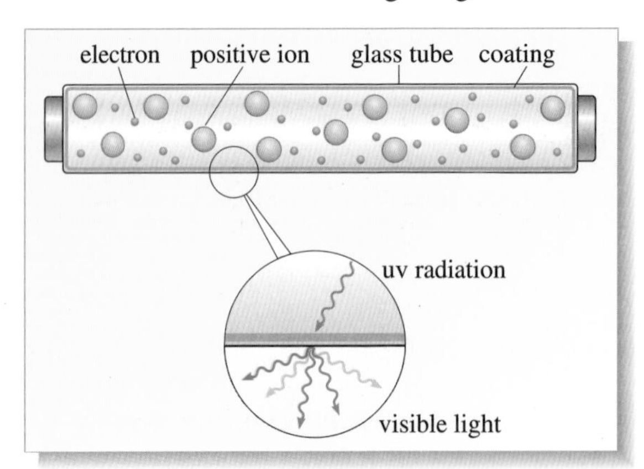

Figure 3.32 A diagram of a fluorescent tube.

To overcome this, the inside of the tube is coated with a substance that emits visible light when it is illuminated with ultraviolet radiation. Such substances are said to fluoresce. In the process of **fluorescence**, a substance absorbs radiation at one wavelength, and subsequently emits radiation at one or more longer wavelengths. Generally, such materials absorb ultraviolet radiation across a wide range of wavelengths, but emit visible light at just a few characteristic wavelengths. The 'white' colour of the tube is achieved by mixing many different compounds with different fluorescent wavelengths.

Fluorescent tubes convert electrical energy into light with an efficiency of around 10%, i.e. roughly ten times better than an incandescent light-bulb and around three times better than a tungsten–halogen bulb.

Question 3.12 *Estimate* the average distance travelled between successive inelastic collisions by the electrons and ions inside a fluorescent light-tube, if the average magnitude of the electric field inside the tube, when it is operating, is $100\,\mathrm{V\,m^{-1}}$. ■

5.3 Currents in the Earth and its atmosphere

The Earth as a conductor

The Earth does not appear at first to be a conductor of electricity. Most of the components of soil — plant debris and minerals — are not good conductors: indeed, rocks are generally excellent insulators. However, despite first appearances, the Earth is in fact a conductor of electricity, and the component of soil responsible for the conductivity is easy to identify: water. Water is present in the gaps between soil particles and dissolves minerals from the soil to form an ionic solution. Water also exists within the pores of many, otherwise insulating, minerals, such as sandstones, and again contains dissolved ions.

The conductivity of the Earth, and its dependence on moisture content, have several applications.

Archaeological mapping

The change in the moisture content of different rocks can be exploited to generate an 'image' of variations in conductivity across an archaeological site (Figure 3.33). Because stones used in construction generally have a lower water content than soil, their presence and approximate position can be determined by systematic measurements of the resistance between probes placed at a given separation.

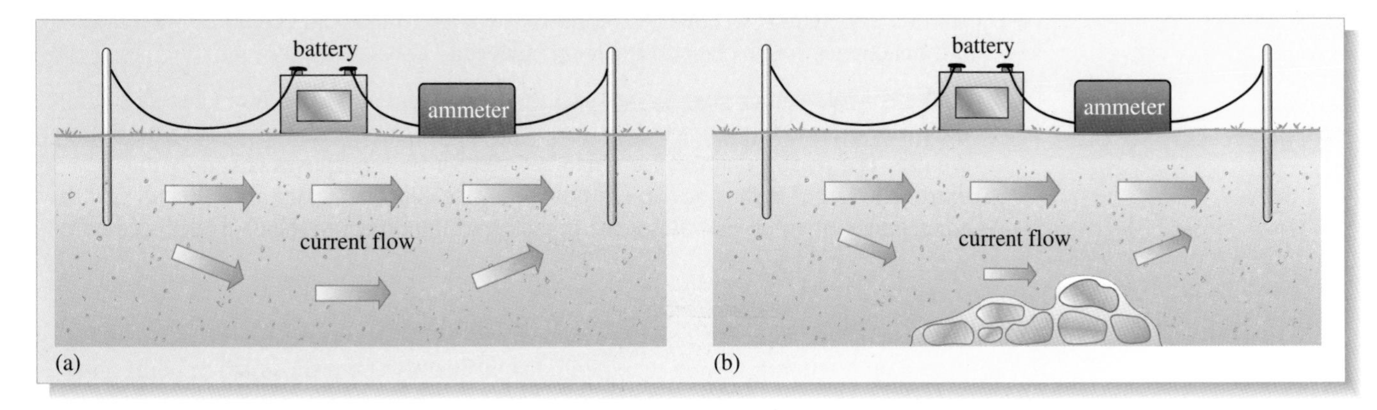

Figure 3.33 Changes in the moisture content below the surface of the Earth can be mapped over an area by systematically measuring the resistance between probes placed at a given separation. The changes in moisture content can be due to natural changes in soil composition, or can reflect human activities. In the example shown, the resistance between the probes is increased in (b) compared with (a) because of a buried structure (a stone wall) with a low moisture content.

Earth potential

The Earth is used as an intrinsic part of the system of electricity distribution. Two related, but distinct, electrical properties of the Earth are used. The first, as we have discussed, is its conductivity, and the second is its capacitance.

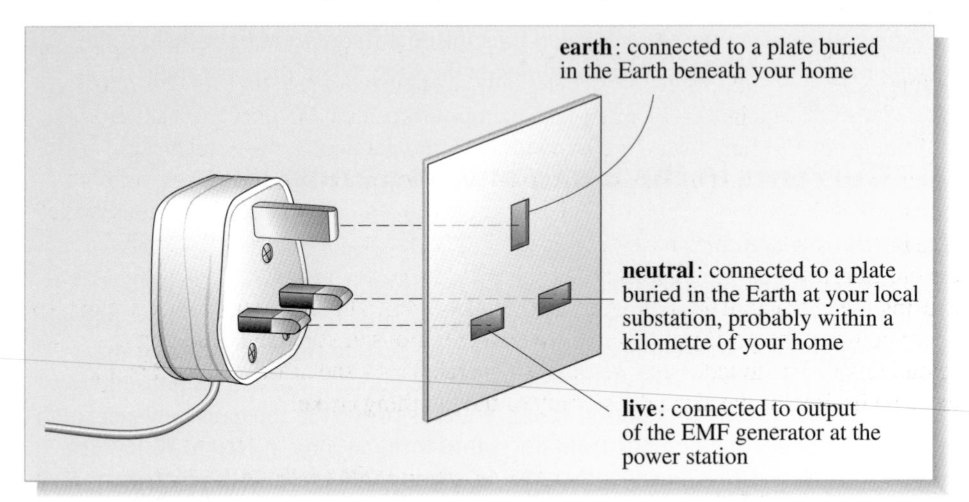

Figure 3.34 The domestic electricity supply uses the conducting properties of the Earth.

Figure 3.34 shows two of the connections between our electricity supply system and the Earth. The first is the connection from the pin marked 'earth' on a domestic plug. The socket is connected directly to a plate buried in the Earth. This connection must have saved thousands of lives during the course of this century. The idea is that all the conducting parts of an electrical device that could conceivably be touched by a person should be connected to this plate through a low resistance connection. In the event that the power supply accidentally makes contact with these parts and applies a lethal potential to these components, then current will flow into the Earth along the low resistance path rather than through our bodies.

The Earth is also used as the return path for the current supplied from the power station. The neutral terminal on domestic sockets is also connected into the Earth, but the connection is close to the local electricity substation.

It might strike you as slightly odd that the Earth can be used both to keep us safe in the event of an electrical accident, and as a return path for all the electric current that we use. The reason that the Earth can be used for both purposes is simply that it is extremely large! It thus has an extremely large electrical capacitance. Recall that the capacitance of an object relates the potential of the object to the charge placed on it:

$$V = \frac{q}{C}. \qquad \text{(Eqn 3.8)}$$

Because of the immense capacitance of the Earth, large amounts of charge may be placed on it without raising its potential significantly. Thus the Earth remains quite safe to touch.

The idea of the Earth as a 'reference' potential into which charge may be returned, or from which charge may be taken, is used widely in electronics. In fact, confusingly, terminals with this property are often referred to as 'earth' even if there is no direct physical connection to the Earth.

The Earth's magnetic field

Conduction is possible not only through the outer part of the Earth but also much deeper within it. Perhaps the most important phenomenon associated with this conduction is the Earth's magnetic field. We discuss this a little more in Section 3.5

of the next chapter, but here we note that the Earth's magnetic field is presently believed to be generated by currents flowing in the outer part of the Earth's core at a depth greater than 2800 km.

Thunderstorms

As we pointed out in Chapter 1, the precise mechanism causing the charge separation in a thundercloud is still a matter of debate, but the net result is that the cloud develops a net negative charge in its lower half and a net positive charge in its upper half (Figure 3.35). The electric field near the ground under a thundercloud can be as great as $5 \times 10^4\,\mathrm{V\,m^{-1}}$ (pointing upwards). Electrical breakdown can begin in regions of high electric field in the lower regions of the cloud, starting an avalanche of electrons, which generates a conducting path called a 'leader'. Once the leader reaches the ground, charge can flow very rapidly (more than 10 000 A) between the cloud and the ground causing the characteristic lightning flash (Figure 3.36). The lightning discharge can also occur between the positive and negative charge regions of the cloud, resulting in a 'cloud flash'. The thunder that we hear is generated by a shock wave caused by the extreme heating of the air in the vicinity of the lightning strike.

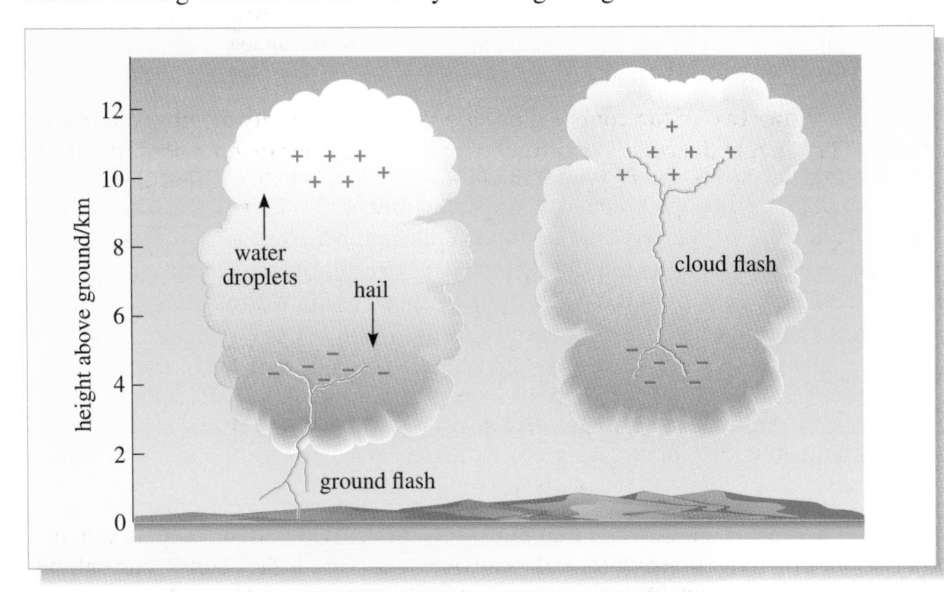

Figure 3.35 Schematic diagram of the charge distribution in a typical, fully developed thundercloud. Two alternative discharge processes are shown.

Figure 3.36 Lightning discharge.

5.4 Currents in living organisms

Living organisms are rich sources of electrical currents. There is a potential difference between the inside and the outside of all living cells, which is maintained by the ion-pumping activity of the cell membrane. In most cells, the potential differences remain essentially constant, but in 'active' cells, notably those of nerves and muscles, the potential difference fluctuates. These fluctuations, known as action potentials, propagate along the cell and so transmit signals from one end to the other. In nerves, the action potentials carry information. They carry sensory information to the brain and provide a means for the richly connected nerves in the brain to process this sensory information, to control the body, to recall memories, to analyse, to experience emotion and so on. In muscles, the action potentials accompany the process of contraction. So, for example, the contraction of the large 'soleus' muscle in your calf is initiated by action potentials. These are transmitted along nerve fibres, which extend all the way from your brain, via your spine, to the long muscle cells that make up the soleus. The action potential is passed from the nerve to the muscle cells, where it spreads out and initiates the contraction. This process is repeated sufficiently often to maintain the required level of contraction.

The currents generated by the active cells are very small but they do provide a window through which it is possible to investigate the working of these cells. You have probably heard of the electrocardiograph. This instrument records the small voltage fluctuations on the skin caused by the beating of the heart. The effectiveness of the heart as a pump depends on the synchronized contraction of the large number of muscle cells that make up the heart. An action potential wave sweeps around the wall of the heart in a precisely choreographed manner, which ensures that the blood is pumped around the body according to need. As a result of this synchronization, the fluctuating currents from the many heart cells add together producing relatively large current pulses, which flow around the chest. These currents generate potential differences, which can be picked up by attaching electrodes to the skin. Such measurements (the electrocardiogram or ecg) can be made very easily with modern instruments and they allow us to monitor the heart's rhythm and detect abnormalities in the working of the heart (Figure 3.37).

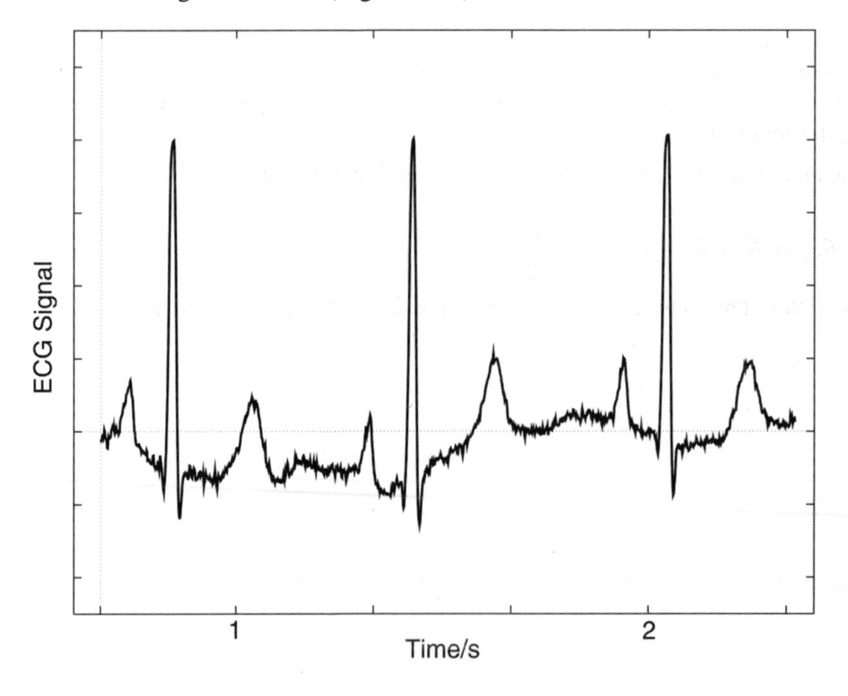

Figure 3.37 The ecg of a normal heart. The large peak is associated with the contraction of the left ventricle, the heart chamber with the thickest wall, which forces blood into the aorta and round the body.

6 Closing items

6.1 Chapter summary

1. An electric current consists of a flow of electrically charged particles. The microscopic nature of the current can be quite different in different states of matter.
2. An electric current is the rate at which charge flows in an assigned direction. The instantaneous current is given by:

$$i = \frac{dq}{dt}. \tag{3.1}$$

The SI unit of current is the ampere (A), where one ampere is a charge flow of one coulomb per second.
3. Electrons flow through a resistor when there is an electrical potential difference across its ends. A battery is a common means of maintaining this potential difference.
4. The resistance of a conductor is defined by:

$$R = \frac{|V_R|}{|i|}. \tag{3.3}$$

The SI unit of resistance is the ohm (Ω) where $1\,\Omega = 1\,\mathrm{V\,A^{-1}}$. Ohm's law ($V = iR$) gives the relationship between the voltage of the source and the current that flows through a conductor. It is obeyed over a large range of values of i by many, but by no means all, components in electrical circuits.
5. The resistivity, ρ, of a material is defined by Equation 3.4

$$R = \frac{\rho L}{A} \tag{3.4}$$

where R is the resistance of a piece of that material of length L and cross-sectional area A.
6. Kirchhoff's first law states that the sum of the potential differences in a given sense around a circuit is zero. In this sum, a potential rise is positive and a potential drop is negative. Kirchhoff's second law states that, at a junction of two or more wires, the current flowing into the junction is equal to the current leaving the junction.
7. When resistors are connected in series, the effective resistance can be found using:

$$R_{\mathrm{eff}} = R_1 + R_2 + R_3 + \ldots = \sum_i R_i. \tag{3.5a}$$

When resistors are connected in parallel, the effective resistance can be found using:

$$\frac{1}{R_{\mathrm{eff}}} = \frac{1}{R_1} + \frac{1}{R_2} + \frac{1}{R_3} + \ldots = \sum_i \frac{1}{R_i}. \tag{3.7a}$$

8. When a capacitor is discharged through a resistance, the charge on the plates decays exponentially according to

$$q = q_0 e^{-t/RC}. \tag{3.12}$$

9. The time constant for the decay of a charged capacitor through a resistor is given by:

$$\tau = RC.$$

This is the time for the current to decay to 1/e (~37%) of its initial value.

10 The current flowing as a capacitor discharges is given by

$$i = i_0 e^{-t/RC}. \qquad (3.13)$$

11 In electric circuits, power is given by:

$$\text{power} = iV. \qquad (3.15)$$

12 When currents flow through circuits, the potential difference across the terminals of the source (battery) drops. This is because the source itself has an internal resistance, r. In general, the current flowing across a resistance R is given by

$$i = \frac{V_{\text{EMF}}}{(r + R)}. \qquad (3.18)$$

In this equation, V_{EMF} is the electromotive force, which is defined as the ratio of the power generated by the battery to the current supplied to the circuit.

6.2 Achievements

After studying this chapter you should be able to:

A1 Explain the meaning of all the newly defined (emboldened) terms introduced in this chapter.

A2 Describe what is meant by 'electric current' and outline some situations in which such a current can flow.

A3 Explain, using force or energy arguments, why and how electrons travel down a metal wire whose ends are held at different electric potentials.

A4 Define the instantaneous current in a wire.

A5 Quote Ohm's law, indicate its range of applicability and use it in simple calculations.

A6 Quote Kirchhoff's laws of circuit analysis and use them in problems. Derive an expression for the effective resistance of resistors in parallel or in series and use these expressions in simple calculations.

A7 Recall how the current provided by a capacitor varies exponentially as a function of time and that the time constant $\tau = RC$ for a capacitor of capacitance C whose stored charge is decaying through a resistance R.

A8 Identify the time constant on a graph representing an exponential decay process.

A9 Express the power dissipated in a resistor in terms of any two of the following parameters: the potential difference across the resistor, the current through the resistor and the resistance of the resistor. Use these relationships in simple calculations.

A10 Understand the difference between the electromotive force of a cell and the effective potential difference across its terminals when a current flows. Thus, recall the constraints on the total energy and the maximum current available from a cell/battery.

A11 Outline some examples of electric currents in the everyday world.

6.3 End-of-chapter questions

Questions 3.13 to 3.15 all refer to the conduction of electricity through a small piece of silver wire.

Question 3.13 The number density of conduction electrons in silver is $5.9 \times 10^{28}\,\text{m}^{-3}$. What is the average drift speed of these electrons in a piece of silver wire of diameter 0.5 mm, which is carrying a current of 0.1 A?

Question 3.14 The resistivity of silver is $1.6 \times 10^{-8}\,\Omega\,\text{m}$. (a) What voltage must be applied across the ends of a piece of silver wire, of diameter 0.5 mm and length 5 cm, to produce a current of 0.1 A? (b) Calculate the power dissipated in the wire under these circumstances.

Question 3.15 (a) Calculate the energy transferred to an electron by the applied voltage when it travels from one end to the other of the wire in Question 3.14.
(b) Calculate the total number of conduction electrons in the piece of wire.
(c) Hence calculate the power dissipated in the wire by first calculating how long it takes for all the conduction electrons in the wire to pass through its entire length. (*You will need your answer to Question 3.13 here.*) How does your result compare with the one you obtained in Question 3.14b?

Question 3.16 Figure 3.38 shows a circuit containing three resistors. If the resistances of the resistors are 100 Ω, 10 Ω and 1 Ω, how should they be combined (i.e. which resistor should be inserted in which position) in order to minimize the current through point A? ■

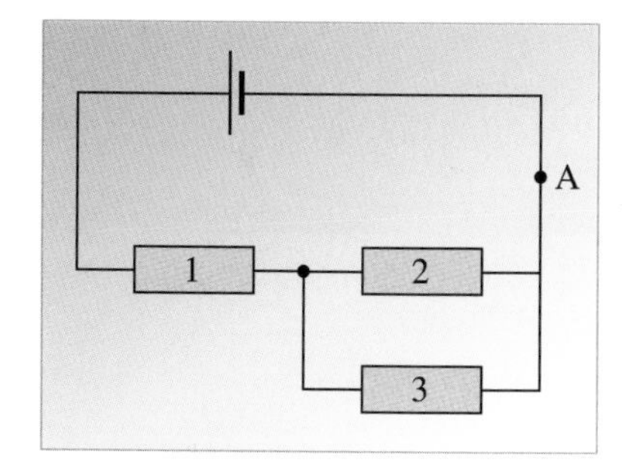

Figure 3.38 For use with Question 3.16.

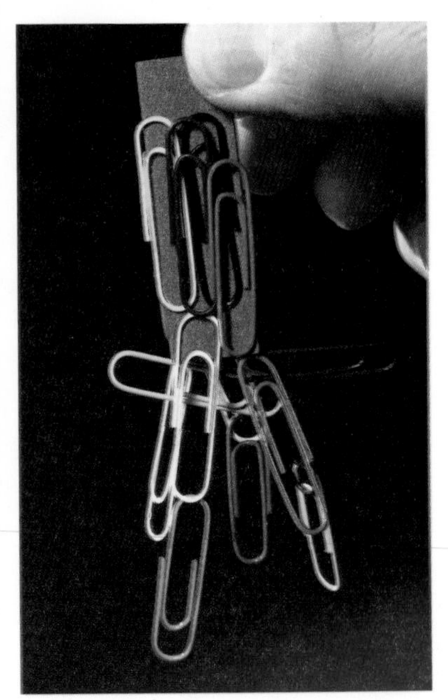

Figure 4.1 Paper clips attracted to a magnet.

Chapter 4 Magnetic fields and the Lorentz force law

1 The magic of magnetism

Even experienced physicists can still be amazed by the phenomenon of magnetism. Holding two bar magnets and feeling their repulsion or attraction is a reminder of what physics is for. It allows us to *describe* succinctly what seems at first indescribable, and it allows us to connect what appear to be unconnected phenomena. In particular, in this chapter, we will make the connection between magnetism and electric currents. This link provides the basis of the subject known as **electromagnetism**.

Figure 4.1 demonstrates a purely magnetic effect. It shows several paper clips adhering to a bar magnet. It is a simple demonstration, and, as such, might be treated with a certain condescension. However, there are many reasons why this would be an inappropriate reaction, and several lessons to be learned.

- The first is that it is hard to devise any explanation of this phenomenon that does not involve the concept of a *field*. The influence of the magnet extends into space away from itself and does not involve the passage of any material particles from the magnet to the paper clips.
- The second is that although the field is not a material object in itself, its sources *are* material and it can only be detected by its effect on matter. Thus, our determination of the properties of a field is, in a sense, always second-hand.
- The third lesson to draw from this demonstration is one of awe. Einstein was in his late sixties when he wrote:

  'I experienced a miracle ... as a child of four or five when my father showed me a compass ... [I realized] there had to be something behind objects that lay deeply hidden.'

The topics covered in this chapter range from the forces between current-carrying wires through to the discovery of the radiation belts surrounding the Earth. However, throughout our exploration of these subjects, one concept will figure prominently — the idea of a *magnetic field*.

The remainder of this chapter is divided into five main sections.

- Section 2 is aimed at defining and describing what we mean by a magnetic field.
- Section 3 is devoted to the generation of magnetic fields by moving charges.
- Section 4 is an examination of the origin of the startling behaviour of the materials that we call 'magnetic'.
- Sections 5 and 6 are aimed at describing the motion of charged particles in uniform and non-uniform fields respectively. These will include descriptions of several technological devices and natural phenomena that can be explained using your new understanding.

2 The magnetic field

2.1 Charged particles in electric and magnetic fields

You will recall from Chapter 1 that the field concept was very useful in describing a force that can act at any point within a specified region. In particular, if there is an electric field $\boldsymbol{\mathscr{E}}(\boldsymbol{r})$ in a certain region such as that shown in Figure 4.2, then the force on a charge q at position $\boldsymbol{r}$ in the region is given by

$$\boldsymbol{F}_{\text{el}}(\boldsymbol{r}) = q\boldsymbol{\mathscr{E}}(\boldsymbol{r}). \tag{4.1}$$

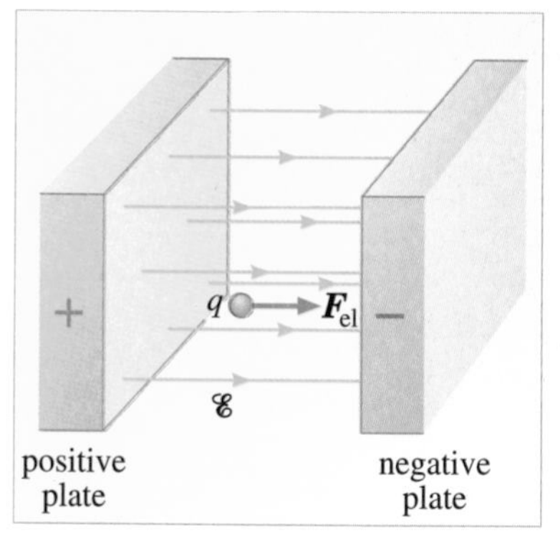

Figure 4.2 The electric force $\boldsymbol{F}_{\text{el}}$ on a positively charged particle located in an electric field between two oppositely charged parallel metal plates. Such an electric field is uniform throughout most of the region between the plates but becomes non-uniform near the edges (not shown).

Equation 4.1 can be used in two ways. A measurement of the force on a known charge serves to define the electric field at that point. Alternatively, if the field is known, it can be used to predict the force on a charge placed in the field. It is important to remember that the electrostatic force acts on all charged particles, whether they are moving or stationary.

Now consider a similar situation — that depicted in Figure 4.3. Instead of being in an electric field as in Figure 4.2, the charge q is now in the region between the poles of a large permanent magnet. By analogy with the electrical case, we may now suppose that the particle is located in a **magnetic field**. Moreover, by extending the analogy, we can suppose (correctly) that the magnetic field is *uniform* throughout most of the region between the pole pieces, so that it can be represented by parallel, uniformly spaced, magnetic field lines. We denote the magnetic field by the symbol $\boldsymbol{B}$ so the magnetic field at a point with position vector $\boldsymbol{r}$ is written as $\boldsymbol{B}(\boldsymbol{r})$. The source of the magnetic field need not concern us here: it is dealt with later in the chapter.

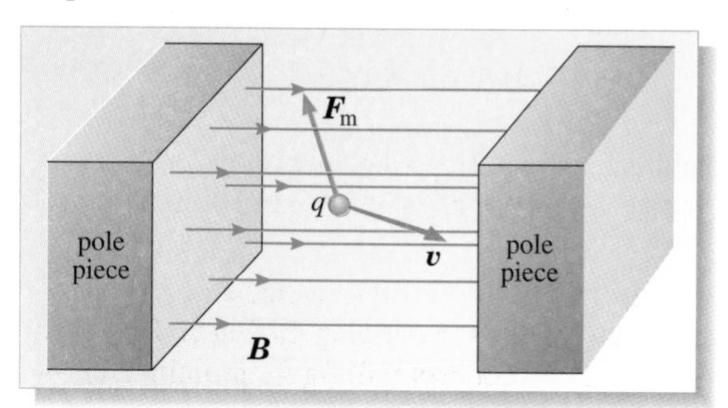

Figure 4.3 The magnetic field between the pole pieces of a large permanent magnet is uniform throughout most of the region, only becoming non-uniform near the edges (not shown). A charged particle (the one shown here is positive), moving with velocity $\boldsymbol{v}$ across the magnetic field in this region, experiences a magnetic force $\boldsymbol{F}_{\text{m}}$ that is perpendicular to both $\boldsymbol{v}$ and $\boldsymbol{B}$.

If there is no electric field in the region between the pole pieces of the magnet, *and* the charge q is stationary, it will experience no force. However, if the charge is *moving across* the magnetic field as shown in Figure 4.3, then it *will* experience a force, even in the absence of any electric field. We call this the **magnetic force** and give it the symbol $\boldsymbol{F}_{\text{m}}$. It is important to note that

For a particle to experience a magnetic force in a uniform magnetic field, it must be (a) charged and (b) moving.

We now want to discover the exact quantitative relationship between the magnetic field and the magnetic force. To do this, we will consider conducting a series of hypothetical experiments in a region such as that shown in Figure 4.3. The experiments consist of firing test particles with a variety of charges, masses and velocities into this region, and then making observations on the changes in motion of the particles. From these experiments we could deduce the acceleration of the particles, and knowing their mass we could then deduce the magnetic force $\boldsymbol{F}_m$ that they experienced. If we imagine carrying out many experiments, then we could determine the dependence of $\boldsymbol{F}_m$ on the test particle's charge q, velocity $\boldsymbol{v}$ and on the relative orientation of $\boldsymbol{v}$ and $\boldsymbol{B}$.

After carrying out a series of experiments and collating the results, we would end up with something like Table 4.1.

Table 4.1 The magnetic force acting on a moving charged particle.

Observation 1:
Particles with charges of the same magnitude but opposite sign that are fired in the same direction and with equal speeds (i.e. with the same velocity) experience magnetic forces of equal magnitude but opposite directions.

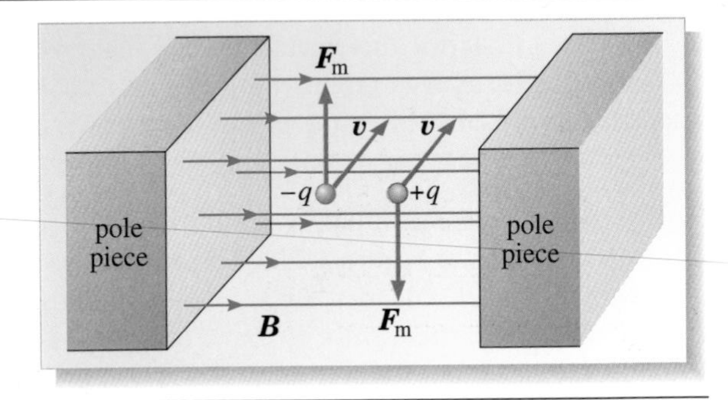

Figure 4.4

Observation 2:
Particles with the same charge, fired with equal but opposite velocities, experience magnetic forces of equal magnitude but opposite directions.

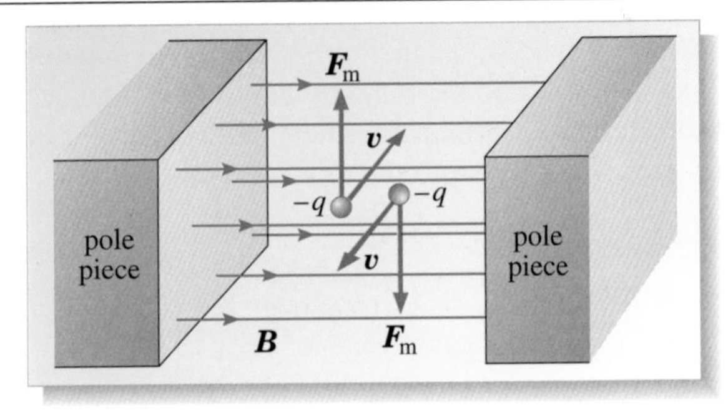

Figure 4.5

Observation 3:
The direction of the magnetic force $\boldsymbol{F}_m$ on a particle at $\boldsymbol{r}$ is invariably at right angles both to its velocity $\boldsymbol{v}$ at $\boldsymbol{r}$ and to the magnetic field $\boldsymbol{B}(\boldsymbol{r})$.

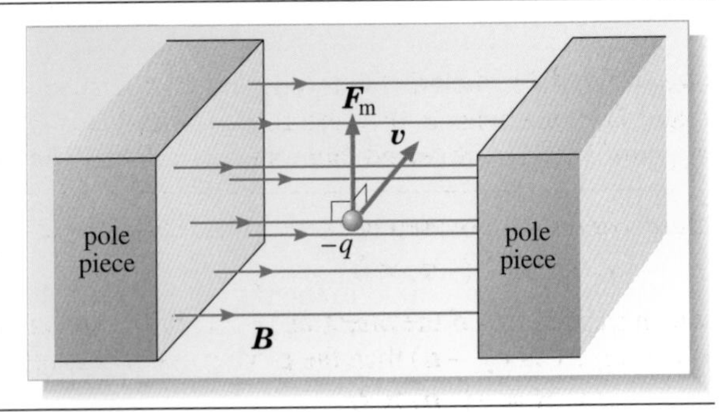

Figure 4.6

Observation 4:
The magnitude of the magnetic force, $|\boldsymbol{F}_m|$, on a particle is always proportional to the magnitude of the particle's charge $|q|$ and the sine of the angle θ between the direction of $\boldsymbol{v}$ and the direction of $\boldsymbol{B}$.

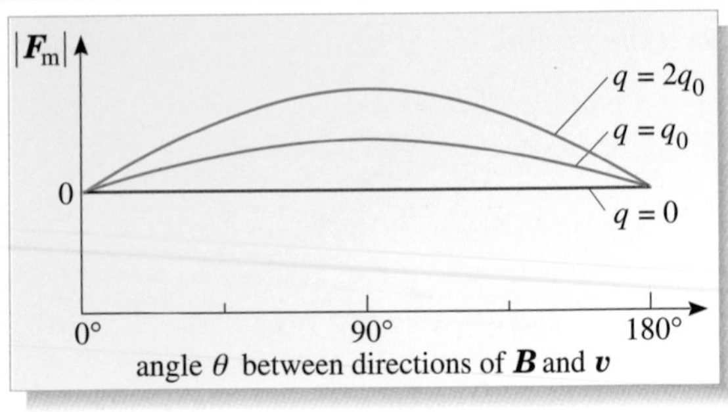

Figure 4.7

As you can see, the results describe quite a complex pattern of behaviour. However, there is a particularly economical way of describing these results. To see this, notice that Observations 3 and 4 describe how one vector $\boldsymbol{F}_\mathrm{m}$ varies as a function of two other vectors $\boldsymbol{v}$ and $\boldsymbol{B}$. This might (I hope) bring to mind the *vector (cross) product* used in the discussion of torque in Chapter 4 of *Predicting motion*. Recall that the vector product of two vectors $\boldsymbol{a}$ and $\boldsymbol{b}$ is written as $\boldsymbol{a} \times \boldsymbol{b}$ and defined to be a vector that is perpendicular to both $\boldsymbol{a}$ and $\boldsymbol{b}$, as given by the right-hand rule. The magnitude of $\boldsymbol{a} \times \boldsymbol{b}$ is given by

$$|\boldsymbol{a} \times \boldsymbol{b}| = ab\sin\theta$$

where θ is the angle between $\boldsymbol{a}$ and $\boldsymbol{b}$.

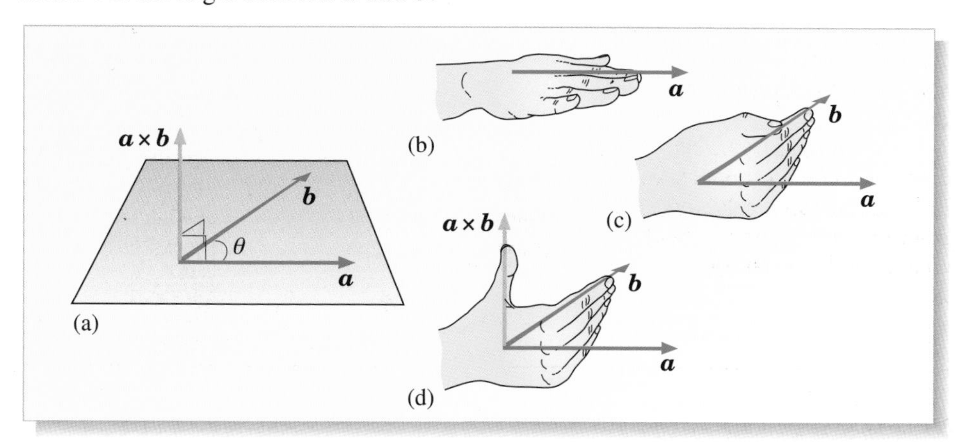

Figure 4.8 (a) The vector product $\boldsymbol{a} \times \boldsymbol{b}$. (b)–(d) The right-hand rule for working out the direction of the vector $\boldsymbol{a} \times \boldsymbol{b}$. *First*, point the straightened fingers of your right hand in the direction of the *first* vector $\boldsymbol{a}$, as in (b). *Secondly*, keep your arm pointing in the direction of the first vector, and turn your wrist until your hand is orientated in such a way that by bending your fingers through less than 180° you can align them with the *second* vector $\boldsymbol{b}$, as in (c). The direction of the product $\boldsymbol{a} \times \boldsymbol{b}$ is then given (approximately) by the direction in which your extended thumb points, as in (d). Vector $\boldsymbol{a} \times \boldsymbol{b}$ is perpendicular to both $\boldsymbol{a}$ and $\boldsymbol{b}$.

Using the cross-product notation, it is possible to incorporate all the observations presented above into a single equation:

$$\boldsymbol{F}_\mathrm{m} = q[\boldsymbol{v} \times \boldsymbol{B}(\boldsymbol{r})]. \tag{4.2}$$

Table 4.2 confirms that Equation 4.2 does indeed encompass the four observations from Table 4.1.

Table 4.2 Confirmation of the observations in Table 4.1.

Observation 1 is included within Equation 4.2 since if we change the sign of q, but keep $\boldsymbol{v}$ and $\boldsymbol{B}(\boldsymbol{r})$ unchanged, then the equation predicts that the direction of the magnetic force $\boldsymbol{F}_\mathrm{m}$ will indeed be reversed, but its magnitude will be unchanged.

Observation 2 is covered because for a cross product

$$\boldsymbol{a} \times \boldsymbol{b} = -(-\boldsymbol{a}) \times \boldsymbol{b}$$

which implies that if the direction of one of the vectors in a cross product is reversed (i.e. if $\boldsymbol{a}$ is replaced by $-\boldsymbol{a}$) then the product itself is reversed in sign. In this case, we have

$$\boldsymbol{v} \times \boldsymbol{B} = -(-\boldsymbol{v}) \times \boldsymbol{B}$$

which implies that

$$\boldsymbol{F}_\mathrm{m}(\boldsymbol{v}) = -\boldsymbol{F}_\mathrm{m}(-\boldsymbol{v}).$$

Observation 3, that $\boldsymbol{F}_\mathrm{m}$ is perpendicular both to $\boldsymbol{v}$ and to $\boldsymbol{B}$, is included since the vector product $\boldsymbol{v} \times \boldsymbol{B}$ is by definition perpendicular to both $\boldsymbol{v}$ and $\boldsymbol{B}$.

Observation 4 is included because the magnitude of $\boldsymbol{v} \times \boldsymbol{B}$ is $vB\sin\theta$ and so according to Equation 4.2 we can write the magnitude of the magnetic force as

$$F_\mathrm{m} = |q|vB\sin\theta. \tag{4.3}$$

The right-hand side of Equation 4.3 includes the speed v, and the sine of the angle θ, in accordance with our observation.

Question 4.1 Use the right-hand rule and Equation 4.2 to convince yourself that the particle in Figure 4.3 is indeed, as stated in the caption, positively charged. ■

The predictions of Equations 4.2 and 4.3 actually go slightly beyond our observations. The third quantity on the right-hand side of Equation 4.3 is the *magnitude of the magnetic field*, B. This did not feature in the list of observations in Table 4.1 because we only considered a single field, and its magnitude was fixed. Equation 4.3 asserts that the magnitude of the force is proportional to the magnitude of the magnetic field. The magnitude of the magnetic field, B, at a point is usually called the **magnetic field strength** at that point. If the point is defined by the position vector $\boldsymbol{r}$, then the magnetic field strength is denoted by $B(\boldsymbol{r})$. Note that although the B is not bold (because it denotes the *magnitude* of a quantity), the $\boldsymbol{r}$ *is* emboldened because it represents the position *vector* of the point at which B is measured.

Question 4.2 (a) It is sometimes stated that 'a charged particle must move *across* a magnetic field in order to experience a magnetic force'. Explain this statement. (*Hint*: What is the value of θ if the particle is moving parallel or anti-parallel to the field?)

(b) In what direction must the particle move if it is to experience the maximum magnetic force? ■

In summary, Equation 4.2, $\boldsymbol{F}_{\rm m} = q[\boldsymbol{v} \times \boldsymbol{B}(\boldsymbol{r})]$, performs the equivalent role for magnetic fields that Equation 4.1, $\boldsymbol{F}_{\rm el}(\boldsymbol{r}) = q\boldsymbol{\mathscr{E}}(\boldsymbol{r})$, performed for electric fields. These equations predict the forces on charged particles, given particular arrangements of electric and magnetic fields.

2.2 Magnetic fields and magnetic field strengths

It is important to appreciate that the general relationship between the magnetic force and the magnetic field

$$\boldsymbol{F}_{\rm m} = q[\boldsymbol{v} \times \boldsymbol{B}(\boldsymbol{r})] \qquad \text{(Eqn 4.2)}$$

can be used in two quite distinct ways.

- First, it can be used to *determine* both the magnitude and the direction of the magnetic field. If we measure the magnetic forces, and know the charges and speeds of the particles involved, we can use Equation 4.2 to determine $\boldsymbol{B}(\boldsymbol{r})$. Since we have been more than a little vague in describing the magnetic field so far, we can be particularly happy that we now have a practical way to determine $\boldsymbol{B}(\boldsymbol{r})$.
- Secondly, if $\boldsymbol{B}(\boldsymbol{r})$ is known, it can be used to determine the magnitude and the direction of the forces on charged particles as they move through the field.

This is directly analogous to the two uses of Equation 4.1 ($\boldsymbol{F}_{\rm el} = q\boldsymbol{\mathscr{E}}$) described at the beginning of Section 2.1. Let's look at each of these two uses of Equation 4.2 in turn.

Determining the magnetic field

Question 4.3 gives you a chance to see how one could determine the magnetic field strength in a particularly simple situation. Try it now before continuing.

Question 4.3 Figure 4.9 shows a situation in which a particle with a positive charge q passes through the point defined by position vector $\boldsymbol{r}$ travelling with speed v in a direction perpendicular to the magnetic field. If the magnitude of the magnetic force on this particle at $\boldsymbol{r}$ is F_m, derive an expression for the magnetic field strength at $\boldsymbol{r}$. ■

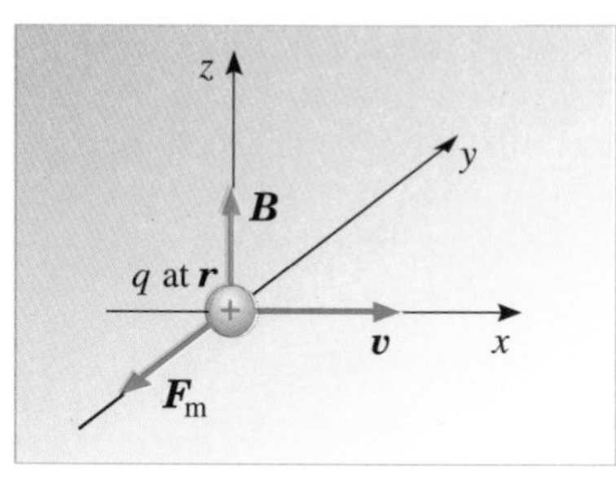

Figure 4.9 For use with Question 4.3.

Units of magnetic field strength

From the equation you have derived in Question 4.3, we can now work out the units in which magnetic field strength is measured. Since F_m is measured in newtons, q in coulombs, and v in metres per second, it *must* be the case that B can be measured in $\mathrm{N\,C^{-1}\,s\,m^{-1}}$. This unit is given a special name — the **tesla** (after the physicist Nikola Tesla) — and is denoted by the symbol T. Another way of stating this is to say that when a particle of charge 1 C moves perpendicularly to a magnetic field of strength 1 T at a speed of $1\,\mathrm{m\,s^{-1}}$, it experiences a force of magnitude 1 N.

$$1 \text{ unit of magnetic field strength} = 1\,\mathrm{N\,C^{-1}\,s\,m^{-1}} = 1\,\mathrm{T}.$$

In fact, a field strength of one tesla represents a fairly substantial magnetic field. Figure 4.10 (overleaf) provides a list of some of the magnetic field strengths of interest to physicists. The key feature of the diagram that you should note is the enormous range of magnetic fields found in nature.

Nikola Tesla (1856–1943)

Nikola Tesla was born in 1856 to a Serbian family in Croatia. He studied engineering at the University of Graz before emigrating to the USA in 1884. For a time he worked with Thomas Edison, but they fell out over money and seemed to argue from then on. Tesla was a brilliant inventor. His main achievement was to develop ways of transforming electricity to and from high voltages, allowing it to be transmitted efficiently. This a.c. technique was opposed by Edison who favoured a d.c. method. Eventually, in spite of Edison's reputation, the advantages of Tesla's method became clear and it was adopted. Tesla refused to share a Nobel Prize with Edison, so great was his resentment of the more famous man. His lasting fame comes from the name given to the unit of magnetic field strength, the tesla. He became at best eccentric and possibly completely deranged in his later years, and died in New York in 1943.

Magnetars are a recently (1999) discovered, special class of neutron star generating the most intense magnetic field ever observed in the Universe.

The gauss (1 gauss = 10^{-4} T) is a unit of field strength in a non-SI system of units. It is in widespread use, so you may even find it in textbooks that otherwise stick to SI units.

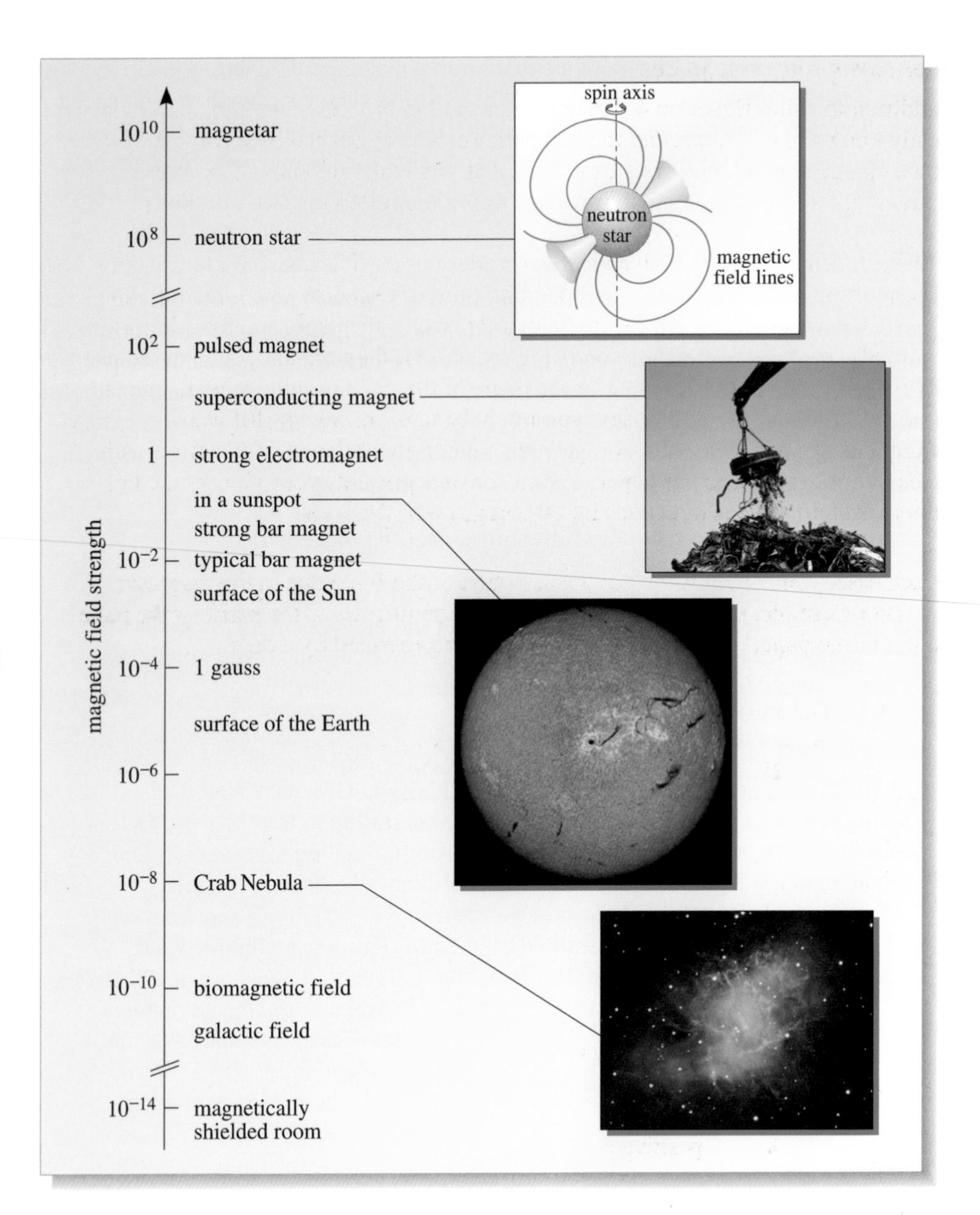

Figure 4.10 Some typical magnetic field strengths.

Magnetic field direction

Using Equation 4.2, we can deduce not only the strength but also the direction of a magnetic field. Try Questions 4.4 and 4.5 to make sure you can see how.

Question 4.4 A particle with a positive charge travelling in the x-direction experiences a force in the $-y$-direction. What is the direction of $\boldsymbol{B}$?

Question 4.5 If the particle in Question 4.4 above has a charge +1 nC and travels along the x-axis of a coordinate system with a speed of $100\,\mathrm{m\,s^{-1}}$, it experiences a force of $1\,\mu\mathrm{N}$. What is the magnitude of the magnetic field? (*Remember, the prefixes* μ *and* n *stand for micro* (10^{-6}) *and nano* (10^{-9}) *respectively.*) ■

Determining magnetic forces

In addition to using Equation 4.2 to determine and *define* $\mathbf{B}(\mathbf{r})$, we can, if $\mathbf{B}(\mathbf{r})$ is already known, use it to *predict* the magnetic force experienced by a particle of known charge and velocity. You can check that you really do know how to use Equation 4.2 to predict magnetic forces by doing the following two questions.

Question 4.6 This question concerns a variety of particles, and Table 4.3 gives the sign of the particle's charge, and the directions of $\boldsymbol{v}$ and $\boldsymbol{B}$. The directions are shown as an arrow if they happen to be in the plane of the page, and by a dot (•) if they are directed out of the page, and by a cross (×) if they are directed into the page. This notation for vectors in or out of the plane of the page is quite general, and is intended to remind you of the view you might have of the vector if it were represented by an arrow. The dot represents your view of the tip of the arrow as it comes out of the page towards you, and the cross represents your view of the tail feathers of the arrow as it recedes into the page away from you.

In each case, you should work out the direction of the force due to the magnetic field. (For example, the answer to part (a) is 'perpendicular to the plane of the page and out of the paper' and so the answer may be represented by a dot).

Table 4.3 For use with Question 4.6.

	$\boldsymbol{v}$	$\boldsymbol{B}$	q	$\boldsymbol{F}_{\rm m} = q[\boldsymbol{v} \times \boldsymbol{B}(\boldsymbol{r})]$
(a)	→	↑	positive	•
(b)	↗	↘	positive	
(c)	↗	↘	negative	
(d)	$= \mathbf{0}$	↖	negative	
(e)	×	↑	negative	
(f)	↓	•	positive	

Question 4.7 What is the magnitude and direction of $\boldsymbol{F}_{\rm m}$ in each of the following cases?

(a) $q = 2\,\text{C}$; $|\boldsymbol{v}| = 10^4\,\text{m}\,\text{s}^{-1}$ and $|\boldsymbol{B}| = 2\,\text{T}$.

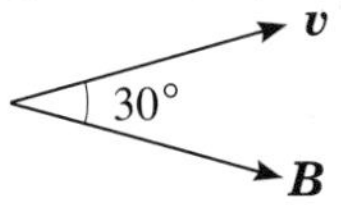

$\boldsymbol{v}$
30°
$\boldsymbol{B}$

(b) $q = -1.6 \times 10^{-19}\,\text{C}$; $|\boldsymbol{v}| = 10^{-1}\,\text{m}\,\text{s}^{-1}$ and $|\boldsymbol{B}| = 10^{-4}\,\text{T}$. ■

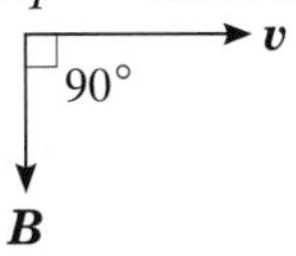

$\boldsymbol{v}$
90°
$\boldsymbol{B}$

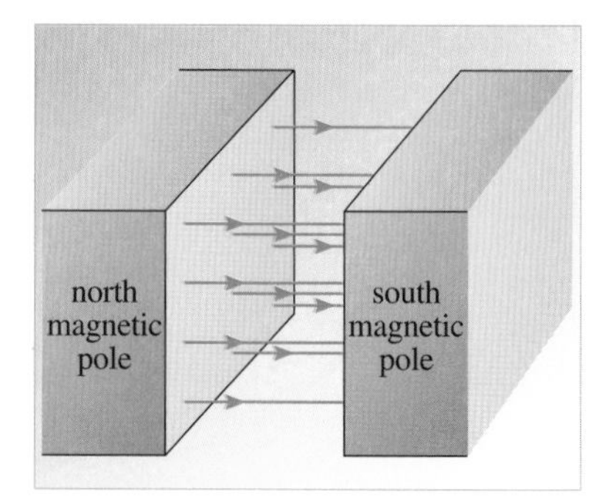

Figure 4.11 The uniform magnetic field between the pole pieces of a large magnet. The magnetic field lines point from the north to the south pole.

2.3 Representations of magnetic fields

So far in this chapter, the only type of magnetic field that we have considered is the uniform magnetic field that exists in the central region between the pole pieces of a large permanent magnet. Such a field is shown again in Figure 4.11, but this time a further piece of information has been added: the magnetic poles have been labelled *north* and *south*. The uniform field between the pole pieces is represented by magnetic field lines which, in this case, are equally spaced parallel lines that point *away* from the north pole and *towards* the south pole.

The rules for drawing magnetic field lines are the same as for electric field lines:

1 the tangent to the field line is parallel to the magnetic field at that point;

2 the lines are close together where the field is strong and further apart where it is weaker.

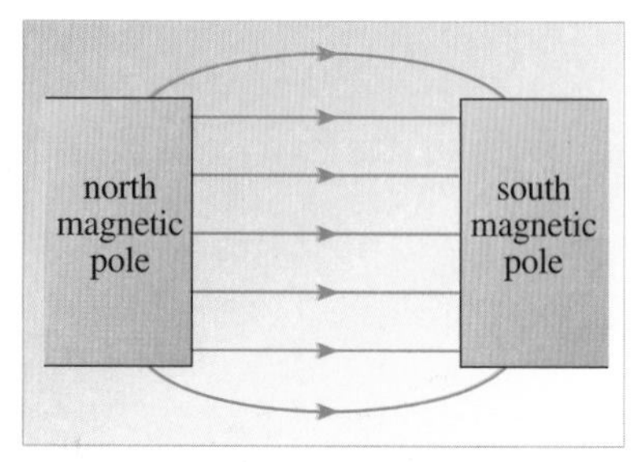

Figure 4.12 The magnetic field in the space between the poles of a large magnet shown as field lines in a two-dimensional cross-section through the region. In this picture, the non-uniform edge effects have been included.

Of course, the magnetic field shown in Figure 4.11 is a particularly simple one. Usually, magnetic fields are much more complex, and, as in the case of electric fields, we show the field lines in planar sections through the system. Then, the field between the pole pieces of the large magnet would be represented by something like Figure 4.12.

Figure 4.13 shows the field of a small permanent magnet such as one might find in a toyshop or laboratory. Note again that the magnetic field lines emerge from the north magnetic pole and converge to the south magnetic pole.

Magnetic fields such as that shown in Figure 4.13 are often revealed by the use of either small compasses or iron filings. If a compass is placed in a magnetic field, the needle (which is just a small, very weak magnet) will line up with, and point in the direction of, the field at that point. Thus, if sufficient compasses are used, the magnetic field can easily be revealed (Figure 4.14a).

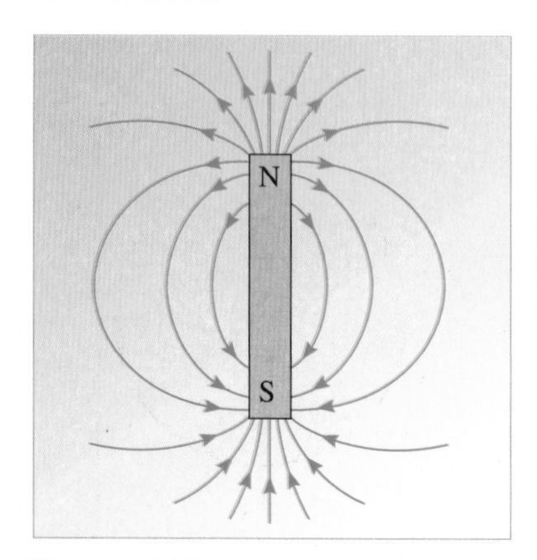

Figure 4.13 The magnetic field of a bar magnet. We show a section through the field, which is, of course, in reality, three-dimensional.

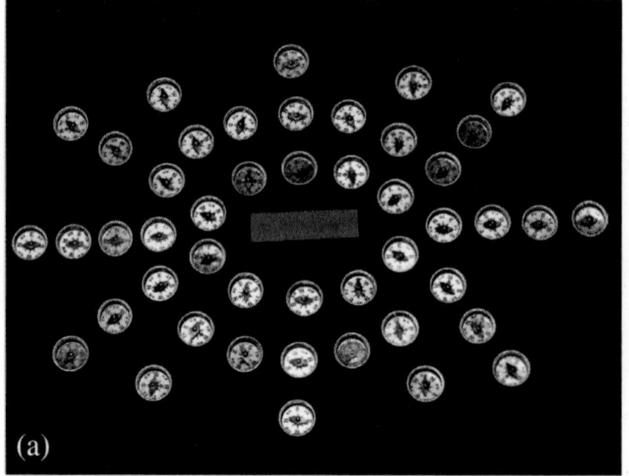

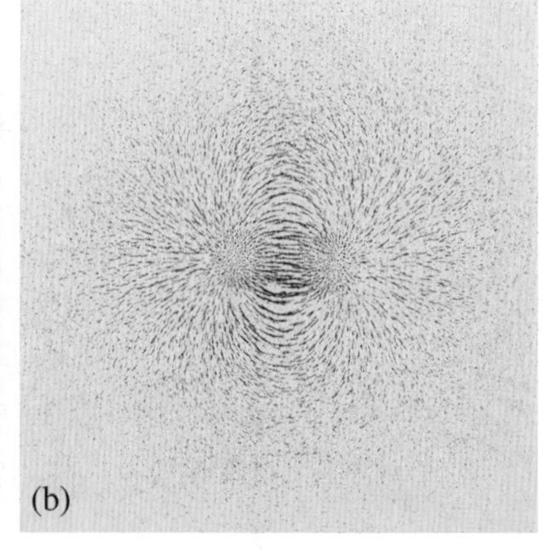

Figure 4.14 (a) The field of a bar magnet revealed using many small compasses. (b) The field of a bar magnet revealed using iron filings.

Iron filings can reveal the shape of a magnetic field if they are sprinkled on a piece of paper placed over the magnet (Figure 4.14b). When the paper is given a few gentle taps, the iron filings line up with the field just as the compass needles did. What happens in this case is that the magnetic field 'magnetizes' the iron filings.

Each individual filing becomes a small magnet, which acts like a miniature compass needle, lining up with the magnetic field in its vicinity and attracting other filings. This method reveals the shape, but not the direction, of the magnetic field.

A fact that is well known to anyone who has played with small bar magnets or with toys that utilize them (Figure 4.15), and the reason that compasses work at all, is that magnets exert forces on other magnets just as they do on moving charged particles. In fact, the two effects are basically the same, but the connection between them is very subtle. Although the physical processes responsible for the forces between permanent magnets are complicated, there is a simple and well-known rule that summarizes a great deal of practical experience:

Like poles repel and unlike poles attract.

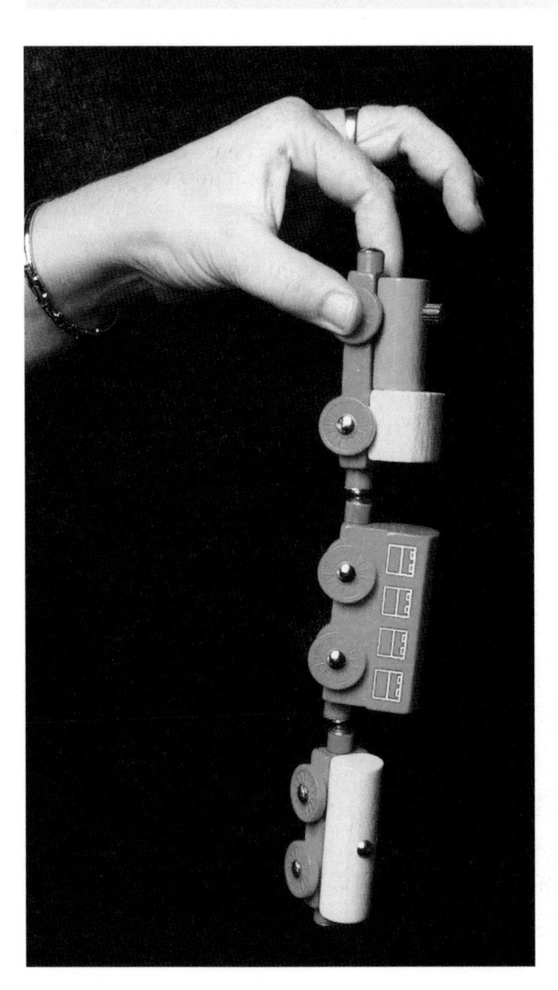

Figure 4.15 A toy train. The engine and carriages are held together by means of small magnets attached at the front and rear of each one.

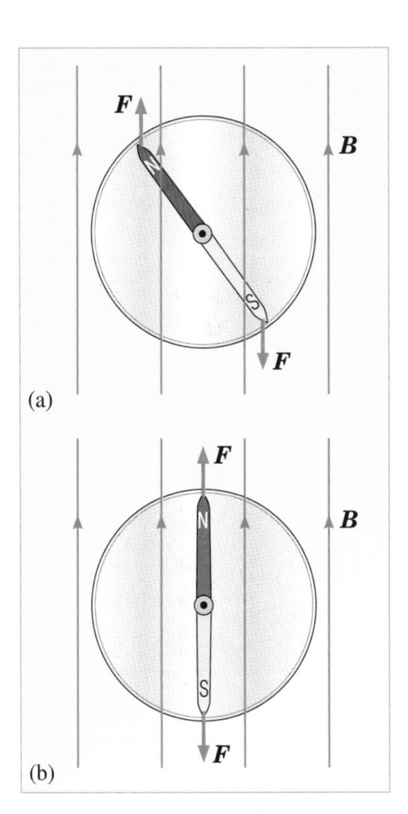

Figure 4.16 (a) The torque on a compass needle in a magnetic field. (b) The torque is zero once the compass needle is aligned with the magnetic field.

This rule explains, though not at any profound level, why a compass needle aligns itself with a magnetic field. The needle, which is really just a tiny bar magnet, experiences a torque due to the magnetic field in which it is located (Figure 4.16). The forces exerted on the needle are such that it lines up with the local magnetic field.

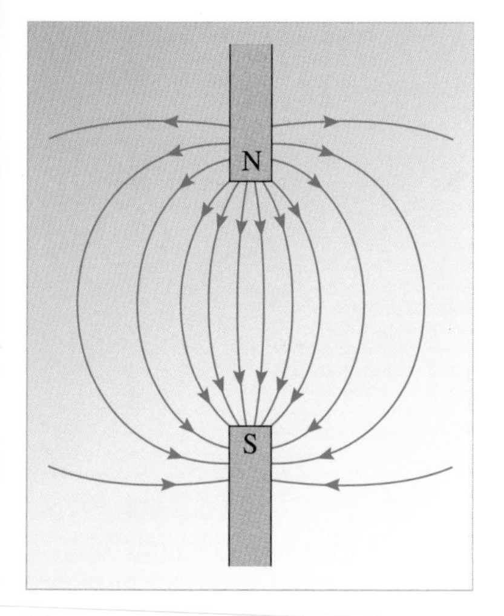

Figure 4.17 One of the many possible field patterns that can be obtained by combining the individual magnetic fields of two bar magnets. (As with Figure 4.13, this is a two-dimensional representation.)

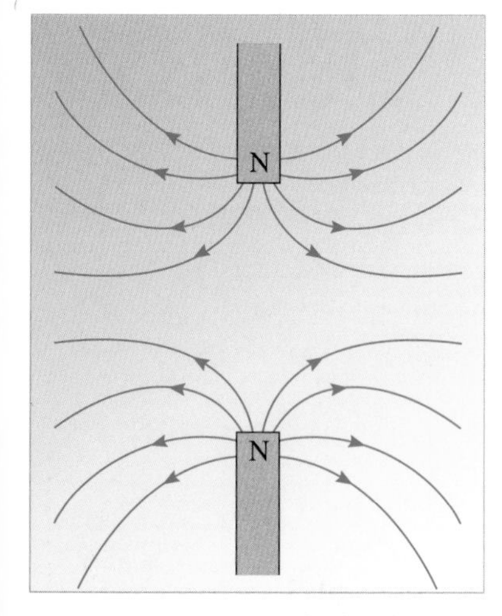

Figure 4.18 A different arrangement of the two magnets shown in Figure 4.17 produces a different total magnetic field.

On the whole, compass needles are such weak magnets that they can be used to determine the direction of a magnetic field without themselves having much influence on the total magnetic field. However, when two magnets of roughly comparable strength are brought together, the total field will be a combination of the fields due to each of the two magnets. In fact, the magnetic fields are combined in exactly the same way as electric fields were combined in Chapter 1, i.e. by means of vector addition.

The resultant magnetic field at any point is equal to the vector sum of each of the individual magnetic fields at that point.

Figures 4.17 and 4.18 show two different arrangements of two identical bar magnets. As you would expect, the fields are different in the two cases.

2.4 Some fundamental questions about magnetic fields

Poles, dipoles and monopoles?

Before proceeding any further with our discussion of magnetic fields, it is important to dispel any idea that might be forming in your mind that there is something fundamental about the poles of a permanent bar magnet. It may well appear to you from the preceding discussion that magnetic field lines begin on north poles and end on south poles, and that if we chopped the magnet up finely enough we would eventually be able to isolate an elementary north or south pole. *This is not the case.* If you tried to isolate the poles of a magnet by cutting it in half, you would simply end up with two smaller bar magnets as shown in Figure 4.19. No matter how finely you chopped up the magnet, each individual fragment would always have both a north and a south magnetic pole: it would be what is called a **magnetic dipole**. Which brings us back to the question of where magnetic field lines start and end.

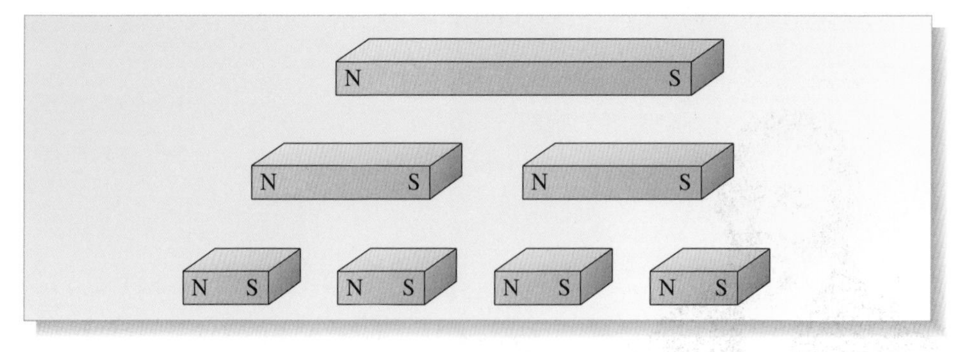

Figure 4.19 By cutting up a magnet, you simply obtain a succession of smaller magnets. The individual poles cannot be isolated.

To solve this problem, let us return to the analogy that we made at the beginning of the chapter. We first recalled that a charge placed in an electric field would experience an electrostatic force and then we pointed out that a *moving* charge in a *magnetic* field would also experience a force — a magnetic force. In the first situation, the electric field originated on the individual charges on the metal plates, but, in the second situation, where does the magnetic field originate? Well, in the same way that magnetic fields are *detected* by moving charges, they are also *created* by moving charges, and this will be the subject of the next section. Let us here consider the simple situation of a point charge moving uniformly in a circular path as shown in Figure 4.20. The moving charge then forms a current loop and the magnetic field is that of a simple magnetic dipole with the field lines as shown on the figure. As you can see, the magnetic field lines *do not start or end anywhere*, but form *closed loops* around the current loop. Permanent magnets are the result of elementary charges such as electrons moving on the atomic scale and lining up with each other to generate large magnetic fields. The magnetic field lines do not start at the poles of the magnet but actually form closed loops passing through the material of the magnet in a manner such as that illustrated in Figure 4.21 (the detailed nature of magnetic materials is addressed in Section 4). Alternatively, macroscopic magnetic fields can be generated by macroscopic currents flowing in wires or other conducting environments. This will be the subject of the next section.

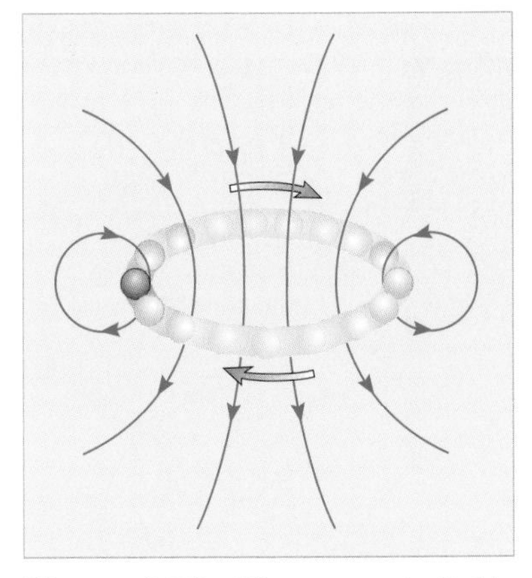

Figure 4.20 The magnetic field produced by a point charge moving in a circular path. You need to think of the charge as if it were spread out over the whole circumference of the circle.

Thus, even by going down to atomic levels it is impossible to isolate a magnetic pole — a **monopole** (Figure 4.22). Indeed, the classical theory of electromagnetism is founded on the belief that magnetic monopoles do not exist. However, physicists are now re-examining this theory in the context of the special circumstances pertaining in the very early Universe. In recent years, detailed studies have been made about the conditions under which magnetic monopoles might be produced. Such studies indicate that, even if magnetic monopoles can exist, their formation requires so much energy that they are unlikely ever to be created in a laboratory experiment. However, it may be possible to detect some of the indirect effects of magnetic monopoles in laboratory experiments, or even to find some monopoles left over from the early stages of the evolution of the Universe, when highly energetic processes were more common than they are now.

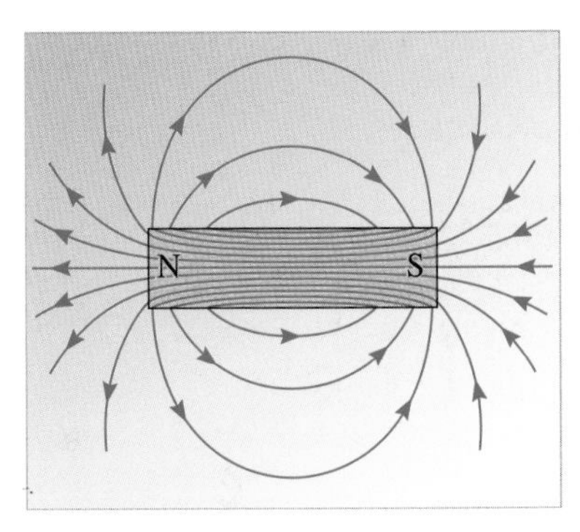

Figure 4.21 The magnetic field lines of a bar magnet.

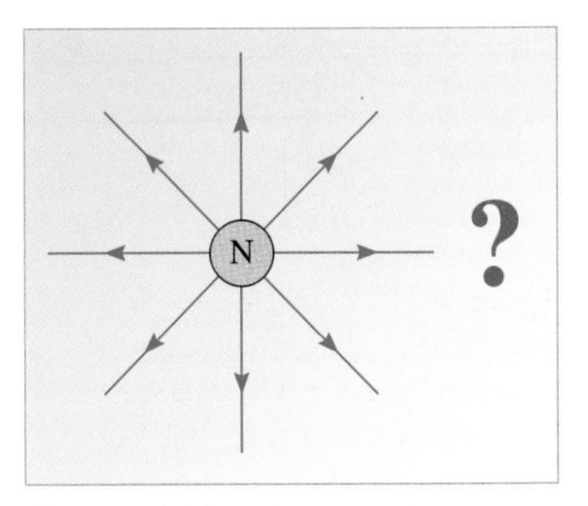

Figure 4.22 A magnetic monopole?

3 The generation of magnetic fields

We have already seen that there is a link between electricity and magnetism. In fact, the first experimental evidence for this relationship came several years before Equation 4.2 was first written down. The evidence arose from the work of the Danish scientist and philosopher, Hans Christian Oersted (Figure 4.23).

Figure 4.23 Hans Christian Oersted (1777–1851) was born at Rudkøbing in Denmark. He began his career as a pharmacist but became professor of physics at Copenhagen in 1806 and, in 1829, director of the Polytechnic Institute in Copenhagen. In 1820, he discovered the link between electric currents and magnetic fields, thus founding the science of electromagnetism.

It could be said that Oersted's inspiration came in a flash, for he was partly prompted to undertake his investigations by phenomena associated with lightning. Oersted knew that lightning sometimes affected compass needles, causing them to swing around. He also knew that lightning was essentially electrical in nature — a kind of

In Figure 4.24, we are using the same convention for currents flowing into or out of the page as was introduced in Question 4.6 for magnetic fields pointing into or out of the page.

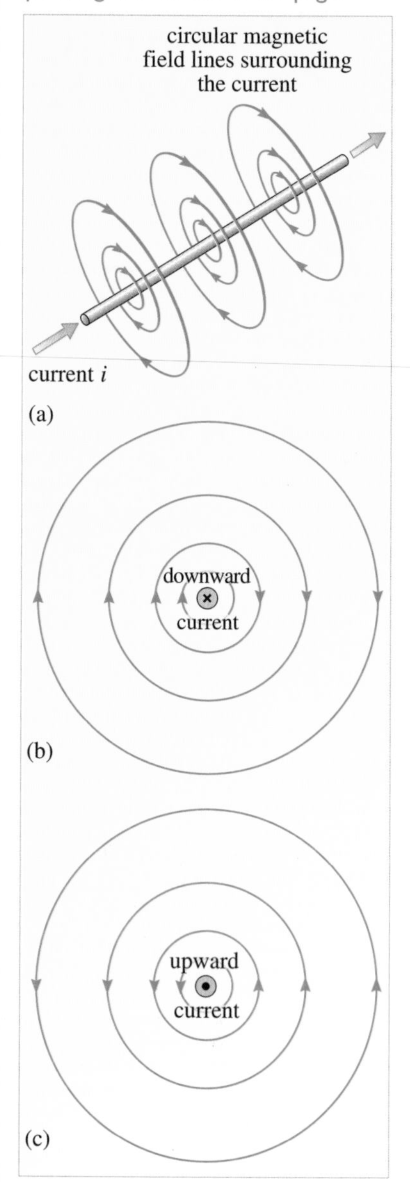

Figure 4.24 The magnetic field of a long straight wire carrying a steady current, i. (a) The field lines form concentric circles around the wire in planes at right angles to the direction of the current flow. The direction in which the field lines point is determined by the direction in which the current flows. (b) When the (conventional) current flows into the page, the field lines form clockwise circles. (c) When the current flows out of the page, the circles are anticlockwise.

electric current. These facts, combined with certain strongly held philosophical convictions, led Oersted to suspect that an electric current might well exhibit a 'magnetical effect' under the right conditions. Using a compass as a magnetic detector, Oersted was able to verify his suspicion and show that electric currents can indeed produce 'magnetical effects' — or, as we would say now, electric currents produce magnetic fields. Oersted's discovery, in 1820, of the link between electricity and magnetism, marked the creation of a new branch of physical science — **electromagnetism**.

When Oersted performed his original experiments, he used a thin platinum wire and a large electric current. The passage of the electric current heated the wire and made it glow, a fact that Oersted believed would increase the similarity between his experiment and a real lightning storm. In fact, as Oersted soon discovered for himself, it was not at all necessary to go to such lengths. Any electric current produces a magnetic field, so the problem is only to make sure the field is strong enough to be detected.

3.1 Magnetic field due to a current in a long straight wire

Figure 4.24a shows what is probably the simplest example of a magnetic field produced by a current. In this case, the field is due to a steady current i flowing through a long straight piece of wire. As you can see from the figure, the field lines form concentric circles around the wire. The direction in which the circular field lines point depends on the direction in which the current flows. If the direction of current flow is reversed, then the direction of the field lines is also reversed; this is shown in Figure 4.24b and c. Fortunately, there is an easy way to remember these directions. Just close the palm of your right hand and point your thumb in the direction of the (conventional) current shown in either Figure 4.24b or c. In either case, you will find that your fingers curl around your thumb in exactly the same way that the magnetic field lines curl around the wire carrying the electric current. This simple technique for remembering the direction of the magnetic field is illustrated in Figure 4.25, and will be referred to as the **right-hand grip rule**. (The same rule was used in *Predicting motion* to associate angular velocity vectors with a given sense of rotation about an axis.)

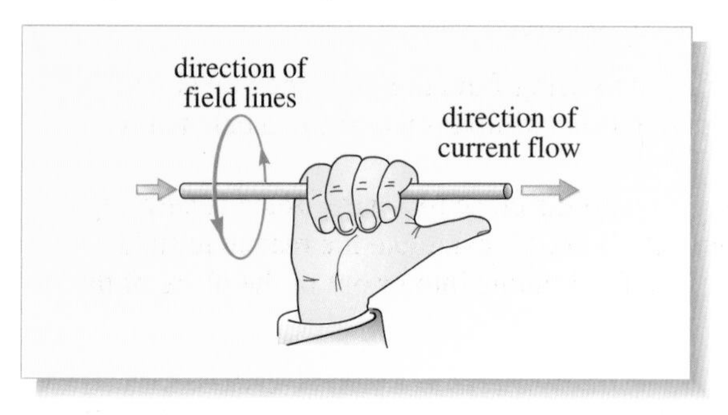

Figure 4.25 If you imagine using your right hand to 'grip' a current-carrying conductor, you can work out the direction of the magnetic field. When your thumb points in the direction of (conventional) current flow, your curled fingers indicate the direction of the magnetic field. Do not confuse this rule with the 'right-hand rule' for vector cross products!

Apart from its direction, the other piece of information that we want to know about the magnetic field is its strength. It can be shown that at a perpendicular distance r from the wire (and *outside* the wire), the field strength $B(r)$ is given by

$$B(r) = \frac{\mu_0 i}{2\pi r} \quad \text{for a long straight current} \tag{4.4}$$

where μ_0 is a constant.

You can see that the form of the equation is sensible. The larger the current in a wire, the stronger the magnetic field that results. The further you are from the wire (larger r), the weaker the magnetic field.

Notice that in Equation 4.4, the field strength has been written as $B(r)$ instead of $B(\mathbf{r})$. This is because in this case the field strength at any point is determined entirely by its perpendicular distance r from the wire. This simplification occurs because the field produced by the wire is *highly symmetrical*. If you imagine a set of coaxial cylinders drawn around the wire, as shown in Figure 4.26, the field strength has the same value at all points on any one of the cylinders. However, the field strength is different on each of the different cylinders. So, for a given current i, the value of B at any point is determined by the cylinder upon which the point lies, and the simplest way of specifying the cylinder is to quote the appropriate value of its radius r. Another notational point concerns the current i. For simplicity, and to avoid the unnecessary use of modulus signs, it will be assumed throughout this chapter that the direction of i is known (or can be chosen) and that i is a positive quantity.

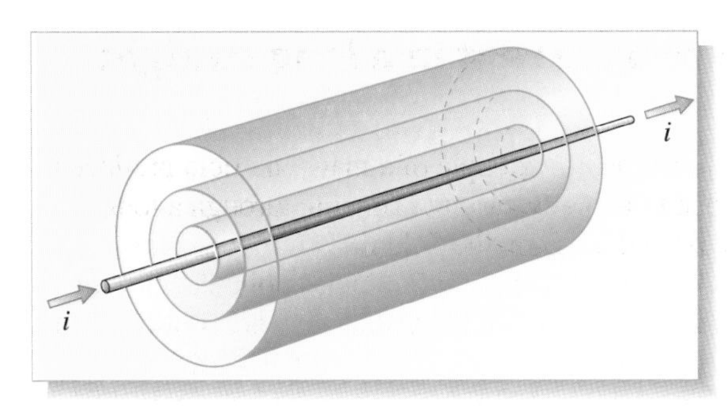

Figure 4.26 Imaginary cylindrical surfaces surrounding the current. The strength of the magnetic field is the same at all points on any given cylinder.

The constant μ_0 is called the **permeability of free space**. It is frequently encountered in electromagnetism just as ε_0 is common in electrostatics. We can see from Equation 4.4 that, if i is measured in amps, B is measured in teslas and r is measured in metres, then the units of μ_0 are $\mathrm{T\,m\,A^{-1}}$. The value of μ_0 is

$$\mu_0 = 4\pi \times 10^{-7}\,\mathrm{T\,m\,A^{-1}}.$$

The surprisingly simple value of μ_0 is a consequence of the way in which the ampere is defined, as you will discover in Section 5.

Question 4.8 will give you some practice in using Equation 4.4, and you should be pleased to note that you should be able to do it without recourse to a calculator!

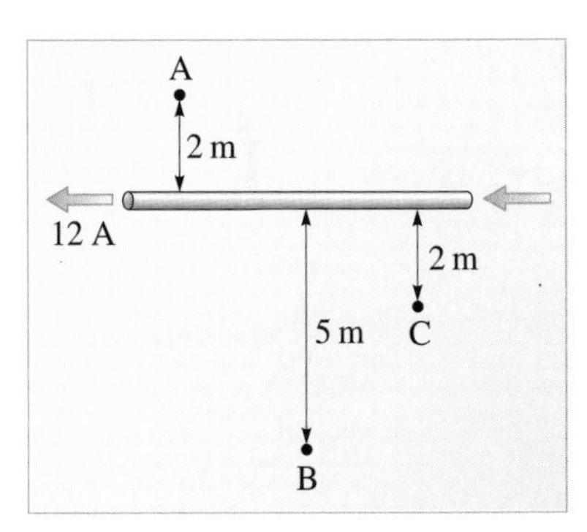

Figure 4.27 The three points referred to in Question 4.8. The current in the wire is flowing in the direction indicated by the arrow. The wire and the three points all lie in the same plane.

Question 4.8 Figure 4.27 shows three points close to a wire which is carrying a current of 12 A. For each of the points A, B and C, calculate the magnetic field strength, and state whether the magnetic field points into or out of the plane of the page. ■

3.2 Magnetic field due to a current in a circular loop of wire

What magnetic field would result from passing a steady current round a circular loop of wire? We have already seen that this current configuration is of fundamental importance in magnetism, and Figure 4.20 illustrated the form of the magnetic field near a current loop. Let's try to see how this field arises using what we already know about the field around a long straight current.

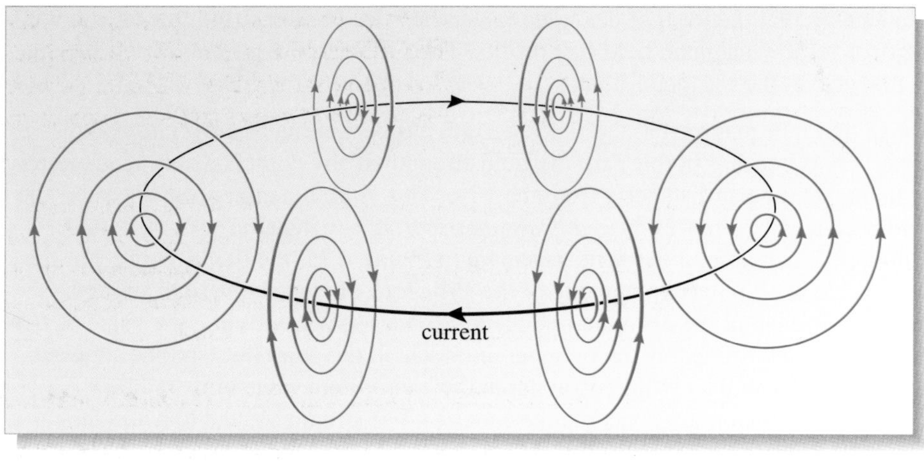

Figure 4.28 A rough approximation to the magnetic field of a circular loop of wire carrying a steady current. A more accurate representation of the field is given in Figure 4.29.

To work out roughly what we might expect, imagine taking the long straight wire of Figure 4.26 and bending it into a circle. At points very close to the wire, the field lines will be concentric circles — at least to the extent that the influence of other parts of the wire can be ignored — as shown in Figure 4.28. Notice that in the centre of the wire loop, the fields due to each individual part of the wire reinforce one another: all the field lines point downwards through the middle of the loop. This is important from a practical point of view because it provides us with a way to generate a relatively large magnetic field in that region.

Of course, in a full calculation the points on the wire cannot be considered in complete isolation, and the pattern of field lines sketched in Figure 4.28 is therefore only an approximate representation. To obtain a more accurate representation, it is necessary to add vectorially all the magnetic fields created by the various parts of the wire, and the true field pattern is shown in Figure 4.29. Calculation predicts that the magnetic field strength at the centre of the loop will be

$$B_{\text{centre}} = \frac{\mu_0 i}{2R} \quad \text{for a circular current loop} \qquad (4.5)$$

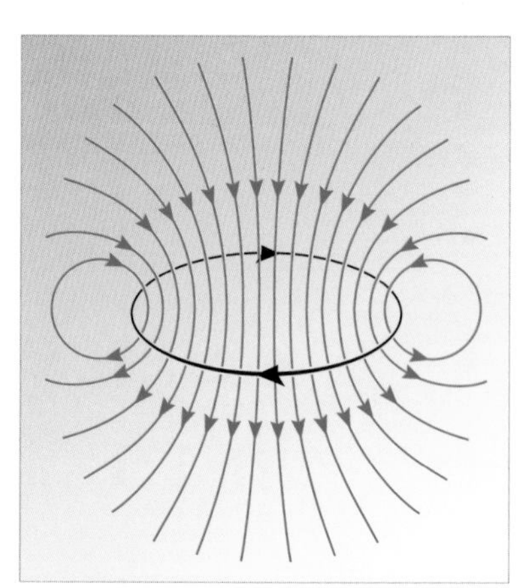

Figure 4.29 An accurate representation of part of the field of a circular current loop. A diagram showing a two-dimensional pattern of field lines in a single plane perpendicular to the loop. In reality, the field lines form a three-dimensional pattern which can be visualized by 'rotating' the pattern around the axis of the loop.

where i is the current and R is the radius of the loop. (Remember, we have, in this context, chosen the directions such that i is positive.) We can see that the form of the equation is plausible: first, the larger the current in a loop of given radius, the stronger the magnetic field that results, and secondly, the larger the radius of the loop (i.e. the further away from the wire is the centre), the weaker the field will be in the middle of the loop. You might also like to check for yourself that the units on the right-hand side of Equation 4.5 do reduce to teslas.

The flat coil

As noted above, bending a current-carrying wire into a loop results in a relatively large magnetic field at the centre of the loop. We can fairly easily increase this field further by looping the wire many times around the same path. Using the *principle of superposition*, the resultant field is the vector sum of the magnetic fields of each loop considered individually. Thus if we have N turns of wire in a loop, the magnetic field at the centre of the loop is just N times the right-hand side of Equation 4.5:

$$B_{\text{centre}} = \frac{\mu_0 N i}{2R} \qquad (4.5a)$$

Question 4.9 Consider a current of 1 A flowing in a wire. (a) If the wire is a long straight wire, what is the magnitude of the field 10 cm from the wire? (b) If the wire is bent into a loop of radius 10 cm, what is the field at the centre of the loop? (c) If the wire is bent into a loop of radius 10 cm and wound around 10 times, what would you expect the field to be at the centre of the loop? ■

The magnetic field associated with a circular current loop is known as a magnetic dipole field. The strength of this field is usually given in terms of the *magnetic dipole moment* of the loop. We will not go into the detailed definition of this quantity, but if an object is described as possessing a magnetic dipole moment, it simply means that it has a magnetic dipole field associated with it.

3.3 Magnetic field due to a current in two coils of wire

For a single coil, calculations show that the magnetic field falls off rapidly as one moves away from the loop along its axis. Figure 4.30 shows how the magnitude of the field varies as one travels along the axis of a coil. In the example of a coil of radius 10 cm given in Question 4.9b, just 5 cm along the axis of the loop, away from the centre, the field has fallen by 30% compared with the field at the centre of the loop. Thus, while we might try to concentrate the magnetic field, by using a loop of wire, or indeed with many loops of wire, this does not provide a *uniform* field. If we wish to achieve a field that is relatively uniform — i.e. with magnitude and direction that do not change much over a specified region — we need to use at least two coils.

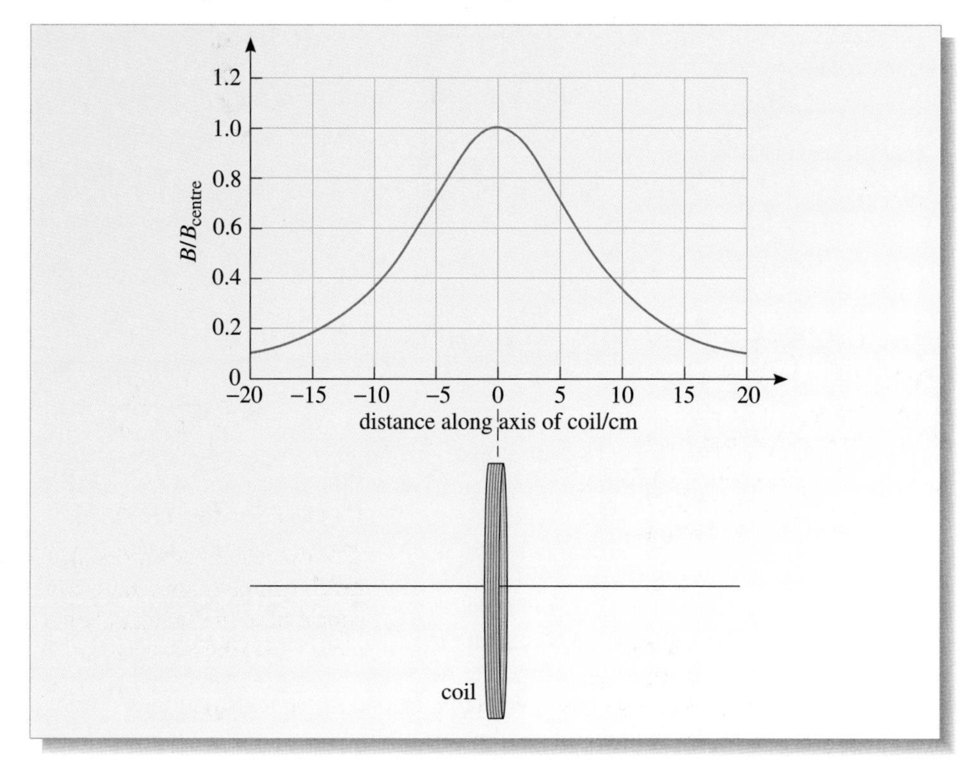

Figure 4.30 A graph showing how the magnetic field strength varies along the axis of a single coil of radius 10 cm. Notice the rapid fall-off in field strength as one leaves the plane of the loop. The magnetic field 5 cm from its centre has fallen by around 30% compared with the value of the field at the centre of the coil.

Figure 4.31 shows how the magnitude of the magnetic field varies along the axis of two similar coils aligned coaxially with the current flowing in the same direction in each, and separated by a distance equal to their radius. If the coils are of radius 10 cm, as in Question 4.10, the magnetic field remains relatively constant (as shown by the flat part of the graph) over a distance of several centimetres along the axis at positions near the midpoint between the coils. A combination of two coils in this configuration is known as a **Helmholtz pair**.

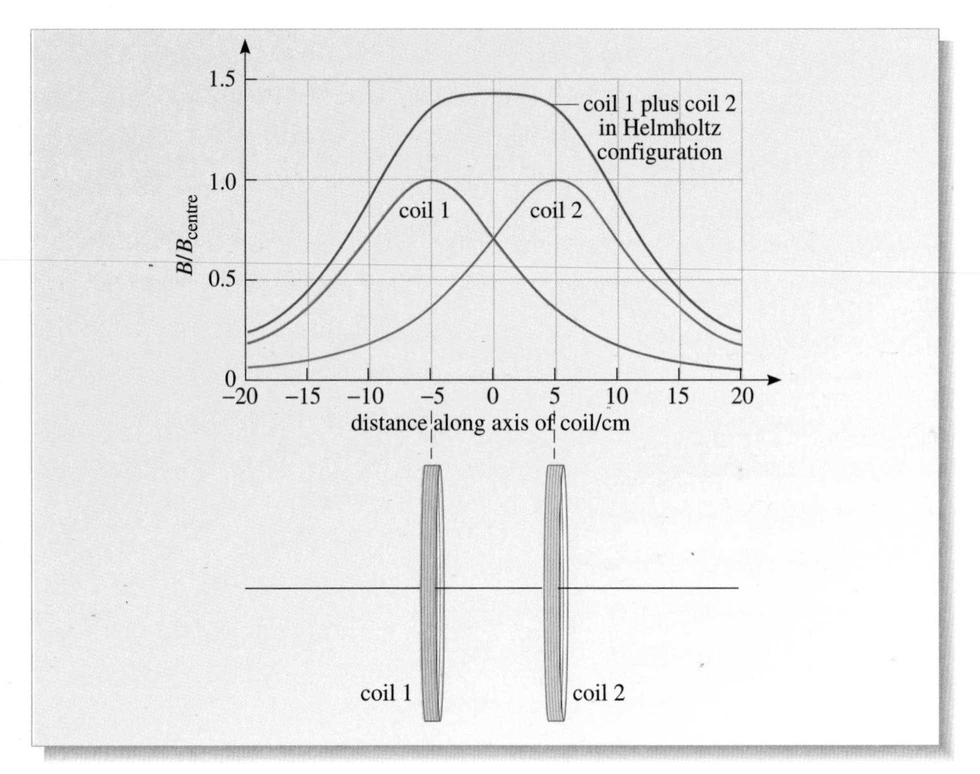

Figure 4.31 A graph showing the magnetic field strength along the axis of two similar coils aligned coaxially, and separated by a distance equal to their radius of 10 cm. The lower curves represent the field strength due to each coil individually. The magnetic field 5 cm from their combined centre has fallen by less than 5% compared with the value of the field midway between the coils. A combination of two coils in this configuration is known as a Helmholtz pair.

3.4 Magnetic field due to a current in a cylindrical solenoid

To generate the largest possible values of magnetic field, while at the same time maintaining reasonable uniformity over a relatively large region, one uses wires in a configuration known as a **cylindrical solenoid**. This consists of coils wound on the outside of a cylinder. If one wishes to increase the field, layers can also be added. The field pattern within a cylindrical solenoid is discussed in Question 4.10.

Question 4.10 (a) Figure 4.32 shows a set of coaxial circular loops, each carrying the same steady current i. Sketch on this figure the fields arising from one individual loop, using the right-hand grip rule to find the field direction. Then, remembering that the fields due to individual loops add vectorially, extend the representation to obtain a rough diagram of the total field due to all the current loops.

(b) Now imagine that the well-separated loops in Figure 4.32 are pushed closer together and connected electrically, so that they form a single continuous coil (or cylindrical solenoid), as illustrated in Figure 4.33a. Figure 4.33b is a cross-section through a cylindrical solenoid which has been tightly wound so that the turns are very close together. Sketch on this diagram the field line representation of the magnetic field produced by the current in the solenoid. Of what does this field pattern remind you?

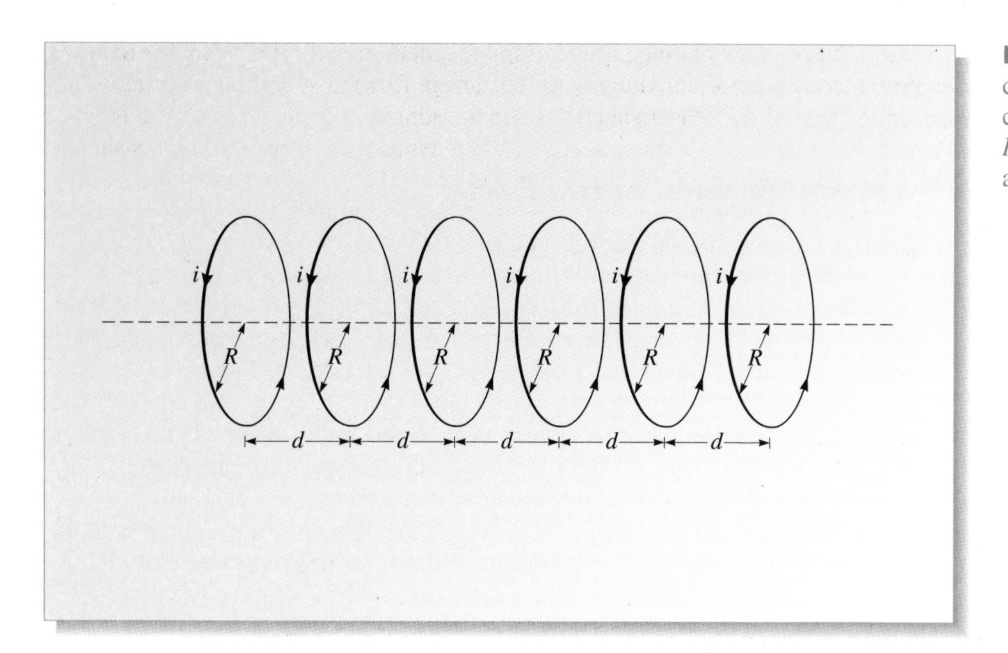

Figure 4.32 A set of parallel coaxial current loops. Each of the circular loops has the same radius R, and the distance d separating adjacent loops is constant.

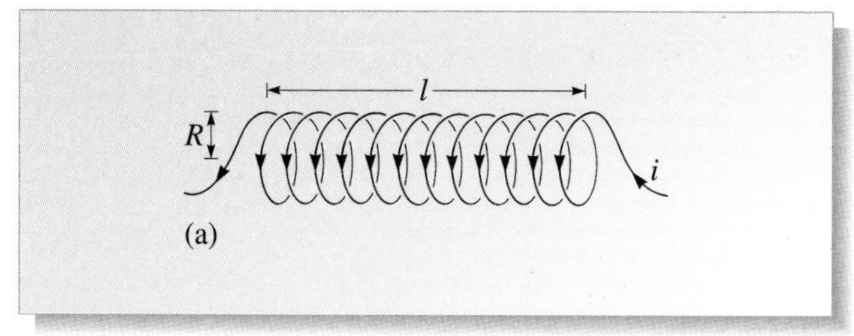

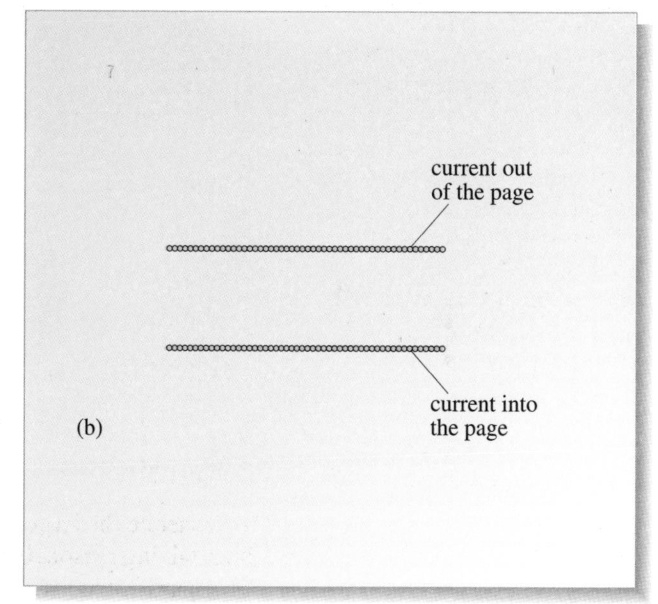

Figure 4.33 (a) A loosely wound cylindrical solenoid. Each loop of wire in the solenoid is called a *turn*. Thus, the diagram shows a solenoid of length l, consisting of 12 turns each of radius R. (b) Cross-section through a cylindrical solenoid.

(c) The magnetic field inside an 'infinitely long' cylindrical solenoid turns out to be parallel to the axis of the solenoid and to have a strength given by $B = \mu_0 Ni/l$, where N is the number of turns in the solenoid and l is its length. Can you suggest the reason for the condition of an 'infinitely long' solenoid , and what this might mean in practice? How would you interpret the quantity (N/l)? Is the form of the equation for B physically reasonable? ■

To summarize, the magnetic field strength inside an infinitely long solenoid with N turns and of length l is

$$B = \frac{\mu_0 Ni}{l} \quad \text{inside an 'infinitely long' cylindrical solenoid} \qquad (4.6)$$

and the field is parallel to the axis of the solenoid. Notice that the magnetic field inside the solenoid is *independent* of its radius.

Question 4.11 An infinitely long cylindrical solenoid with 500 turns per metre carries a current of 4 A. What is the magnetic field strength at a point halfway between the axis of the solenoid and the windings? ■

3.5 Magnetic field of the Earth

The approximate shape of the Earth's magnetic field has been known for hundreds of years, since the pioneering work of William Gilbert in *De Magnete*. Working with little more than needles of iron floating on corks, he deduced that the magnetic field of the Earth must be of the shape that we would now describe as essentially dipolar, i.e. similar to that of a (rather large!) bar magnet or current loop (Figure 4.34). Notice that the field lines converge on a point close to the geographical North Pole. This means that (confusingly) the magnetic pole closest to the geographical North Pole (the so-called 'magnetic North Pole') is actually a south magnetic pole according to our conventional definition. The end of a compass needle which is labelled 'North' is indeed the north pole of the magnet that constitutes the needle. It thus points geographically north as it lines up with the Earth's magnetic field. For reasons such as this, it is important to take great care in the use of the terminology 'north' and 'south' in discussion of magnetic fields.

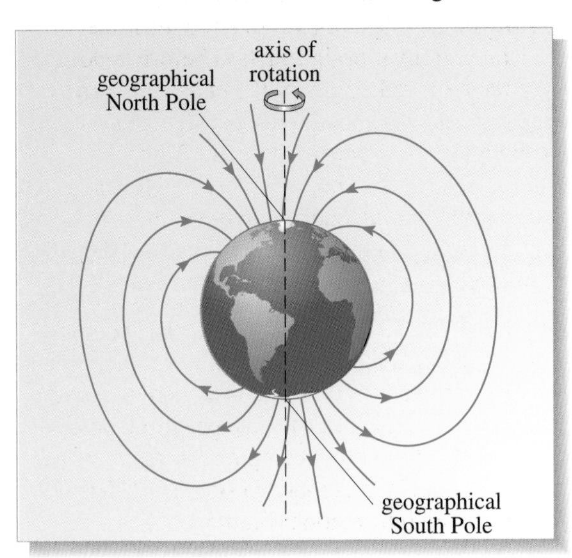

Figure 4.34 Two-dimensional representation of the magnetic field of the Earth. Note that a south magnetic pole seems to be located near the geographical North Pole. The magnetic axis of the Earth is currently (2000) roughly 11° away from the rotational axis.

The Earth's magnetic field does not arise from permanent magnetism, but from currents flowing in the molten 'outer core' of the Earth. The two most important factors giving rise to the so-called 'geo-dynamo' are thought to be the Earth's rotation, which induces some circulation in the outer core, and heat flow from the centre of the Earth, which induces convective mixing. However, the core of the Earth is inaccessible to most probes, and calculations involving magnetic fluids are exceedingly complex. These factors have so far conspired to deny us a detailed explanation of the origin of the Earth's magnetic field.

3.6 Biological magnetic fields

At the end of Chapter 3, we learnt how electric currents and potentials abound in human tissue. The tiny electrical currents in the muscles and nerves of our bodies produce magnetic fields, which, if they can be measured, provide medically useful information. The currents are typically at the microamp (10^{-6} A) level and the associated magnetic fields are extremely weak. You might like to calculate the magnetic field at a distance of 10 cm from a long wire carrying a current of 1 microamp. This should help to explain the fact that biomagnetic fields are measured in picotesla (10^{-12} T) or even femtotesla (10^{-15} T).

Fortunately, there are detectors capable of measuring such small fields. Superconducting Quantum Interference Devices (SQUIDs) rely on the extraordinary quantum properties of superconducting materials to achieve a sensitivity to magnetic field changes of just a few femtotesla. They lie at the heart of a family of new large medical imaging instruments that measure the magnetic fields produced by the brain and the heart and allow us to follow the evolving activity millisecond by millisecond.

Greatest interest is focused on the brain where magnetoencephalography (MEG), that is, the measurement of the magnetic fields around the head, is being used to provide diagnostic information about epilepsy, ischaemia, neural degradation and sensory malfunction, as well as disorders such as dyslexia and autism. An interesting application is in the planning for neurosurgery. A major problem in such surgery is the need for precise knowledge about the functional role of the parts of the brain that the surgeon is investigating. It is essential to know the effect of any excision or transection but the brain landmarks are far from clear. There are some general similarities between the brains of different individuals, but the detailed patterns of folds are as different as fingerprints. How then is the surgeon to know how the operation will affect the patient? One way is to map brain function before making any incision by electrically stimulating the exposed cortex of the conscious patient during the operation and observing the patient's response. This is a slow and potentially distressing process. A non-invasive alternative is to make MEG measurements of the fields produced in standard situations (e.g. hearing tones, moving fingers, watching a screen etc.) and then map back from the data to produce an image of brain function which can be overlaid onto a structural image. Trials of such procedures are now in progress in large hospitals in several countries.

The long-term future of MEG probably lies in its ability to follow the activity of the brain in time. There has been a growing realization that, to understand the more complex functions of the brain, it is necessary to know not only *where* there is activity but also how the activity develops in *time*. The brain isn't just a computer that processes linearly. There is massive use of feedback in the build up of complex processes such as perception, memory retrieval, language production and comprehension. MEG has unique capabilities to provide such time-resolved information (Figure 4.35).

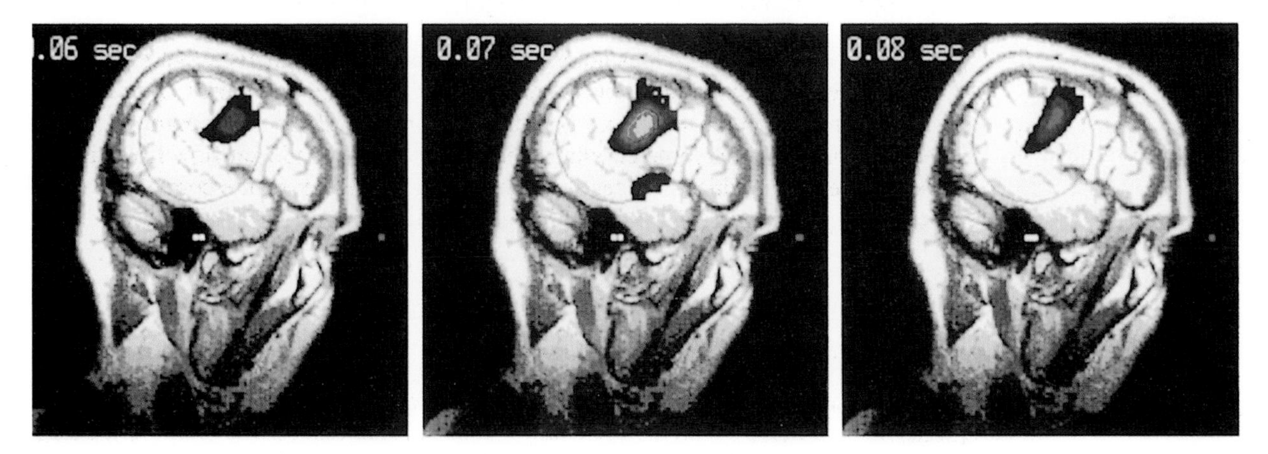

Figure 4.35 Time-resolved MEG image. The black and white picture shows an MRI image of a 'slice' through the brain. Superimposed on this is a coloured image showing the electric current activity in the brain (yellow indicates very strong activity) at various times after the subject heard a tone in one ear.

4 Magnetic materials

4.1 Permanent magnets

Figure 4.36 Two boxes, A and B, one containing a permanent magnet and the other a coil of wire and a battery. Is it possible to tell which box is which simply by investigating the nature of the magnetic field outside the box?

So far in this chapter, we have described in detail only how magnetic fields arise from electric currents. However, for most of us, our first encounter with magnetic phenomena is when we play with permanent magnets. It is not immediately obvious that there are any electric currents flowing in such devices. You may therefore (in spite of Section 2.4!) have in the back of your mind a lingering thought that somehow there are two types of magnetic field: those due to permanent magnets and those arising from movements of charged particles. Let's dispel that lingering thought once and for all: there is only one type of magnetic field and *all* magnetic fields arise from the behaviour of charged particles.

We have delayed detailed discussion of permanent magnets to this point because it is important to emphasize that permanent magnetism really is a phenomenon fundamentally linked to the *motion* of electric charges. In permanent magnets, the moving charges are not taking part in macroscopic currents, such as those described in Chapter 3 and in Section 3 of this chapter. Rather, it is the motion of charges *within each atom* of the magnet itself, and in the very nature of those charges.

Question 4.12 Consider an experiment such as that illustrated in Figure 4.36. A permanent magnet is placed into a case, and a solenoid with a battery placed in a second identical case. A researcher is asked to distinguish between the two on the basis only of the magnetic fields that they produce. Could this be done? ■

4.2 The response of substances to applied magnetic fields

When *any* substance is placed in a magnetic field, it acquires a magnetic dipole moment similar to that of a current loop or a bar magnet. The substance is said to be **magnetized**. For most substances, this is a very weak effect and, when removed from the field, the substance loses its magnetization and retains no history of its exposure to the field. Sometimes, the magnetization is such that the magnetic dipole field increases the magnetic field within the substance, and sometimes it is such that it reduces the magnetic field within the substance. These two types of response are termed **paramagnetic** and **diamagnetic** respectively (Figure 4.37).

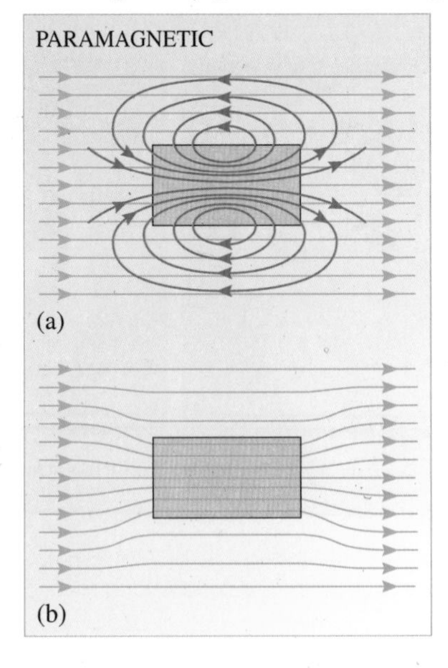

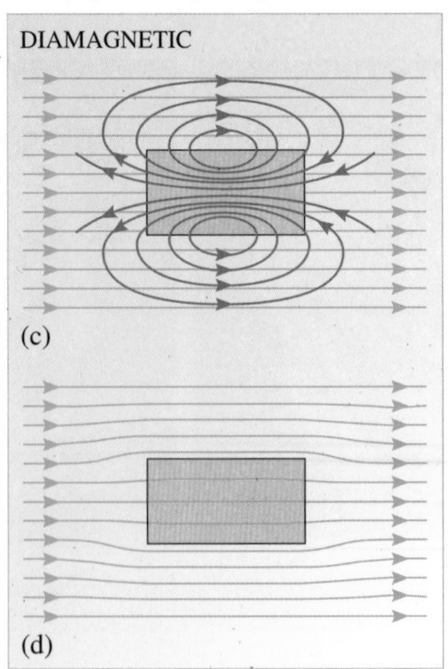

Figure 4.37 The response of paramagnetic (a and b) and diamagnetic (c and d) materials to an applied magnetic field. The *applied* magnetic field is indicated by the pale green lines. The magnetic field induced in the sample (indicated by the dark green lines) has the form of the field due to a bar magnet which, in a paramagnetic material (a), is in such a direction as to *increase* the field inside the material, and *decrease* the field on either side of the material as shown in (b): the sample has 'concentrated' field lines. In a diamagnetic material (c), the magnetic field induced in the sample (dark green lines) is in such a direction as to *decrease* the field inside the material, and *increase* the field on either side of the material as shown in (d): the sample has 'repelled' field lines.

However, a few substances behave in a dramatically different manner. Exposure to even a weak magnetic field causes them to become strongly magnetized. Furthermore, after exposure to a magnetic field they sometimes remain permanently magnetized, even in the absence of an applied magnetic field. Such substances are termed **ferromagnets**, named after the Latin *ferro* for iron, the most common substance to exhibit this behaviour. Nickel, cobalt, a couple of the more exotic elements and a number of compounds and alloys are also ferromagnets. These are the materials from which permanent magnets are made. In the next section, we will see how we can understand the behaviour of ferromagnets.

4.3 Understanding ferromagnets

There are three steps to understanding how ferromagnetism arises.

1. In ferromagnets, each atom of the ferromagnet behaves like a small magnetic dipole as shown in Figure 4.38a. Our first step will be to see how this behaviour can arise.
2. Many materials exhibit this first property, but in ferromagnets the individual atomic magnetic moments align themselves so that, as shown in Figure 4.38b, their magnetic fields add together. Our second step will be to see what causes this to happen.
3. Although neighbouring atomic magnetic dipoles are aligned in a ferromagnet, in pieces of ferromagnet larger than a few microns across, different regions of the ferromagnet — known as **domains** — align themselves in different directions as shown in Figure 4.38c.

One micron (μm) is equal to 10^{-6} m.

It is Step 2 that is the key to understanding the fundamental origin of ferromagnetic behaviour. However, in order to understand the actual properties of, say, a piece of iron, it is essential to understand the origin and properties of domains. Our third step will be to see how understanding domains allows us to explain how one piece of iron can behave as a permanent magnet whereas another, similar piece, easily loses its magnetization.

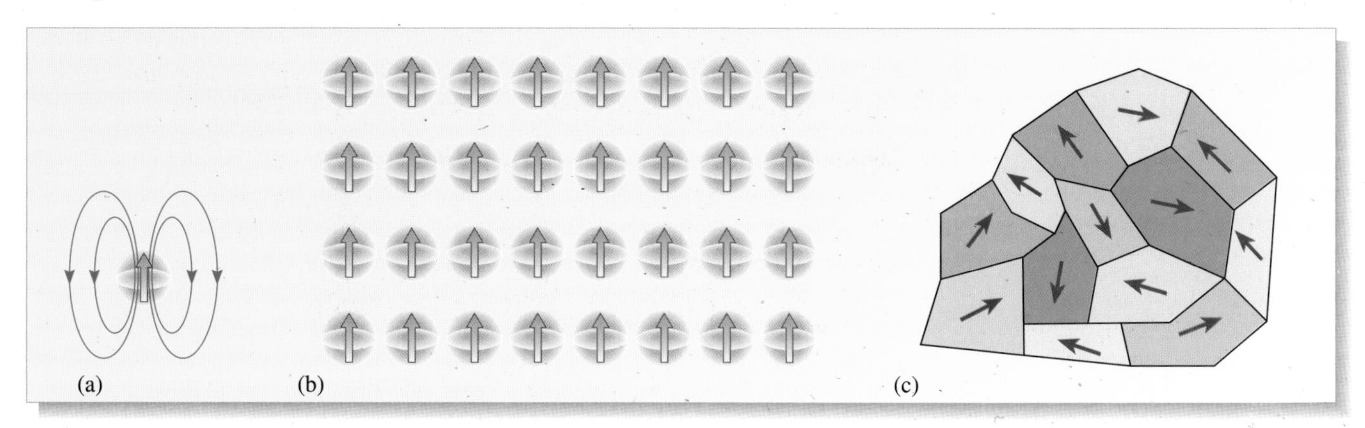

Figure 4.38 Schematic illustration of ferromagnetism. (a) The magnetic dipole field associated with each atom in a ferromagnetic material. The short green arrow through the centre of the atom indicates the magnetic dipole moment of the atom. (b) The spontaneous alignment of atomic magnetic dipoles. (c) The domain structure, which dominates the properties of ferromagnets that are larger than a few microns in size. Each arrow in (c) represents the overall magnetization of a single domain.

Let's look at these three steps in turn and then briefly review some simple applications of magnetic materials.

Step 1: The origin of atomic magnetic dipoles

First, a note of caution. You will learn, when you come to *Quantum physics: an introduction*, that the behaviour of particles on an atomic scale can be properly discussed only by applying the theory of quantum mechanics, and that concepts such as orbits of electrons in atoms must be treated with some scepticism. However, for the purposes of this discussion the following rather 'classical' treatment provides valuable insight into the origins of permanent magnetism.

There are two quite distinct ways in which atoms can come to behave like magnetic dipoles.

- The first mechanism arises because some orbits of an electron about an atom take place in one sense only. Their orbits can thus be thought of as small current loops, which give rise to a magnetic dipole field similar to that due to a macroscopic current loop (Section 3.2 and Figure 4.39a). The atom is then said to possess a magnetic dipole moment.
- The second mechanism is quite different and arises because electrons, like most elementary particles, themselves behave as tiny magnetic dipoles. In other words, in addition to possessing an intrinsic charge of $-e$, electrons also possess what is termed an *intrinsic magnetic dipole moment* (Figure 4.39b).

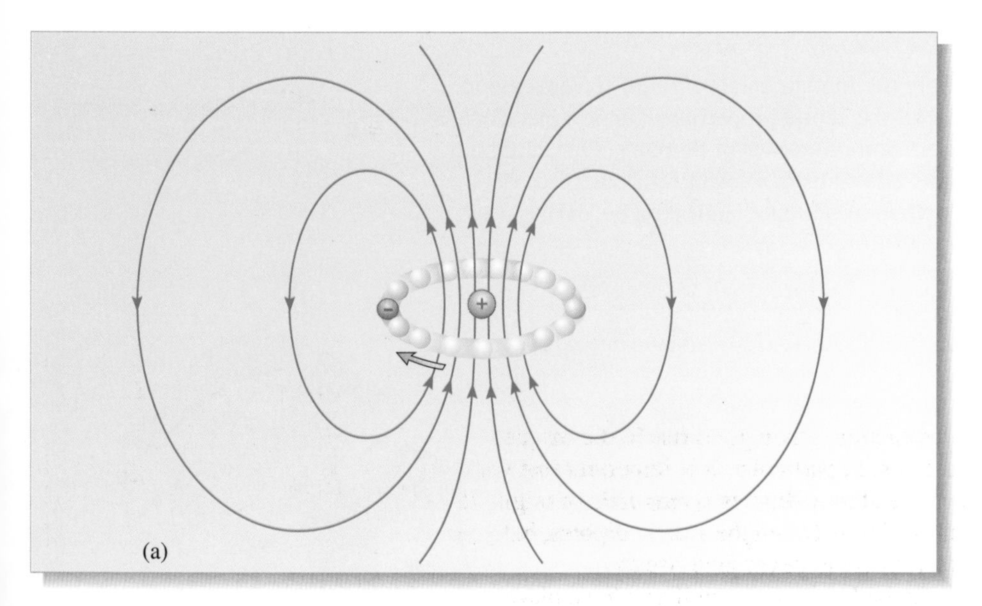

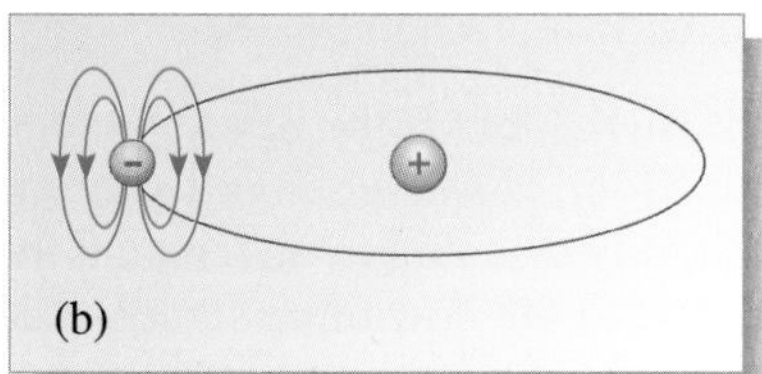

Figure 4.39 (a) Illustration of the origin of an atomic magnetic dipole moment due to the orbital motion of an electron in an atom. (b) In addition to the orbital magnetic dipole moment, each electron in an atom also possesses an *intrinsic* magnetic dipole moment.

These two phenomena might lead one to expect that all atoms should have a magnetic dipole moment. However, this is not so. In many atoms, the number of electrons orbiting in one sense in an atom is matched by an equal number orbiting in the opposite sense, and their dipole fields cancel. Similarly, the *intrinsic* magnetic dipole moments of electrons often cancel each other if the dipoles of pairs of electrons are aligned opposite to each other. In atoms that *do* possess magnetic dipole moments, it is because the cancellation from different sources is incomplete.

In fact, there are quantum mechanical reasons why electrons within the same atom should actually align their dipoles (both orbital and intrinsic). These reasons are associated with the fact that two electrons orbiting in the same sense within an atom are actually physically further apart, on average, than electrons orbiting in the opposite sense. This means that the *electrostatic* potential energy due to the mutual repulsion of the electrons is *lower* in the former case, which is thus favoured. You will learn more about the nature of electrons in atoms in *Quantum physics: an introduction*.

Step 2: Ordering of atomic dipoles

In most substances whose constituent atoms do possess a net magnetic dipole moment, the dipoles orient themselves randomly and so the substance as a whole has no net magnetic dipole moment. However, in some substances, the atoms are so close that electrons on one atom interact electrically with electrons on immediately neighbouring atoms. In some of these substances, the mutual electric repulsion of the electrons causes electrons on neighbouring atoms to orbit in the same sense. This is exactly the same mechanism that caused electrons in the same atom to orbit the atom in the same sense. If electrons in each atom orbit in the same sense as electrons in their immediate neighbours, then the atoms in the entire sample will orbit in the same sense and all the atomic magnetic dipoles will be aligned (Figure 4.40). Thus the magnetic dipole moments from each atom add together to give rise to a macroscopic magnetic dipole moment.

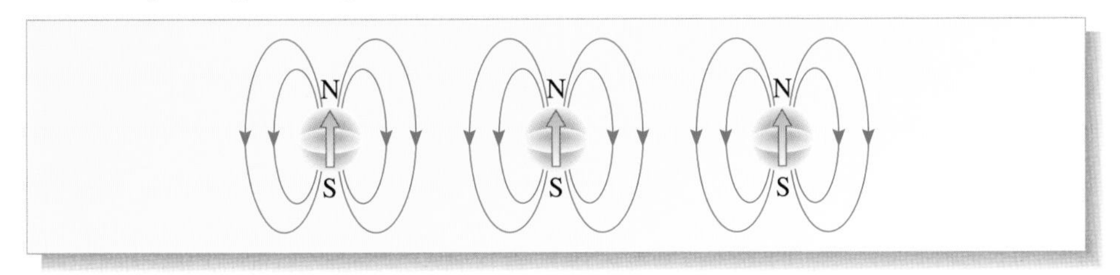

Figure 4.40 Electrons on neighbouring atoms can minimize their Coulomb repulsion by correlating their orbital motions. Even though this interaction is very short range, acting only between neighbouring atoms, it can cause large numbers of atoms to align their magnetic dipoles because each atom prefers to orient its magnetic dipole parallel to its neighbour's.

It is important that you notice that the interaction which gives rise to the magnetic ordering is the *electric* repulsion of electrons. In particular, it is important that you do not think that the interaction between the atomic dipoles is *magnetic* in origin. It is true that there is a weak magnetic interaction between the atomic dipoles, but you should be able to see that such an interaction could never give rise to ferromagnetism. This is because — as may be familiar to you if you have played with a pair of bar magnets — magnetic interactions cause the dipoles to align in opposite directions: colloquially, north pole to south pole. This is illustrated in Figure 4.41.

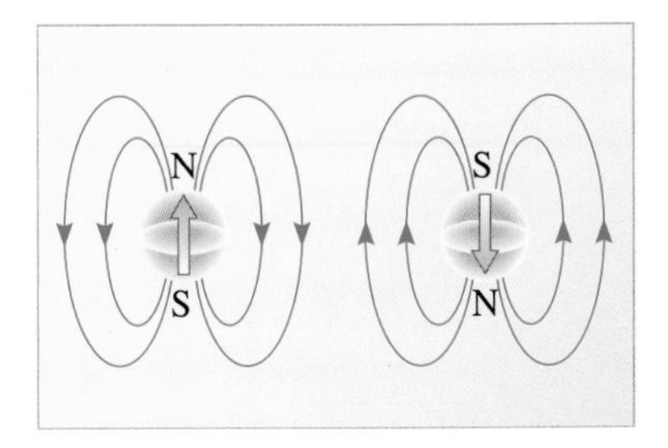

Figure 4.41 Ferromagnetism does *not* arise from the magnetic interaction of atomic magnetic dipoles. To see this, recall that the magnetic interaction between dipoles will cause them to align so that neighbouring atomic dipoles point in opposite directions.

Thus *magnetic* interactions alone could never give rise to the phenomenon of ferromagnetism. However, magnetic interactions *do* become important when we consider the behaviour of magnetic domains.

Step 3: Magnetic domains

The underlying reason for the magnetic properties of ferromagnets is that they are the only substances in which neighbouring atomic dipoles tend spontaneously to line up parallel to each other. In practice, it is found that the alignment is perfect only within small areas, called domains, whose typical dimensions are of the order of a few microns. Although this may seem physically quite a small space, it nevertheless contains around 10^{16} atoms. The boundaries between domains, known as **domain walls**, are just a few atoms thick. The total field outside a sample of material is just the vector sum of the fields of all the domains. In an unmagnetized iron bar, the domains are randomly distributed, as shown in Figure 4.42a, and there is no net field around the bar.

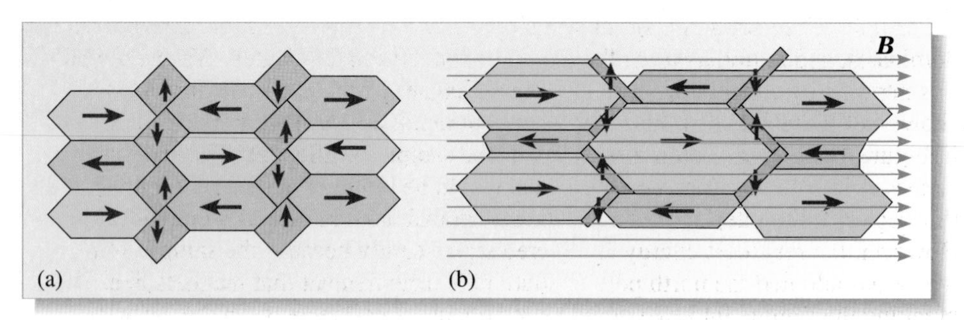

Figure 4.42 (a) An unmagnetized sample of a ferromagnet with randomly oriented domains. (b) In the presence of an applied magnetic field $\boldsymbol{B}$ pointing to the right, the domains in which the net magnetic dipole is parallel to $\boldsymbol{B}$ (pale green) grow at the expense of domains oriented in other directions. The ease with which domains can grow or shrink depends on the ease with which a domain's wall can move. It is this property — the ease or difficulty of domain wall motion — that dominates the macroscopic properties of ferromagnets.

When an external magnetic field is applied, it is sometimes possible for the domains in which the atomic dipoles are aligned parallel to the applied field to grow in size, while other domains shrink. If the external field is strong, the domains also tend to rotate in order to align with it. This is illustrated schematically in Figure 4.42b. As a result of this process, the field of the whole bar becomes dipolar: a macroscopic magnet has been created. Depending on the material involved, the new domain arrangement sometimes persists even after the external magnetic field has been removed — in other words, a permanent bar magnet has been made.

The two paragraphs above have described roughly what happens inside a ferromagnet, but there are many unanswered questions. In particular,

- what causes the formation of magnetic domains?
- what factors determine whether a material will become permanently magnetized when an applied field is removed?

Let's consider these two questions in turn.

Why do domains form?

Domains arise in ferromagnets because there is a competition between the two kinds of interaction between the magnetic dipole moments in atoms:

- The first interaction is the basic interaction that gives rise to *alignment* of atomic magnetic dipole moments: that is, the electrostatic interaction of electrons in neighbouring atoms. This is a rather powerful interaction, but it is only effective over a very short range, acting only between atoms that are actually next to one another.
- The second interaction is the magnetic interaction between dipoles that tends to cause dipoles to align in *opposite* directions. If we consider just two neighbouring atoms, then the electrostatic interaction is generally very much stronger than the magnetic interaction. However, despite the weakness of the interaction on the scale of an individual atom, the magnetic interaction has a long range.

In a large piece of ferromagnet, competition between these two interactions causes the sample to break up into small domains *within which* the electrical interaction dominates, and *between which* the magnetic interaction dominates. We can see how this happens by considering a piece of ferromagnet which is all one domain, as shown in Figure 4.43a. If this sample were to break up into two domains, as shown in Figure 4.43b, how would the energy of the sample be affected? The short range electrical interactions between the atoms would be largely unaffected, except in the thin region at the interface between the two domains — known as a domain wall. However, the magnetic energy is lowered significantly because the sample as a whole has adopted the north pole to south pole arrangement that magnets like. Thus, breaking the sample into two domains results in a small increase in the electrical energy and a large loss in the magnetic energy. So, such a change will occur spontaneously. However, if the sample breaks into two domains spontaneously, then why not four or eight? In this way, a large sample of ferromagnet breaks up into smaller and smaller regions, until the magnetic energy gained in this process is offset by the cost in electrical energy required to form domain walls in which atomic dipole moments are not optimally aligned. In practice, this results in domains of the order a few tens of microns in size, and domain walls around 10 nm thick.

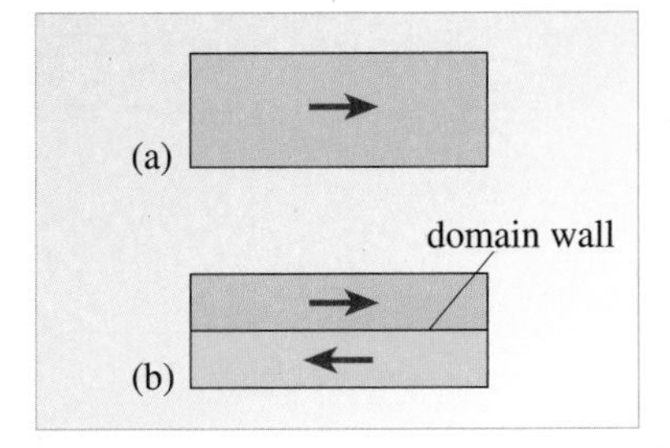

Figure 4.43 A piece of ferromagnet in which all atomic magnetic moments are aligned as in (a) can lower its energy by splitting into two oppositely oriented domains (b). This lowers the magnetic energy of most atomic dipole moments, but raises the electrical energy of a small number of atoms in the region between the two domains known as the domain wall.

What happens when an external field is applied?

The application of an external magnetic field alters the balance between the magnetic and electric forces that establishes the patchwork of domains throughout a sample. In an applied magnetic field, it becomes energetically preferable for domains whose magnetic dipoles are parallel to the applied field to grow at the expense of domains oriented in the opposite sense. However, the ease of motion of the domain walls critically affects the magnetic properties of the material.

If the domain walls can move freely, then the material will be easily magnetized in an applied magnetic field. However, the material will also lose its magnetization easily when the applied magnetic field is removed. Such behaviour is referred to as **soft magnetic behaviour**. Examples are silicon steel and soft iron.

If the domain walls require some work to make them move, then the material will be harder to magnetize in an applied magnetic field. Indeed, it may appear to be non-magnetic. However, if such a material *does* become magnetized, it will then tend to keep its magnetization when the applied magnetic field is removed. Such behaviour is referred to as **hard magnetic behaviour**. Examples are cobalt steel and some alloys of nickel, aluminium and cobalt.

Both behaviours are common and both are useful in different circumstances.

4.4 Magnetic materials in action

Soft magnetic materials

An electromagnet is a device for cheaply and easily producing a strong magnetic field, usually up to around 1 T, but occasionally as large as 3 T. One is illustrated in Figure 4.44. A current flowing through a coil of wire generates a small magnetic field, typically only of the order of 10^{-3} T. This is sufficient to strongly magnetize some iron in the centre of the coil, and it is the magnetic field of the magnetized iron which is utilized. As the current in the coil is changed, the magnetization of the iron changes and the field produced is roughly proportional to the current in the coil, up to some maximum current at which the iron is fully magnetized — i.e. all the domains are aligned.

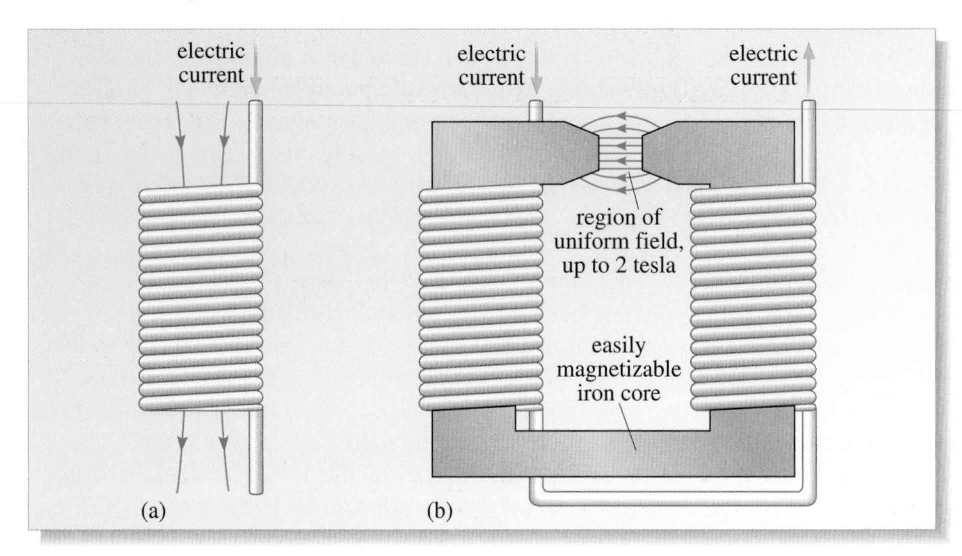

Figure 4.44 Illustration of the operation of an electromagnet. (a) A current through a solenoidal coil of wire generates a small magnetic field. (b) If a solenoidal coil is wrapped around a soft magnetic material, then the material becomes strongly magnetized even by the small applied field.

Hard magnetic materials

Permanent magnets are examples of hard magnetic materials. The fabrication of strong permanent magnets is a complex process, but from our point of view we need to notice only one thing: the key requirement is for a material in which domain wall motion is very difficult. Thus when a parallel arrangement of domains is created — generally by the application of strong magnetic fields and high temperatures — the arrangement is unable to change.

Although permanent magnets are common in all kinds of devices, perhaps the most important application is as a component in small powerful motors. These are now commonplace in many technological devices. For example, many modern cameras have an autofocus lens driven by a motor, and another motor automatically winds on the film after each exposure and finally rewinds it. Motors are essential whenever a low-power device does mechanical work of any kind. The physical principles underlying the operation of such motors are discussed in *Dynamic fields and waves*.

Magnetic media

Cassette tapes and computer disks are examples of magnetic media. The requirements for such media are that they should become fairly easily magnetized, and yet retain their magnetization so as to preserve the information recorded. Indeed, signals recorded onto magnetic media are stable for periods as long as a few decades. The design of such materials involves the use of particles that are so small that within each particle there is room for only one magnetic domain. In such circumstances, the magnetization process does not involve the motion of domain walls, but the rotation of the magnetization of the individual particles of the magnetic powder.

When the medium is read, a soft magnetic material is passed over the recording medium, and becomes magnetized by the weak magnetic field above the medium. This magnetization can be detected electrically and converted into a form suitable for electronic amplification (Figure 4.45).

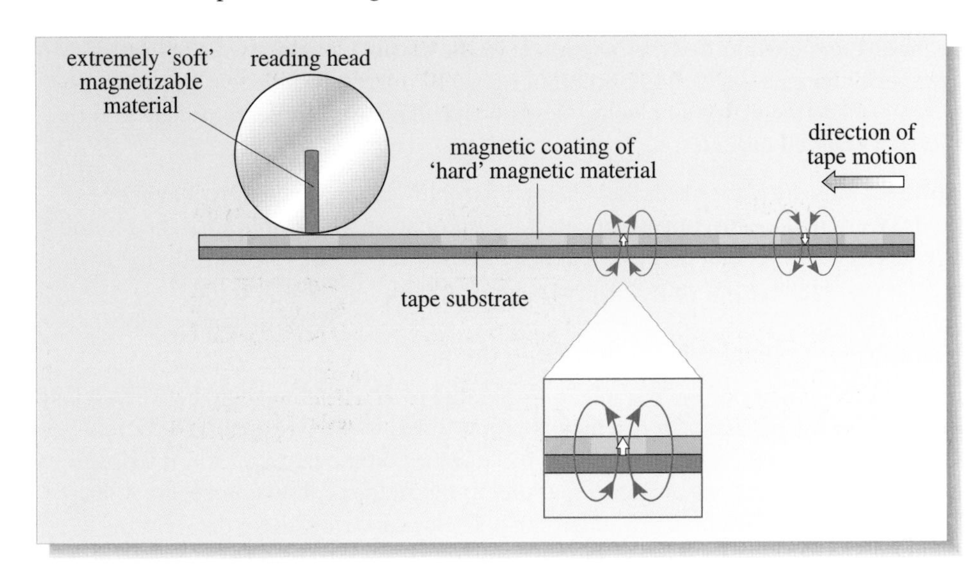

Figure 4.45 Schematic illustration of a device for reading magnetic tape. The flexible tape is coated with a hard magnetic material. As the tape is drawn past a reading head, the small fields due to differently magnetized regions of the tape magnetize a small sample of soft magnetic material. This material is designed so that it retains almost no magnetization when applied fields are removed. The magnetized reading head effectively amplifies the field of the tape and allows it to be registered electronically and electrically filtered and amplified.

5 The motion of charged particles in uniform magnetic fields

5.1 Motion of a charged particle in a uniform magnetic field

The simplest kind of field is a *uniform* one — in other words, a field that has the same magnitude and (for vector fields) the same direction at each point. In a uniform magnetic field we can rewrite Equation 4.2 as

$$\boldsymbol{F}_{\rm m} = q(\boldsymbol{v} \times \boldsymbol{B}). \qquad \text{(Eqn 4.2a)}$$

Here, we have just written $\boldsymbol{B}$ instead of $\boldsymbol{B}(\boldsymbol{r})$ because, in a uniform field, $\boldsymbol{B}$ has the same value for all points $\boldsymbol{r}$. We shall see that even with this simplification, the motion of charged particles through such a field can be quite complex. However, this is an important topic and we will be able to arrive at a quantitative understanding of a number of interesting phenomena and devices. In Section 6, we will relax the restriction of a uniform field and qualitatively extend some of the results from this section.

The kinetic energy of a particle in a uniform magnetic field

It is a curious property of the motion of charged particles in uniform magnetic fields that a particle's speed cannot be changed by the field. To see how this comes about, let's think about the kinetic energy of a particle as it moves through a uniform magnetic field. The work that is done on the particle by the field as it moves through a small vector displacement $\Delta\boldsymbol{s}$ is given in *Predicting motion* Chapter 2 Section 2 as

$$\Delta W = \boldsymbol{F} \cdot \Delta\boldsymbol{s} \,. \qquad (4.7)$$

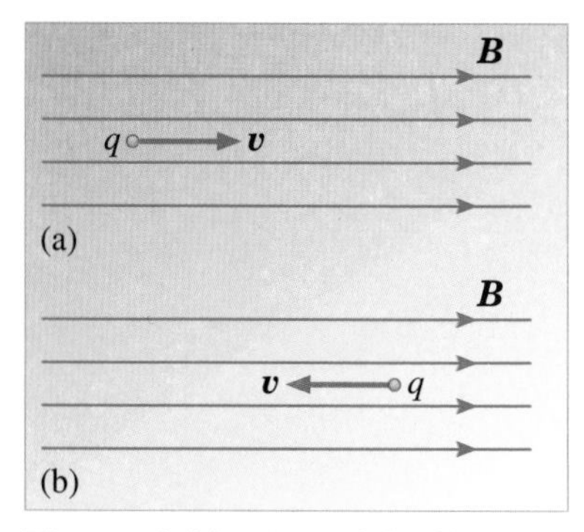

Figure 4.46 A particle with charge q moving either parallel, as in (a), or antiparallel, as in (b), to a uniform magnetic field does not change its speed or direction of motion.

This states that the work done is given by the scalar product of the force — which in this case is the magnetic force $\boldsymbol{F}_{\mathrm{m}}$ — and the displacement $\Delta\boldsymbol{s}$ along the path of the particle over which the force acts.

For small displacements, $\Delta\boldsymbol{s}$ is a vector pointing in the same direction as $\boldsymbol{v}$. Equation 4.2a tells us that $\boldsymbol{F}_{\mathrm{m}}$ is always perpendicular to $\boldsymbol{v}$ and so we can conclude that $\boldsymbol{F}_{\mathrm{m}}$ is also always perpendicular to $\Delta\boldsymbol{s}$. Hence the scalar product in Equation 4.7, $\boldsymbol{F}_{\mathrm{m}} \cdot \Delta\boldsymbol{s}$, is always zero and so the magnetic force is not able to change (either to increase or to decrease) the kinetic energy of the particle. We thus draw the, perhaps rather surprising, conclusion that the magnitude of the particle's velocity, i.e. its speed, cannot be changed through the action of magnetic forces. Notice that this is quite different from the action of electric fields which can easily increase or decrease the speed of charged particles.

However, this does not mean that charged particles are unaffected by magnetic fields. We will now investigate in detail what Equation 4.2a implies for the motion of charged particles in a uniform magnetic field. We will start by considering two special cases before arriving at a general description.

Special case 1: initial velocity parallel to the field

If the initial velocity of the particle, $\boldsymbol{v}$, is parallel to $\boldsymbol{B}$, then the angle between $\boldsymbol{v}$ and $\boldsymbol{B}$ is 0°. The magnitude of $\boldsymbol{F}_{\mathrm{m}}$, which is proportional to $\sin\theta$ (Equation 4.3), is therefore zero. So there is no magnetic force acting on the particle, and it will simply continue to move at constant speed parallel to the field, as shown in Figure 4.46. Note that this analysis holds, whether the particle is moving along the field (as in Figure 4.46a) or against it (as in Figure 4.46b).

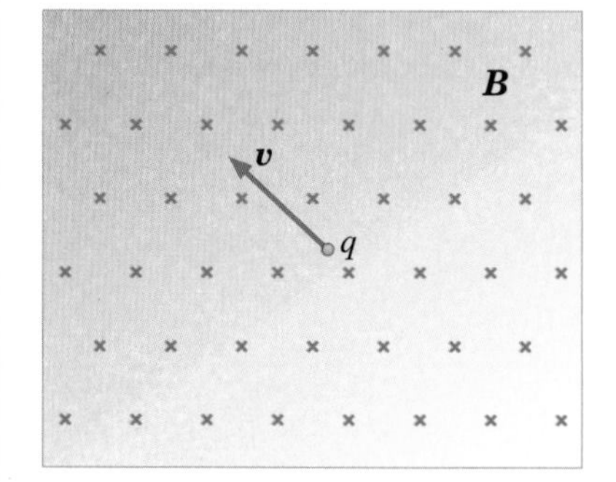

Figure 4.47 A uniform magnetic field directed into the page. $\boldsymbol{v}$ is the initial velocity vector of a particle of charge q moving in the plane of the page.

Special case 2: initial velocity perpendicular to the field

A situation in which a charged particle is moving at right angles to a uniform magnetic field is illustrated in Figure 4.47. The particle has a velocity $\boldsymbol{v}$ in the plane of the page. The magnetic field is perpendicular to the paper, and directed into the page. Thus, a uniform magnetic field, perpendicular to the paper and directed into the page, is represented by a regular pattern of small crosses.

Question 4.13 Assuming the charge q to be positive, draw an arrow onto Figure 4.47 showing the direction of the acceleration experienced by the particle. ■

Now imagine that $\boldsymbol{v}$ remains in the plane of the paper, but changes its direction slightly. You will realize from your answer to Question 4.13 that, as the direction of the velocity changes, so does the direction of the acceleration; the velocity and the acceleration are *always* perpendicular in this case.

You have already met situations in which a particle has acceleration $\boldsymbol{a}$ of constant magnitude that is always directed at right angles to the particle's velocity (*Predicting motion* Chapter 1 Section 5). This is the situation in circular motion: the particle moves in a circular path at constant speed. The instantaneous velocity $\boldsymbol{v}$ is tangential to the path and the acceleration $\boldsymbol{a}$ acts towards the centre of the circle. In exactly the same way, the particle shown in Figure 4.47 will move at constant speed v, following a circular path in the plane of the paper. Such a path is shown in Figure 4.48, for the case in which the charge q is positive.

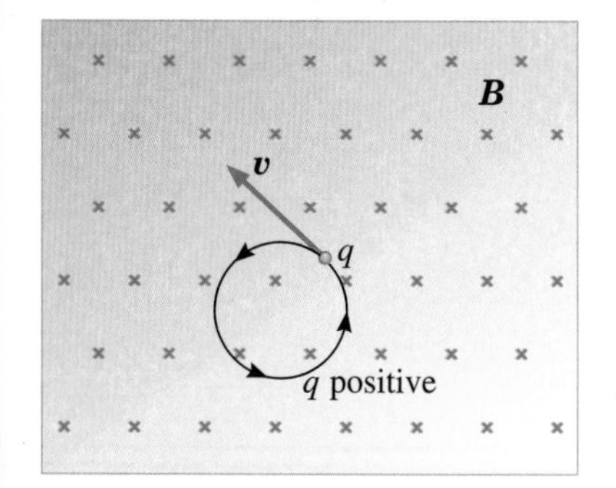

Figure 4.48 The circular orbit of a positively charged particle, initially moving in a plane perpendicular to the direction of a uniform magnetic field.

Question 4.14 If the sign of the charge of the particle shown in Figure 4.48 were reversed, so that q was negative, what would the orbit look like? Sketch onto Figure 4.48 the orbit of the negatively charged particle. ■

The circular motion of a particle travelling at right angles to a magnetic field is known in general as **cyclotron motion**. For a complete description of the cyclotron motion of the particle shown in Figure 4.48, we need to work out both the radius of the circular trajectory of the particle, and the frequency with which the particle orbits on this trajectory. To do this, remember that the magnitude of the acceleration of a particle with speed v in a circular orbit of radius R is:

$$a = \frac{v^2}{R}.$$

It is the magnetic force that maintains this acceleration, and, since $\boldsymbol{v}$ and $\boldsymbol{B}$ are always perpendicular to each other, the magnitude of the force is

$$F_\mathrm{m} = |q|vB\sin 90^\circ = |q|vB.$$

If the particle has mass m, then by Newton's second law ($\boldsymbol{F} = m\boldsymbol{a}$)

$$a = \frac{F_\mathrm{m}}{m} = \frac{|q|vB}{m} = \frac{v^2}{R}$$

which can be rearranged to give an expression for the radius of the circular path, known as the **cyclotron radius**, R_C:

$$R_\mathrm{C} = \frac{mv}{|q|B}. \tag{4.8}$$

Since the speed of the particle is v and the length of an orbit is $2\pi R_\mathrm{C}$, the time (period) for an orbit is given by

$$T = \frac{2\pi R_\mathrm{C}}{v} = \frac{2\pi mv}{v|q|B} = \frac{2\pi m}{|q|B}.$$

The time for a single orbit depends only on the intrinsic properties of the particle, its charge and its mass, and on the applied field. Perhaps surprisingly, it does not depend at all on the particle's speed (Figure 4.49). One can see why by looking at Equation 4.8: a fast-moving particle will have a larger cyclotron radius than a slow-moving particle, but, of course, it will travel around the perimeter more quickly. These two effects exactly cancel because the magnitude of the magnetic force F_m is directly proportional to v. The rate at which a particle executes such orbits is known as the **cyclotron frequency** f_C and is just the reciprocal of the orbital period:

$$f_\mathrm{C} = \frac{1}{T}.$$

$$f_\mathrm{C} = \frac{1}{2\pi}\frac{|q|B}{m}. \tag{4.9}$$

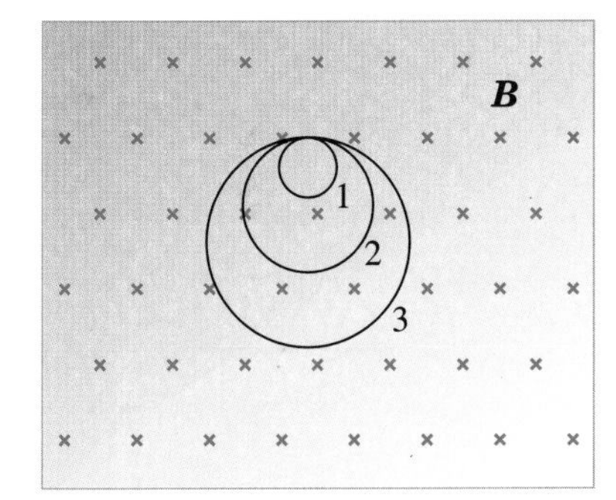

Figure 4.49 Three cyclotron trajectories corresponding to three particles with identical mass and charge, but different speeds ($v_1 < v_2 < v_3$). Although all three orbits have different radii, they are all completed in the same time, i.e. the cyclotron frequency does not depend on a particle's speed.

Question 4.15 What are the cyclotron frequency and radius for protons with kinetic energy 1 MeV, in a magnetic field of 1 T? (Take care to quote the appropriate units.) ■

General case: initial velocity in an arbitrary direction

We now know how a charged particle will move if its initial velocity is either parallel or perpendicular to a uniform magnetic field. It is a comparatively simple matter to generalize to the case in which the particle moves in an arbitrary direction with respect to the field. To do this, we recall that the velocity vector at any instant can always be resolved into two components at right angles, and if one of those components is chosen to be parallel (or antiparallel) to the magnetic field direction, the other must lie in a plane perpendicular to the field. Alternatively, we can resolve the field into two components, one parallel to the velocity (which has no effect on the particle's motion) and one perpendicular to its velocity (which gives rise to cyclotron motion).

Thus, in the general case, the path of the particle is the result of combining the circular motion in a plane perpendicular to the field with a steady movement parallel to the field. Such a path is called a *helix*, and is shown in Figure 4.50a. A helix describes the shape of a coiled spring, or the path of the screw thread on a bolt. In the case of the charged particle in Figure 4.50, the cyclotron radius of the helix is related only to v_{perp}, the component of velocity perpendicular to the field:

$$R_C = \frac{m\,|\,v_{\text{perp}}\,|}{|\,q\,|\,B}. \qquad (4.8a)$$

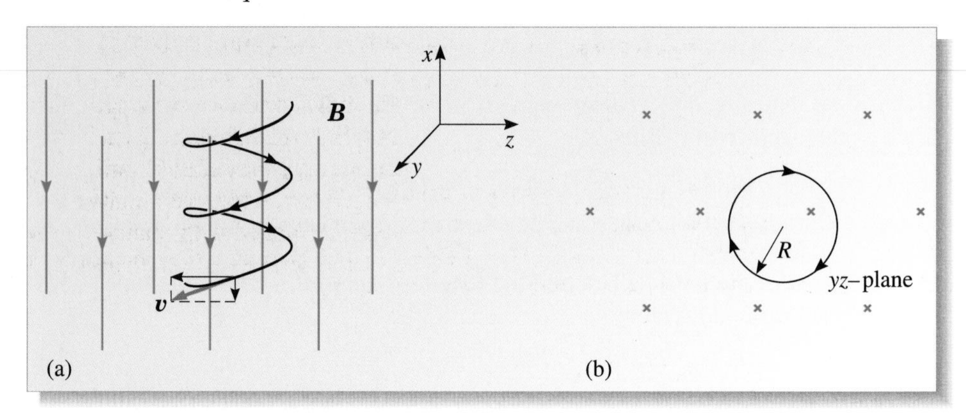

Figure 4.50 Helical path of a particle in a uniform magnetic field. (a) shows a perspective view of the helix. The velocity vector $\boldsymbol{v}$ is shown resolved parallel and perpendicular to the field. (b) shows the same helix viewed looking along the $-x$-direction. The diagrams assume that the particle is negatively charged.

Since here the field is in the $-x$-direction, then $v_{\text{perp}} = \sqrt{v_y^2 + v_z^2}$ and the axis of the helix is parallel to the direction of the magnetic field.

When we considered the special case of a particle moving perpendicular to the magnetic field, we saw that the cyclotron frequency is not affected by the speed of the particle. Similarly, the cyclotron frequency is not affected by the component of velocity perpendicular to the magnetic field and remains as before:

$$f_C = \frac{1}{2\pi}\frac{|\,q\,|\,B}{m}. \qquad \text{(Eqn 4.9)}$$

A particle detector

The curved paths followed by charged particles moving through magnetic fields can be recorded on film using a device called a *bubble chamber*. At the centre of a bubble chamber is a large, liquid-filled vessel situated in a magnetic field. The liquid is usually something exotic, like liquid hydrogen, which has a very low boiling point (20.4 K). A charged elementary particle, moving through the chamber, leaves behind a trail of vapour bubbles that can be photographed, thus providing a visual record of the curved path of the particle as it travelled through the magnetic field. Figure 4.51 shows a typical example of a bubble chamber photograph, involving just a few charged particles. Electrically neutral particles leave no tracks, but their presence can often be deduced if they undergo a process known as 'decay'. This involves the disappearance of the original particle, and the creation of one of more other particles, some of which may be electrically charged. Such a decay is shown in the lower half of Figure 4.51: a neutral particle has decayed into two oppositely charged particles. By analysing pictures such as these, it is possible to identify the particles concerned and determine the speeds of the particles from the curvature of their trajectories.

Figure 4.51 Tracks left by the passage of charged particles through a bubble chamber.

Question 4.16 Suppose the magnetic field in Figure 4.51 points into the page.

(a) Which particle tracks correspond to positively charged particles and which to negatively charged particles?

(b) Explain why the paths of the two particles spiral inwards. (*Hint*: Think about changes in energy or speed.) ■

The following example shows the quantitative use of the concepts and equations developed in this section.

Example 4.1

Figure 4.52 shows a uniform magnetic field of magnitude 1.00 T, pointing in the $-x$-direction. An electron with speed $1.95 \times 10^7\,\mathrm{m\,s^{-1}}$ moves at 45° to the field lines. What is the vertical distance d through which the electron moves during the time it takes to execute one 'twist' of its helical path?

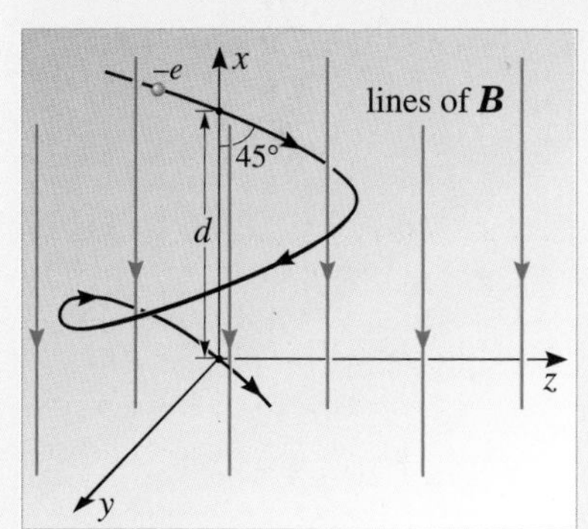

Figure 4.52 Helical path of a (negative) electron in a uniform magnetic field.

Solution

Preparation Our approach to this problem is to take it in two parts.

First, we need to work out the time T taken for an electron to make a single orbit. Recall that this does not depend on the particle's speed, but only on its charge, mass and the value of the magnetic field. The time T for a single orbit is given by the reciprocal of the cyclotron frequency (Equation 4.9):

$$f_{\mathrm{C}} = \frac{1}{2\pi}\frac{|q|B}{m}. \qquad \text{(Eqn 4.9)}$$

Secondly, recalling that the component of velocity parallel to the magnetic field is unchanged by the field, we need to work out how far the particle has travelled *along* the field direction in the time T. To do this we need to find the component of the velocity in the direction parallel to $\boldsymbol{B}$ (i.e. along the $-x$-direction).

Working We start on the first step. The time T taken for a single orbit is given by the reciprocal of the cyclotron frequency:

$$T = \frac{1}{f_C} = \frac{2\pi m}{|q|B}.$$

For an electron, we have $m = m_e$ and $|q| = e$ and so this becomes:

$$T = \frac{2\pi m_e}{eB}.$$

Now we do step 2.

The electron moves parallel to the field with speed $|v_x|$, where $|v_x| = v\cos 45°$ so the distance d it travels in time T is

$$d = |v_x|T = v\cos 45°\,T.$$

Now we combine the results of steps 1 and 2 to get our final expression for d:

$$d = v\cos 45° \frac{2\pi m_e}{eB}.$$

Now all we need to do is substitute the relevant values for the quantities in the expression for d.

Recalling the properties of an electron:

$$m_e = 9.1 \times 10^{-31}\,\text{kg}$$

$$e = 1.60 \times 10^{-19}\,\text{C}$$

and the values of B and v given in the problem:

$$v = 1.95 \times 10^7\,\text{m s}^{-1}$$

$$B = 1.00\,\text{T}$$

and $\cos 45° \approx 0.707$,

we can finally substitute for d:

Notice that we have left this step right until the end of the problem. This reduces the possibility of arithmetical mistakes and makes checking the logic of a calculation more straightforward.

$$d = \left[\frac{1.95 \times 10^7 \times 0.707 \times 2 \times \pi \times 9.1 \times 10^{-31}}{1.60 \times 10^{-19} \times 1.00}\right]\text{m}$$

$$= 4.93 \times 10^{-4}\ \text{m}$$

$$\approx 0.5\ \text{mm}.$$

Checking Apart from checking the algebra and the units, there is no real way of knowing if this is a sensible answer. Since the electron is travelling at such a high speed (nearly one-tenth of the speed of light), half a millimetre might seem a rather short distance for it to have travelled during one orbit. We can check the time taken for one orbit:

$$T = \frac{2\pi m_e}{eB} = \frac{2 \times \pi \times 9.1 \times 10^{-31}}{1.60 \times 10^{-19} \times 1.00}\text{s} = 3.6 \times 10^{-11}\,\text{s}.$$

This is an extremely short time, so our answer for d would seem, after all, to be quite reasonable.

5.2 The Lorentz force law

We have seen, in the last few pages, how a magnetic field affects the motion of a charged particle. As you will recall from Chapter 1, an electric field will also affect the motion of a charged particle according to the equation

$$\boldsymbol{F}_{\text{el}}(\boldsymbol{r}) = q\boldsymbol{\mathscr{E}}(\boldsymbol{r}). \qquad \text{(Eqn 4.1)}$$

Equation 4.1 gives the electrical force on any charge q whether it is moving or stationary. The following question provides some revision of the most important aspects of the motion of a charged particle in a uniform electric field.

Question 4.17 (*Revision of Chapter 1 Section 4*) Figure 4.53 shows a situation in which a uniform electric field $\boldsymbol{\mathscr{E}}$ acts in the $-y$-direction, with the plane of the page being defined as the xy-plane. A particle of positive charge q and mass m moving with initial velocity $\boldsymbol{u}$ in the xy-plane is fired into this field from point A. Throughout this problem, neglect any possible gravitational effects.

(a) Obtain an expression for the acceleration $\boldsymbol{a}$ of the particle, and state whether or not it is dependent on the position and initial velocity of the particle. What is the direction of this acceleration?

(b) Hence, sketch the trajectory of the particle through the field. (*Hint*: Think about the analogy with a projectile in a gravitational field.)

(c) Find expressions for the x- and y-components of the particle's velocity after a time t has elapsed. ■

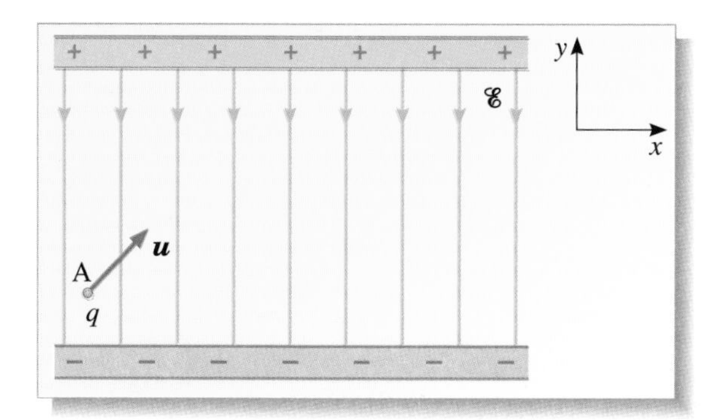

Figure 4.53 A positively charged particle moving with an initial velocity $\boldsymbol{u}$ into a uniform electric field.

As we stated at the end of Section 2.1, the interaction between charged particles has two components: an electrical part and a magnetic part. In uniform magnetic fields, the magnetic force is given by $\boldsymbol{F}_{\text{m}} = q(\boldsymbol{v} \times \boldsymbol{B})$, and in uniform electric fields, the electrostatic force is given by $\boldsymbol{F}_{\text{el}} = q\boldsymbol{\mathscr{E}}$. If both forces act simultaneously on a particle, then it experiences a resultant force $\boldsymbol{F}$ that is just the vector sum of $\boldsymbol{F}_{\text{el}} + \boldsymbol{F}_{\text{m}}$, that is

$$\boldsymbol{F} = q\boldsymbol{\mathscr{E}} + q\,(\boldsymbol{v} \times \boldsymbol{B})$$

$$\boldsymbol{F} = q[\boldsymbol{\mathscr{E}} + \boldsymbol{v} \times \boldsymbol{B}]\,. \qquad (4.10)$$

In fact, nothing in the way this last equation has been derived restricts it to uniform fields. Were the fields to vary with position, then the resultant force on the particle at the point $\boldsymbol{r}$ would be given by:

$$\boldsymbol{F}(\boldsymbol{r}) = q\left[\boldsymbol{\mathscr{E}}(\boldsymbol{r}) + \boldsymbol{v} \times \boldsymbol{B}(\boldsymbol{r})\right]\,. \qquad (4.10\text{a})$$

This general form of the rule giving the force on a charged particle in combined electric and magnetic fields is called the **Lorentz force law**. Because it describes both magnetic and electric forces, the Lorentz force law describes the physics underlying an enormous range of phenomena. In the rest of this section, we will consider three applications of the law in a wide variety of situations.

Before going on to tackle these various applications, this is an appropriate point to review the essential results obtained so far.

Revision Box: Charged particles in electric and magnetic fields

Magnetic fields

A particle with mass m and charge q moving with velocity $\boldsymbol{v}$ *in a uniform magnetic field* $\boldsymbol{B}$ experiences a magnetic force $\boldsymbol{F}_\mathrm{m}$ given by

$$\boldsymbol{F}_\mathrm{m} = q[\boldsymbol{v} \times \boldsymbol{B}]. \qquad \text{(Eqn 4.2a)}$$

This force may change the direction of motion of the particle, but it cannot change its speed.

If $\boldsymbol{v}$ is parallel (or antiparallel) to $\boldsymbol{B}$, then $|\boldsymbol{F}_\mathrm{m}| = 0$, and the particle's motion is completely unaffected by the presence of the field.

If $\boldsymbol{v}$ is perpendicular to $\boldsymbol{B}$, then the particle will move in a circle of radius

$$R_\mathrm{C} = \frac{mv}{|q|B}. \qquad \text{(Eqn 4.8)}$$

If $\boldsymbol{v}$ is at any other arbitrary angle to $\boldsymbol{B}$, then the particle moves in a helical path.

The frequency with which a particle executes its cyclotron motion is independent of the angle between $\boldsymbol{v}$ and $\boldsymbol{B}$, and is given by

$$f_\mathrm{C} = \frac{1}{2\pi}\frac{|q|B}{m} \qquad \text{(Eqn 4.9)}$$

Electric fields

A particle of charge q, whether moving or stationary, *in an electric field* $\boldsymbol{\mathscr{E}}(\boldsymbol{r})$ will experience a force

$$\boldsymbol{F}_\mathrm{el} = q\boldsymbol{\mathscr{E}}(\boldsymbol{r}). \qquad \text{(Eqn 4.1)}$$

A positively charged particle will be accelerated in the direction of the field; a negatively charged particle will be accelerated in a direction antiparallel to the field.

Electric and magnetic fields

If a particle of charge q is moving with velocity $\boldsymbol{v}$ under the influence of both an electric field $\boldsymbol{\mathscr{E}}(\boldsymbol{r})$ and a magnetic field $\boldsymbol{B}(\boldsymbol{r})$, then the resultant force it experiences at $\boldsymbol{r}$ is given by the Lorentz force law:

$$\boldsymbol{F}(\boldsymbol{r}) = q[\boldsymbol{\mathscr{E}}(\boldsymbol{r}) + \boldsymbol{v} \times \boldsymbol{B}(\boldsymbol{r})]. \qquad \text{(Eqn 4.10a)}$$

In a uniform field, the relationship may simply be written as

$$\boldsymbol{F} = q[\boldsymbol{\mathscr{E}} + \boldsymbol{v} \times \boldsymbol{B}]. \qquad \text{(Eqn 4.10)}$$

5.3 Devices involving uniform fields

The Hall effect

The **Hall effect** may be described as follows: when a magnetic field is applied perpendicular to a wafer of conducting material already carrying a current along its length, then a potential difference appears across its width. This is illustrated schematically in Figure 4.54.

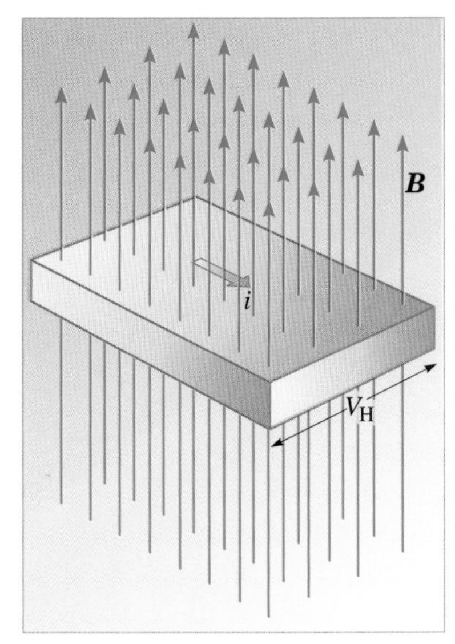

Figure 4.54 The Hall effect in a thin gold strip. When a current i flows through a thin metal strip in a magnetic field $\boldsymbol{B}$, then a transverse voltage V_H appears, known as the Hall voltage.

Figure 4.54 shows a strip of gold leaf, subject to an electric field $\mathscr{E}$ (not shown) and a magnetic field $\boldsymbol{B}$. The electric field $\mathscr{E}$ gives rise to a current i through the sample and a potential difference V along the length of the sample, as described by Ohm's law, $V = iR$ (Chapter 3 Section 2). Application of the magnetic field $\boldsymbol{B}$ gives rise to a *transverse* potential difference V_H which we can understand by considering the following three points:

- First, consider the directions in which electrons move in the absence of an applied magnetic field. On Figure 4.55, we have drawn one arrow representing the drift velocity $\boldsymbol{v}$ of an electron in the metal, and another indicating the direction of conventional current flow.
- Secondly, in an applied magnetic field $\boldsymbol{B}$, each electron experiences an additional magnetic force $\boldsymbol{F}_m = q(\boldsymbol{v} \times \boldsymbol{B})$. Notice (Figure 4.55) that because of the negative charge of the electron, $\boldsymbol{F}_m$ points in the opposite direction to the vector $(\boldsymbol{v} \times \boldsymbol{B})$.
- Thirdly, the magnetic force acts to push electrons to the left-hand side of the sample, that is, in the $-x$-direction. Since the electrons cannot leave the sample, this side becomes negatively charged. The charging process builds up until the transverse electric field created is sufficient to prevent any more electrons being pushed to the left-hand side. In other words, in the steady state — which is established a few picoseconds (10^{-12} s) after the current begins to flow — the transverse magnetic and electric forces are balanced and are of equal magnitude but opposite sign.

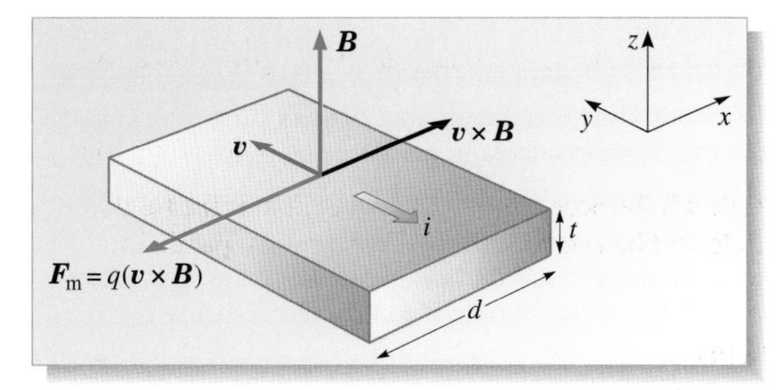

Figure 4.55 The figure shows an arrow representing the drift velocity $\boldsymbol{v}$ of an electron in the metal, and another representing the direction of conventional current flow i. In an applied magnetic field $\boldsymbol{B}$, each electron experiences an additional magnetic force $\boldsymbol{F}_m = q\,(\boldsymbol{v} \times \boldsymbol{B})$ but notice that because of the negative charge of the electron, $\boldsymbol{F}_m$ points in the opposite direction to the vector $(\boldsymbol{v} \times \boldsymbol{B})$. The magnetic force acts to push electrons to the left-hand side of the sample, which becomes negatively charged. Here, d and t are the width and thickness of the wafer respectively.

Mathematically, we can write the condition that the Lorentz force on the electrons be zero as

$$\boldsymbol{F} = \boldsymbol{F}_{\text{el}} + \boldsymbol{F}_{\text{m}} = \boldsymbol{0}$$

which implies that $\boldsymbol{F}_{\text{el}} = -\boldsymbol{F}_{\text{m}}$. In this geometry, both the Hall electric force, $\boldsymbol{F}_{\text{el}}$, and $\boldsymbol{F}_{\text{m}}$, act along the x-axis and so their y- and z-components are both zero. Considering only the x-component of the vectors, we can then write

$$q\mathscr{E}_{\text{H}} = qvB$$

or

$$\mathscr{E}_{\text{H}} = vB$$

where $\mathscr{E}_{\text{H}}$ is the magnitude of the Hall electric field. In order to work out the potential difference V_{H} measured across the sample, we need to recall two results from previous chapters:

- First, from Chapter 2, recall that the magnitude of the electric field $\mathscr{E}_{\text{H}}$ is just V_{H}/d where d is the width of the sample.
- Secondly, from Chapter 3, Section 2 recall that the current can be written as $i = nAqv$, where n is the number density of charge carriers, A the cross-sectional area of the wire, q the charge on an individual carrier and v the average (drift) speed of the carriers.

With these two pieces of information, we can rewrite the previous expression as

$$\frac{V_{\text{H}}}{d} = \frac{i}{nAq}B$$

and hence the Hall voltage may be written as

$$V_{\text{H}} = \frac{id}{nAq}B.$$

Since the cross-sectional area A is equal to the width d of the sample times its thickness t, we have finally

$$V_{\text{H}} = \frac{i}{nqt}B.$$

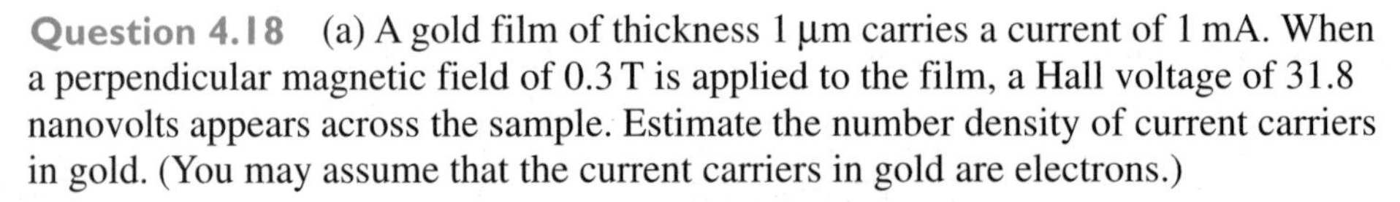

Question 4.18 (a) A gold film of thickness 1 μm carries a current of 1 mA. When a perpendicular magnetic field of 0.3 T is applied to the film, a Hall voltage of 31.8 nanovolts appears across the sample. Estimate the number density of current carriers in gold. (You may assume that the current carriers in gold are electrons.)

(b) The number density n of current carriers in a semiconductor can be in the region of $10^{20}\,\text{m}^{-3}$. A sample of such a material, 1 mm thick, is used in a device to measure magnetic field strengths perpendicular to the sample. Estimate how many volts per tesla would be generated across the sample if a current of 1 mA was flowing along it. ■

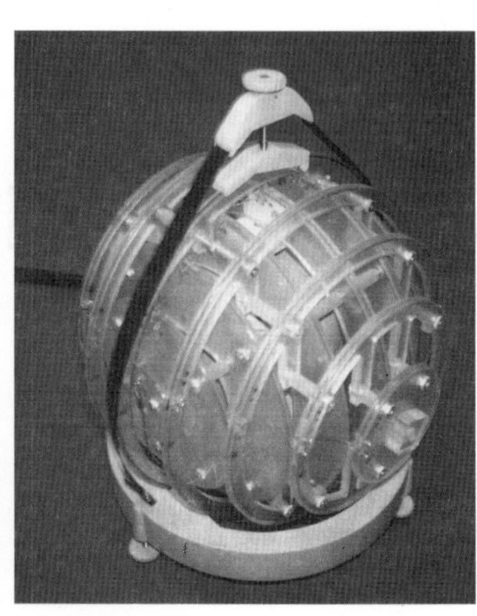

Figure 4.56 A commercial device for simultaneous determination of all components of a magnetic field.

As Question 4.18 brings out, using a material with a relatively low number density of carriers, it is possible to make a device that produces an appreciable Hall voltage that is directly proportional to the applied magnetic field. The device is only sensitive to the component of magnetic field perpendicular to the plane of the wafer. Thus, using three wafers at right angles, it is possible to build a device which will measure all three components of the magnetic field. A commercial instrument with three mutually perpendicular sensors (but operating on a slightly different principle) is shown in Figure 4.56.

The Hall effect is used extensively to study conduction in materials, particularly in semiconductors. We have assumed so far that the mobile charge carriers within solids are electrons and this is in agreement with the sign of the Hall voltage, as experimentally observed, for most materials. However, anomalous results can be obtained for metals such as zinc or lead, and some semiconducting materials behave as if there were positive charge carriers in the materials.

Question 4.19 A small, hand-held magnetometer is being developed that will use a sample of silicon of thickness 0.1 mm. Silicon has a current carrier density of $n = 1 \times 10^{23}\,\mathrm{m}^{-3}$. The current flowing along the sample is chosen to be 1 mA and the smallest voltage that can be reliably detected is about 10 μV. What is the smallest magnetic field that can be measured using the device? ■

The mass spectrometer

A **mass spectrometer** is an instrument used to separate different kinds of atom from one another. Figure 4.57 is a schematic diagram of one common type. Particles passing through traverse three distinct stages.

In the preparatory stage, the atoms of the substance under examination are ionized — usually by stripping them of one or more electrons, leaving them with a net positive charge. This is achieved by a combination of heating and an intense electric field.

The ions are then accelerated by a second electric field and enter a *velocity selector*, which allows through only those ions travelling at a specific speed v. The velocity selector is a device which exploits the fact that, whereas the magnitude of the *magnetic* force on a charged particle is directly proportional to its speed, the *electric* force does not depend at all on the particle's speed.

The essential components of a velocity selector are shown in Figure 4.58. The ions enter the selector travelling in the $+x$-direction, and are acted upon by a third electric field $\mathscr{E}$ in the $-y$-direction and a magnetic field $\boldsymbol{B}_1$ acting in the $-z$-direction. Question 4.20 leads you through the physics of the velocity selector, but the key point to note is that in order to get through the velocity selector, there must be a balance between the electric and magnetic components of the Lorentz force.

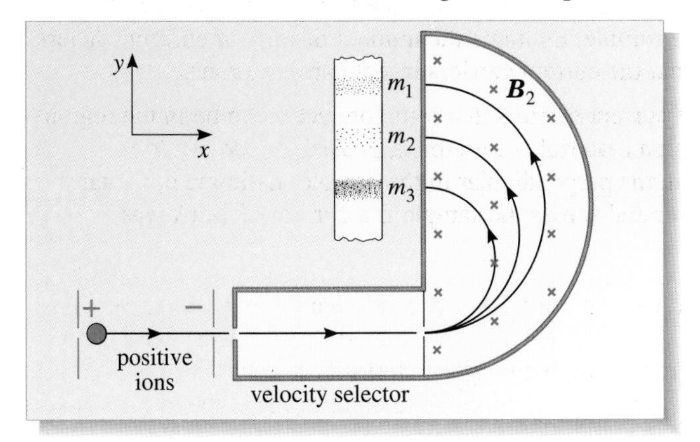

Figure 4.57 A mass spectrometer. The uniform magnetic field $\boldsymbol{B}_2$ points into the page (i.e. along the $-z$-direction). In the example shown here, the spectrometer is being used to separate three types of ions, all carrying the same positive charge, but with different masses, such that $m_1 > m_2 > m_3$.

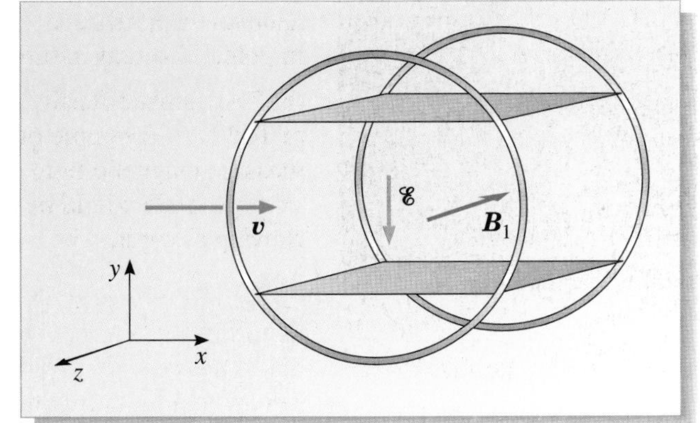

Figure 4.58 Schematic diagram of a velocity selector, consisting of two coils of wire each carrying the same current i in the same direction, and two charged metal plates.

Question 4.20 (a) Mark on Figure 4.58 the direction in which the current must be flowing in the two coils in order to generate the magnetic field $\boldsymbol{B}_1$.

(b) Mark on Figure 4.58 the way in which the two plates must be charged in order to establish the electric field $\boldsymbol{\mathscr{E}}$.

(c) Write down an expression for the resultant force $\boldsymbol{F}$ acting on a positive ion of charge q and velocity $\boldsymbol{v}$ that enters the crossed field region. Hence show that an ion, initially travelling in the $+x$-direction, will pass undeflected through the fields if its speed v is such that $v = \mathscr{E}/B_1$. (*Hint*: Remember Newton's first law!) ■

The device shown in Figure 4.58 thus acts as a velocity selector by allowing only those ions with initial speed v to pass through it without deflection. The actual values of $\mathscr{E}$ and B_1 can be adjusted until their ratio corresponds to whatever value of v is required for a particular experiment. Ions travelling at other speeds will not travel in a straight line through the selector, and therefore will not enter the main part of the spectrometer.

The third stage of the mass spectrometer is the D-shaped chamber shown in Figure 4.57. In this chamber, a magnetic field $\boldsymbol{B}_2$ is applied. Notice that $\boldsymbol{B}_2$ is different in magnitude from $\boldsymbol{B}_1$ (the field applied in the velocity selector) but parallel to it. The field $\boldsymbol{B}_2$ is thus perpendicular to the velocity $\boldsymbol{v}$ of the ions. Under the influence of $\boldsymbol{B}_2$ therefore, the ions move in a semicircular path with radius R_C, the cyclotron radius calculated from Equation 4.8:

$$R_C = \frac{mv}{|q| B_2}.$$

Since we already know that $v = \mathscr{E}/B_1$, this can be written as

$$R_C = \frac{m\mathscr{E}}{|q| B_1 B_2}.$$

In the type of mass spectrometer illustrated in Figure 4.57, a detector plate records the arrival of the ions and allows the radius R_C to be measured (Figure 4.59). Knowing the magnitudes of the electric and magnetic fields, the *charge to mass ratio* of the ions can then be inferred:

$$\frac{|q|}{m} = \frac{\mathscr{E}}{R_C B_1 B_2}.$$

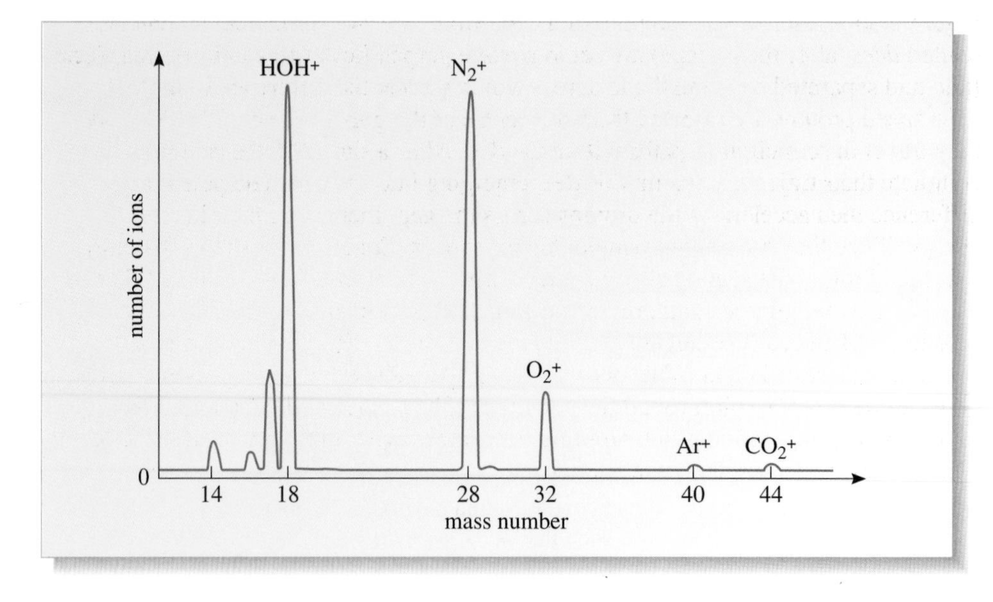

Figure 4.59 The output of a modern mass spectrometer displaying the composition of the residual air left in a vacuum chamber after evacuation. It is clearly possible to identify peaks corresponding to H_2O (mass number 18), N_2 (mass number 28), O_2 (mass number 32) and CO_2 (mass number 44).

Strictly speaking, therefore, a mass spectrometer separates ions according to their charge to mass ratio: the larger $|q|/m$, then the smaller R_C. Only if all the ions have the same charge does the instrument separate all the masses precisely. In that case, the lowest-mass ions travel in the tightest semicircle, as shown in Figure 4.57. In the instrument as a whole, there are two quite different separation processes. The velocity selector (involving field $\boldsymbol{B}_1$) selects ions on the basis of their speed, but irrespective of their mass. In the D-shaped chamber (where $\boldsymbol{B}_2$ is applied), the ions are separated according to their charge to mass ratio.

Mass spectrometers are sensitive instruments, and can be used to detect mass differences of about 0.01%. They can separate *isotopes* which are atoms of a particular element (and which therefore cannot be separated chemically) that differ only in the number of neutrons in their nucleus, and thus have slightly different masses. Because of their sensitivity, mass spectrometers are also useful in the detection of small quantities of impurities or pollutants.

Question 4.21 (a) Suppose that in the velocity selector shown in Figure 4.58 the coil currents are adjusted to give a magnetic field strength of 0.5 T, and that the plate separation is 10 cm. If ions travelling at $2 \times 10^5\,\mathrm{m\,s^{-1}}$ are to be selected, what potential difference should be applied across the plates? (*Note*: *You should not need to use a calculator*!)

(b) The ions selected in (a) then enter the mass analyser region where the field strength $B_2 = 1$ T. They travel in a semicircle, striking the detector plate 16.2 cm from the point at which they enter the D-shaped chamber. Assuming the ions have each lost only a single electron, identify the ions. ■

The cyclotron

A **cyclotron** is a particle accelerator, most commonly used to produce a stream of high energy protons. The operating principle of the cyclotron depends crucially on the fact that (as shown in Section 4.1) the orbital (cyclotron) frequency f_C of a charged particle moving in a plane perpendicular to a magnetic field does not depend on the particle's speed:

$$f_C = \frac{1}{2\pi}\frac{|q|B}{m}. \qquad \text{(Eqn 4.9)}$$

Figure 4.60 is a schematic diagram of a cyclotron. The two semicircular chambers (called *dees*, after their shape) are set in a plane perpendicular to a uniform magnetic field and separated by a small gap across which a potential difference is applied. Low speed protons are injected from the centre of the gap into one of the dees, so they travel in semicircular paths within the dee. After a time $T/2$, the protons complete their trajectory within this dee, emerging into the gap. The potential difference then accelerates the protons across the gap, increasing their kinetic energy. Since they are now moving at higher speeds, Equation 4.8 shows that they will travel round the second dee in a larger orbit. But, after a further time $T/2$ has elapsed, they will, once again, reach the gap. At this moment, the polarity of the potential difference between the dees is reversed, so the protons are once more accelerated across the gap. This periodic reversal of the voltage is repeated many times as the protons spiral round the chambers. The particles speed up each time they cross the gap, moving in a progressively larger orbit, until eventually a deflector plate guides them out of the accelerator.

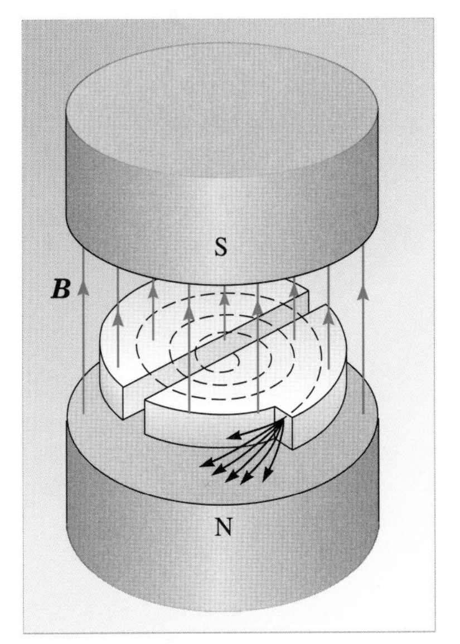

Figure 4.60 A cyclotron, used to accelerate protons.

The cyclotron is an ingenious device, but in fact we have nearly all taken part in an 'accelerator' using a similar principle but in a quite different context. If you give someone on a swing a push at regular intervals matched to the natural frequency of the swing, the amplitude of their motion will increase systematically. This is just like the regular 'pushes' given to a proton in a cyclotron, except that in that device the protons are 'pushed' twice in each cycle. In the swing analogy, this is equivalent to having another person on the far side of the swing pushing in the other direction at alternate half cycles (Figure 4.61).

Figure 4.61 Regular pushing on a swing in time with its natural frequency of oscillation, increases its amplitude of oscillation. The rate of acceleration can be increased if there are two pushes in each cycle, one in one direction and one in the other. This is analogous to the two 'kicks' per cycle received by a proton in its orbit around a cyclotron.

Protons in a cyclotron can attain kinetic energies of up to about 25 MeV. The limit on the acceleration is imposed by relativistic effects. As you will see in *Dynamic fields and waves* Chapter 4, when the protons speed up, they become harder to accelerate and they take longer to complete each successive semicircular orbit. Then the regular voltage reversals get out of step with the protons' travel across the gap and no further acceleration of the protons can occur. This problem is overcome in a *synchrocyclotron* or *synchrotron* by reducing the frequency of the voltage reversals to compensate for the change in crossing frequency. A synchrocyclotron can accelerate protons to energies of several hundred MeV.

Question 4.22 (a) In a cyclotron with a magnetic field in the dee region of 0.5 T, at what frequency should the accelerating potential between the dees be reversed in order to accelerate protons?

(b) If the accelerating voltage between the dees has a magnitude of 1000 V, how many orbits must a proton make to reach an energy of 25 MeV? ■

5.4 Magnetic forces on current-carrying conductors

Most types of electric motor, and many other electrical devices, involve parts that move as a result of a simple physical phenomenon:

> When a wire carrying an electric current is located at an angle to a magnetic field, the wire experiences a force.

We can understand how this happens by considering the force on the individual electrons within the wire.

Magnetic force on a wire

Figure 4.62 shows a length of wire carrying a current i in a uniform magnetic field $\boldsymbol{B}$. Each moving electron in the current has an average velocity $\boldsymbol{v}$ and experiences a force $\boldsymbol{F}_{\text{m}} = q[\boldsymbol{v} \times \boldsymbol{B}]$. The forces on *all* of the electrons moving along a given short length of wire add together vectorially and the resultant force may be sufficient to have an observable effect on the wire. From Figure 4.62 we can see that the resultant force is just the magnetic force on a single electron multiplied by the number N of electrons in the length of wire.

Let's consider a length l of wire of cross-sectional area A. Notice that, even though electrons are moving through the wire, the same number of electrons leave the length l per unit time as enter it. Thus, the average number in the length l is just N, the same as it was when no current flowed. If the number density of electrons in the wire is n, then the total number N of electrons in the length l is nAl. Then the total magnetic force on the length l of wire is given by

$$\boldsymbol{F}_{\text{m}} = (nAl)\,q[\boldsymbol{v} \times \boldsymbol{B}].$$

The current i is defined as the charge passing a point in the wire per unit time. Using the analysis of Chapter 3 Section 2, we recall that $|i| = nAv|q|$, where v is the average drift speed of the electrons. Defining a vector current $\boldsymbol{i} = nAq\boldsymbol{v}$ with magnitude i and directed along the wire, we can rewrite the total magnetic force as:

$$\boldsymbol{F}_{\text{m}} = l[\boldsymbol{i} \times \boldsymbol{B}]. \tag{4.11}$$

Notice that, because electrons are negatively charged, the vector $\boldsymbol{i}$, which represents conventional current, is generally in the opposite direction to $\boldsymbol{v}$. If the wire and the magnetic field are perpendicular to one another, Equation 4.11 implies that the magnitude of the force is

$$F_{\text{m}} = Bil. \tag{4.12}$$

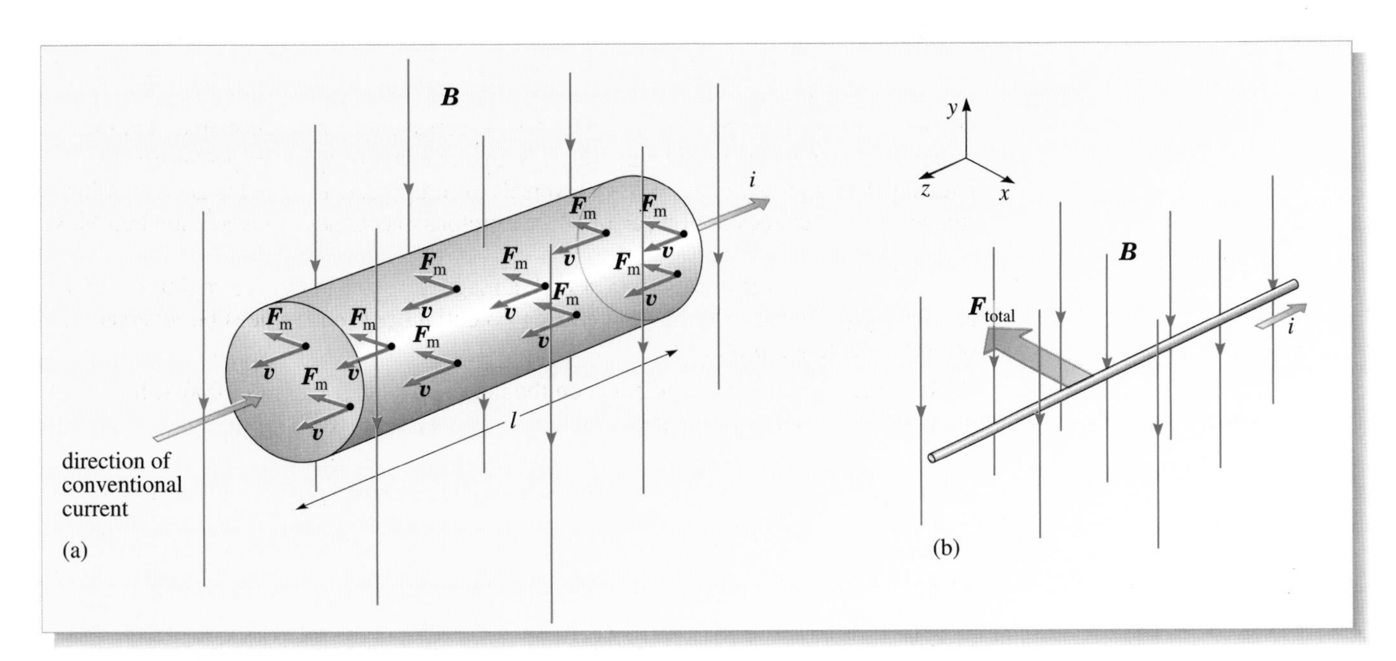

Figure 4.62 (a) A segment of wire carrying a current i in a magnetic field $\boldsymbol{B}$. Each electron has velocity $\boldsymbol{v}$ and is subject to the force $\boldsymbol{F}_{\text{m}}$. (b) The force $\boldsymbol{F}_{\text{total}}$ on a current-carrying wire in a magnetic field.

Question 4.23 A cable at an electricity generating station carries a current of 10^4 A over a distance of 100 m. Estimate the magnitude of the magnetic force on the cable assuming that it is perpendicular to the Earth's magnetic field which has a magnitude of 10^{-4} T. ■

Magnetic forces between current-carrying wires

In general, the magnetic force between two current-carrying wires depends in a complicated way on their shape and relative position. However, there is one particular arrangement of wires for which it is especially simple to calculate both the magnitude and the direction of the force between currents. That arrangement is shown in Figure 4.63. It consists of two long straight wires that are parallel to one another and separated by a distance d. In discussing this set-up, it is very important to distinguish the two wires, so we will refer to them as wire 1 and wire 2, and their respective steady currents will be called i_1 and i_2. The magnetic field lines arising from these currents form concentric circles around each of the wires, the direction of the field lines being given by the right-hand grip rule.

In seeking to apply the analysis of Figure 4.62 to this situation, we note that, although the magnetic fields $\boldsymbol{B}_1$ and $\boldsymbol{B}_2$ in Figure 4.63 are not uniform, when wire 2 is positioned parallel to wire 1, the field $\boldsymbol{B}_1$ has the same magnitude and direction *at all points along wire 2*. At all points along wire 2, the field arising from the current in wire 1 points downwards (i.e. perpendicular to wire 2), and is of magnitude

$$B_1 = \frac{\mu_0 i_1}{2\pi d}.$$

(To remind yourself of the origin of this expression, look back at Equation 4.4. Because Equation 4.4 assumed a positive value of i, the final expression we derive in this section will have the same assumption built into it.)

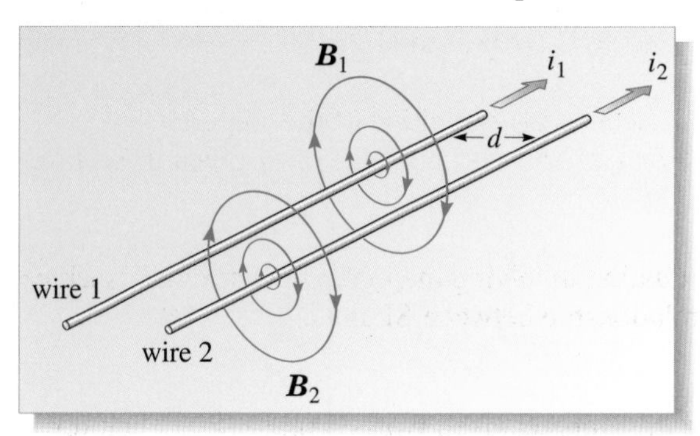

Figure 4.63 Two long straight parallel wires separated by a distance d. Wire 1 carries a current i_1, generating a magnetic field $\boldsymbol{B}_1$. Wire 2 carries a current i_2, generating a magnetic field $\boldsymbol{B}_2$.

Question 4.24 (a) Using similar arguments to those employed in analysing Figure 4.62, work out the direction of the force on wire 2 due to wire 1 in Figure 4.63. Repeat the process to find the direction of the force on wire 1 due to wire 2. (b) Convince yourself that the use of Equation 4.11 gives the same answer. ■

As you will have found by answering Question 4.24, when the currents i_1 and i_2 are flowing in the same direction, the force on each of the wires is directed *towards* the other wire. In other words, the wires attract each other. We now want to derive an expression for the *magnitude* of the force between the two wires. To do this, we apply Equation 4.12 to the situation depicted in Figure 4.63. The wires are separated by a distance d. The current i_1 (in wire 1) gives rise to a magnetic field, which all along wire 2 is of strength

$$B_1 = \frac{\mu_0 i_1}{2\pi d}.$$

Wire 2 is carrying a current i_2 in this field. The force F_2 on a length l of wire 2 is therefore of magnitude

$$F_2 = B_1 i_2 l = \frac{\mu_0 i_1 i_2 l}{2\pi d}.$$

We would of course have obtained exactly the same final expression had we worked out B_2 and F_1. Thus, the magnitude of the force per unit length on *either* of the wires may be written:

$$\frac{F}{l} = \frac{\mu_0 i_1 i_2}{2\pi d}. \tag{4.13}$$

The definition of the ampere

Equation 4.13 is especially important because it is through this equation that the SI unit of the ampere is defined:

One **ampere** is the steady current that, flowing in each of two straight, parallel, infinitely long wires set one metre apart in a vacuum, causes each wire to experience a force of magnitude 2×10^{-7} newtons per metre of its length.

Because Equation 4.13 is at the heart of the whole system of electrical units, it is worth pausing to reflect on it for a moment. The reason for making the ampere the fundamental unit, rather than, say, the coulomb, is a practical one: because the ampere is defined entirely in terms of mechanical quantities (forces and distances), it is easier to make accurate measurements of current than of charge. So it is much more convenient to define the ampere first, and then to define the coulomb in terms of it:

When a steady current of one ampere flows along a wire, then one coulomb of net charge passes a fixed point in the wire every second.

Finally, Question 4.25 allows you to see how the definition of the ampere, encapsulated in Equation 4.13, accords with the defined value of μ_0 given in Section 3.1.

Question 4.25 (a) Use a basic equation involving magnetic field strength, such as $F = |q|vB$, to show the following relationship between SI units:

$$1\,\text{A} = 1\,\text{N}\,\text{m}^{-1}\,\text{T}^{-1}.$$

(b) Then show that the definition of the ampere, taken with Equation 4.13, requires that $\mu_0 = 4\pi \times 10^{-7}\,\text{T}\,\text{m}\,\text{A}^{-1}$ as stated in Section 2.4. ■

To practise the use of Equation 4.12, and some of the fundamental principles of this chapter, try the following question. (*Hint:* It would be a good idea to use our *Problem-solving technique* for this question.)

Question 4.26 Figure 4.64 shows a non-magnetic cylinder of length $l = 20\,\text{cm}$ and mass 0.15 kg, on which have been wound 25 turns of copper wire. (The axis of the cylinder lies within the plane of the coil.) The cylinder is placed on a plane inclined to the horizontal, with the plane of the coil parallel to this slope. A uniform magnetic field of strength 0.3 T points vertically upwards. What is the current required to keep the cylinder stationary and in what direction must this current flow? (Assume the coefficient of friction is just sufficient to ensure that the cylinder does not slide, and take $g = 10\,\text{m}\,\text{s}^{-2}$.) ■

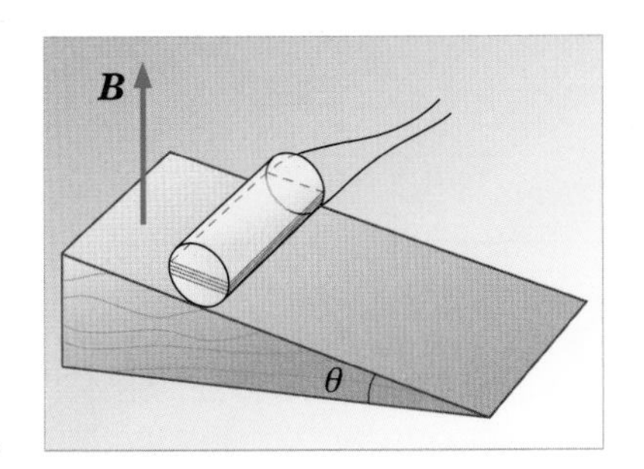

Figure 4.64 For use with Question 4.26.

Torque on a loop of wire

Study Note: If you need reminding of the concept of torque then this would be a good point at which to revise Predicting motion, *Chapter 4, Section 2.*

We have seen that the force on a length l of wire carrying a current i in a uniform magnetic field is given by $F_m = Bil$ when the wire is perpendicular to the magnetic field. We can use this result to analyse the magnetic torque on a rectangular, current-carrying loop of wire (Figure 4.65).

We can see that the forces on sides BC and AD of the coil are zero because in these sections the current flows parallel, or anti-parallel, to $\boldsymbol{B}$. Using Equation 4.11 we determine that the vector product $\boldsymbol{i} \times \boldsymbol{B}$, and therefore the force on side AB, points upwards. The magnitude of the force is

$$F_{AB} = il_1B.$$

Similarly, the force on side CD points downwards and is of the same magnitude

$$F_{CD} = il_1B.$$

Thus the magnitude of the torque on the loop about the axis KK′ in Figure 4.65 is

$$\begin{aligned} \Gamma &= F_{AB} \times [l_2/2] + F_{CD} \times [l_2/2] \\ &= il_1B[l_2/2] + il_1B[l_2/2] \\ &= il_1l_2B. \end{aligned}$$

In fact, this expression simplifies even more if we note that product l_1l_2 is just the area A of the loop, giving

$$\Gamma = iAB.$$

The torque will cause the loop to rotate as shown in Figure 4.66. The magnitudes of the forces F_{AB} and F_{CD} remain the same as the loop rotates, but the torque diminishes until, at the point when the plane of the loop is perpendicular to the magnetic field, the torque is zero because the forces act through the axis KK′.

It is this torque on a current-carrying loop in a magnetic field that is the basis of the operation of numerous devices including motors and moving-coil current meters.

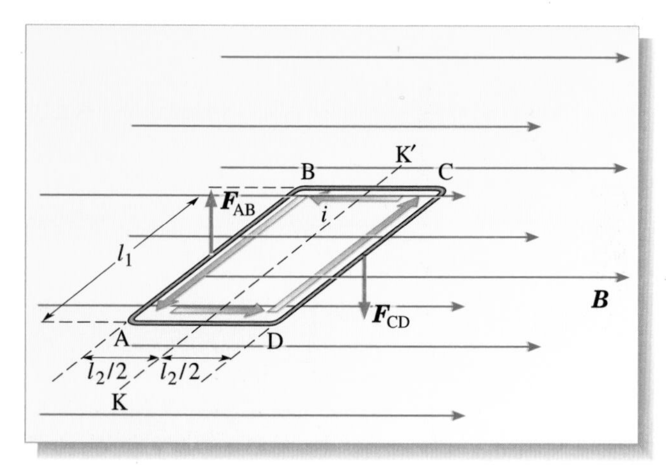

Figure 4.65 Analysis of the torque on a current-carrying loop in a uniform magnetic field. The total *force* on a current-carrying loop in a uniform field is zero. Notice that the sides BC and AD do not experience a force because $\boldsymbol{i}$ flows parallel (or anti-parallel) to $\boldsymbol{B}$.

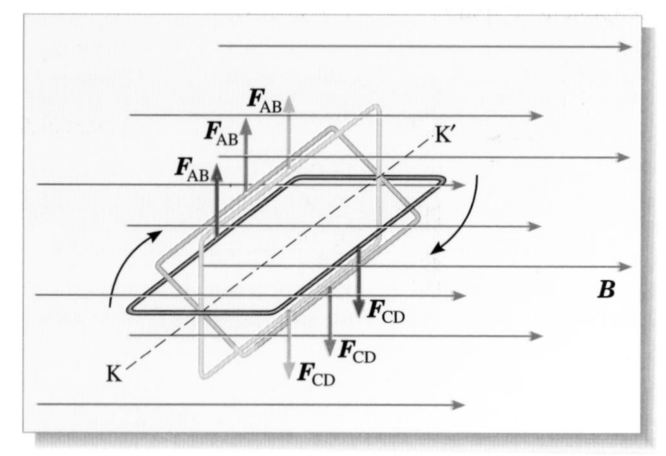

Figure 4.66 Illustration of the effect of the torque on the current-carrying loop in Figure 4.65. The loop rotates until the torque on the loop is zero. Notice that forces on the sides AB and CD do not change with position, but the torque they exert does.

6 The motion of charged particles in non-uniform magnetic fields

In the previous section, we considered a variety of devices which operated in essentially uniform magnetic fields. Even where the fields were not quite uniform, such as in the case of the force between two long, straight current-carrying wires, it did not affect the analysis to any great extent. In this section, however, we will consider phenomena that can *only* occur in *non-uniform* magnetic fields.

The paths followed by charged particles in non-uniform fields tend to be even more complicated than in a uniform field. However, we can use the understanding we gained from analysing the uniform-field examples to gain some insight into these more complicated examples. We will still use the Lorentz force law except that now we have to use the most general form of the law:

$$\boldsymbol{F}(\boldsymbol{r}) = q[\boldsymbol{\mathscr{E}}(\boldsymbol{r}) + \boldsymbol{v} \times \boldsymbol{B}(\boldsymbol{r})] \qquad \text{(Eqn 4.10a)}$$

since the fields depend on position. We also need to be careful about how we denote the magnitude of the magnetic field. Just writing B would not give a full description of the situation: if the field is non-uniform, the field strength is still a scalar quantity, but it varies from point to point, and should therefore be written as $B(\boldsymbol{r})$.

In Section 6.1, we will explore two particular configurations of non-uniform magnetic fields and see how charged particles may actually become *trapped* by a field. Then, Section 6.2 will show how such trapping mechanisms may be set up for use in technological projects, and how charged particles trapped by the Earth's field at altitudes of many thousands of kilometres can lead to the magnificent auroral displays visible from the Earth's surface at high latitudes.

6.1 Magnetic mirrors and magnetic bottles

Figure 4.67 shows a situation in which the magnetic field is stronger on the right of the figure than on the left. We illustrate this by drawing the field lines closer together on the right than on the left. A *positively* charged particle of mass m is shown entering from the left with a velocity $\boldsymbol{v}$, which may be resolved into two velocity components v_x parallel to the axis of the field and v_y perpendicular to it.

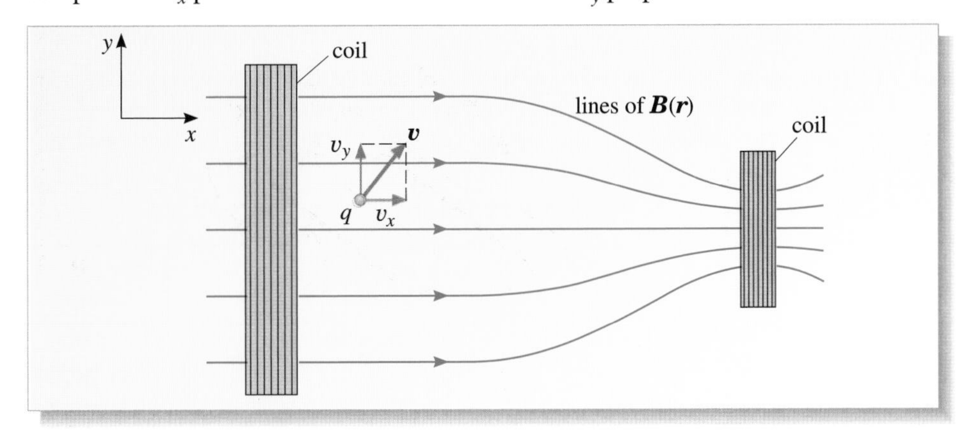

Figure 4.67 A magnetic mirror — a magnetic field that increases in strength towards the right. Such a field can be created by two coils.

The particle will initially find itself in a region in which the magnetic field is almost uniform and will start to follow a helical path. The radius of the helix, R_C, will be given by:

$$R_C = \frac{m\,|v_{\text{perp}}|}{|q|\,B(\boldsymbol{r})} \qquad (4.8b)$$

where v_{perp} is the component of the particle's velocity perpendicular to the field.

This equation is very similar to Equation 4.8a, for a helical path in a uniform field, except that the field strength is now written $B(\boldsymbol{r})$, and not just B.

What will happen to the path of the particle as it enters the region of increasing field strength? The answer is shown in Figure 4.68. As you can see, the stronger field produces two effects: the radius of the helix gradually decreases, and the angle between the velocity vector of the particle and the magnetic field increases, approaching a right angle. In other words, the helix becomes tighter and flatter (more compressed).

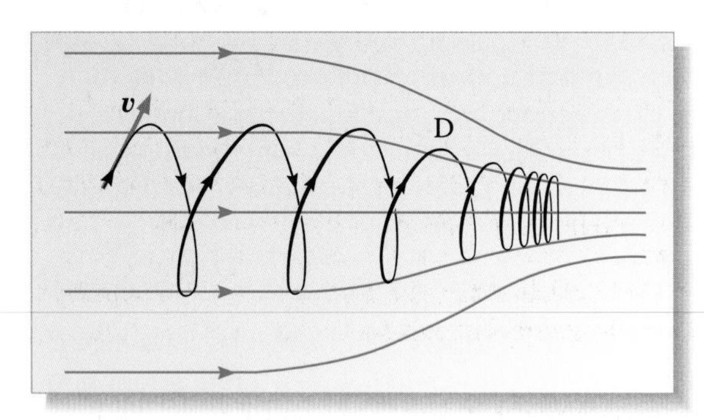

Figure 4.68 Motion of a positively charged particle in the field shown in Figure 4.67. Note that at point D the velocity is mainly directed into the page, and that the magnetic field has a component that points towards the bottom of the page, as well as a component that points to the right.

The first of the two effects, the shrinking radius, is to be expected because the stronger field turns the particle in a 'tighter' circle. Equation 4.8b predicts that an increasing field strength $B(\boldsymbol{r})$ implies a decreasing value of R_C. The second effect, the flattening of the helix, is in accord with Example 4.1 in Section 5.1 in which we saw that the distance moved during one cycle of the helix in a uniform field is given by

$$d = \frac{2\pi m v \cos\theta}{qB}.$$

This shows that d is also inversely proportional to $B(\boldsymbol{r})$. One can understand the physics underlying this effect by considering the force experienced by a positively charged particle as it moves through the point marked D on Figure 4.68. You should be able to see that because of the direction of the magnetic field, the Lorentz force has a part that is pointing towards the left. This part of the force is shown in Figure 4.69, where it is called $\boldsymbol{F}_{\text{left}}$. In fact, if you use the right-hand rule to work out the direction of the Lorentz force at *any* point on the helix, you will find that the gathering together of the field lines always gives rise to a leftward-pointing force that tends to push the particle to the left, away from the region in which the field is strongest. The effect of this force is to slow down the motion of the particle on the axis of its helix, and the particle makes less and less progress towards the right during each successive twist of the helix, so the orbit becomes flatter.

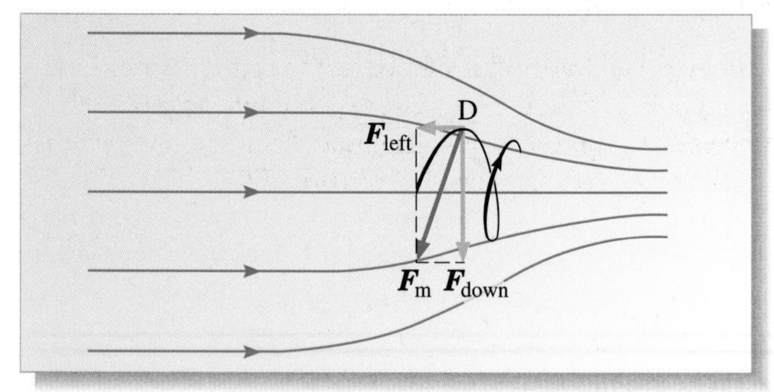

Figure 4.69 The Lorentz force $\boldsymbol{F}_{\text{m}}$ at the point D can be resolved into two mutually perpendicular forces $\boldsymbol{F}_{\text{left}}$ and $\boldsymbol{F}_{\text{down}}$.

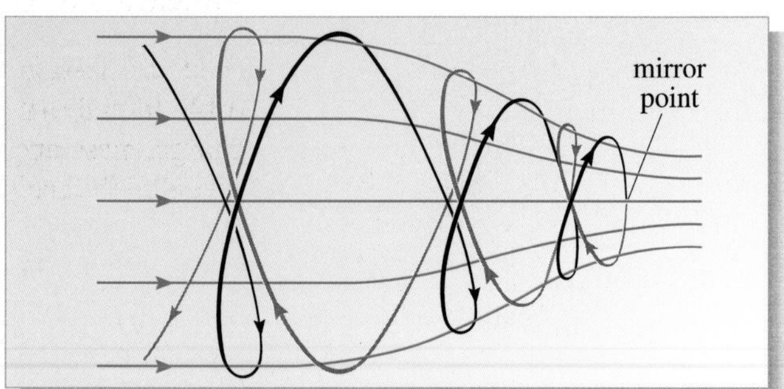

Figure 4.70 Reflection of motion at a mirror point. The (positive) particle enters from the left, travelling towards the region of high field strength. The helical trajectory becomes 'tighter and flatter' until the mirror point is reached. At the mirror point, the motion is 'reflected', and the particle travels back towards the left.

Because of the force $\boldsymbol{F}_{\text{left}}$ pointing towards the left in Figure 4.69, there will be a point on the trajectory of the particle at which motion to the right ceases altogether. This is called the *mirror point*. The exact location of the mirror point on any given trajectory will depend on the strength of the magnetic field, the mass and charge of the particle, and the velocity with which the particle enters the region of non-uniform field. But notice that even when the particle of Figure 4.69 reaches its mirror point, it is still acted on by a force with a component that points towards the left, so after passing through the mirror point the particle will start to move towards the left, returning to the region whence it came, as shown in Figure 4.70.

In a sense, the mirror point 'reflects' the motion of the particle — which, of course, is how it gets its name. A field of the kind we have been discussing is similarly known as a *magnetic mirror*. A particularly interesting aspect of magnetic mirrors is that they can be assembled in pairs to form a device that traps charged particles by means of successive reflections from each mirror in turn. Such traps, as illustrated schematically in Figure 4.71, are called **magnetic bottles**. They are used in devices such as fusion reactors to contain the plasma particles, but they can also arise in nature as the next section indicates.

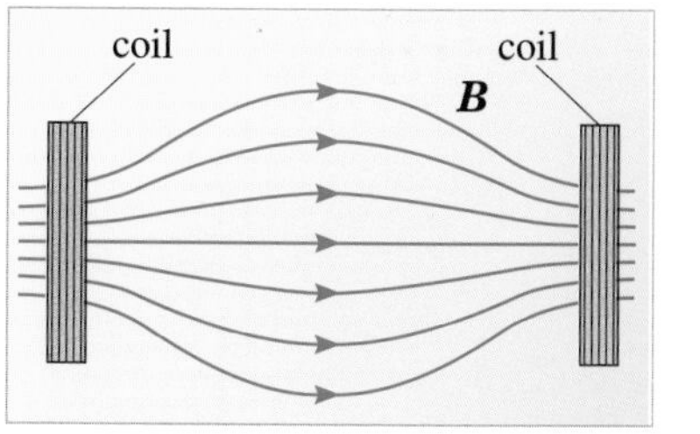

Figure 4.71 A magnetic bottle consisting of two magnetic mirrors facing one another. Charged particles will move back and forth between the ends of the bottle.

6.2 Trapped particles in the Earth's magnetic field

The Earth's magnetic field, being to a good approximation dipolar, is highly non-uniform. The field lines come closer together near the poles, creating exactly the kind of field configuration described in the previous section. When a charged particle, spiralling around a field line, approaches one of the poles, it encounters an increasingly strong magnetic field. Eventually, the particle reaches its mirror point and is reflected back towards the other pole where the same process is repeated. In this way, particles are caught in a magnetic trap; they bounce back and forth between mirror points until liberated by some kind of collision. The trapping mechanism is illustrated schematically in Figure 4.72. The field lines in this diagram are shown in cross-section, so only two 'traps' are shown, but of course the Earth's field extends three-dimensionally into space. In three dimensions, the trapped particles actually form two 'ring doughnut-shaped' belts around the Earth, as shown in Figure 4.73. These concentrations of particles, discovered by Van Allen using the early American space satellites *Explorer 1* and *Pioneer 3* in 1958, are now called the **Van Allen belts**.

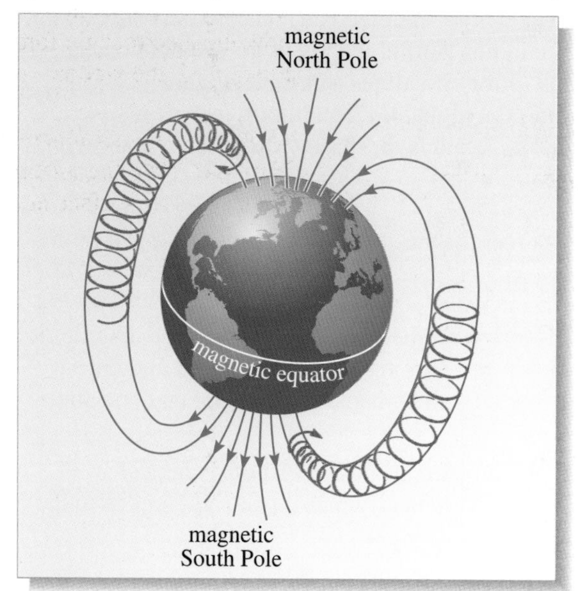

Figure 4.72 The convergence of magnetic field lines near the Earth's magnetic poles creates the conditions of a natural magnetic bottle and traps charged particles.

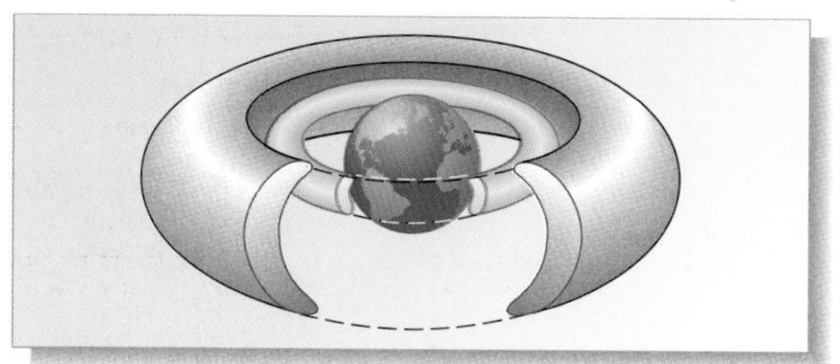

Figure 4.73 The Van Allen belts contain a high density of charged particles trapped by the Earth's magnetic field. The inner and outer belts have mean radii respectively about 1.6 and 3.5 times that of the Earth. The inner belt is now recognized as the region in which the density of protons with energy greater than 30 MeV attains its maximum, while the outer belt is the region in which electrons with energy greater than 1.6 MeV are most concentrated. Particles with lower energies are spread over much broader regions, and cannot be assigned to distinct belts.

The aurora

The supply of particles in the Earth's upper atmosphere is constantly replenished by the **solar wind**. The solar wind is a stream of plasma particles (mostly electrons and protons) that are emitted from the Sun's outermost layer. The mean speed of the solar wind relative to the Earth is of the order of $500\,\text{km}\,\text{s}^{-1}$ (roughly a million miles an hour!), and the average number density of positive particles (balanced by an equal number density of electrons) is about $5\,\text{cm}^{-3}$.

Figure 4.74 The aurora.

The magnetic bottles that give rise to the Van Allen belts are 'leaky'. The **aurora** visible in extreme northern and southern latitudes is associated with the leakage of charged particles into the upper atmosphere near the poles, where the Van Allen belts approach closest to the Earth. The high energy, charged particles then interact with atoms or molecules (mostly oxygen and nitrogen) in the upper atmosphere causing them to emit light. The auroral display often resembles a curtain (Figure 4.74) of pink and green light hanging down into the night sky. The importance of magnetic effects is shown by pictures of the aurora taken from space, which show it as a ring round the *magnetic* pole (Figure 4.75).

This next question will give you some feel for the speeds of particles trapped in the Earth's field.

Question 4.27 At an altitude of 2000 km on the magnetic equator, the magnitude of the Earth's magnetic field is roughly $1.5 \times 10^{-5}\,\text{T}$. Consider a proton and an electron, each of energy 1 MeV, twisting around a magnetic field line and crossing the magnetic equator at this altitude.

(a) How long does each particle take to complete a single twist of its helical trajectory? (*Hint*: Assume the particles are very nearly moving at right angles to the field, and remember that for circular motion $v/r = \omega$.)

(b) What is the radius of the proton's helical trajectory? ■

Figure 4.75 Extent of the 'auroral oval', as seen from a position in space directly above the geographic North Pole. It is clear that the auroral oval is centred not on the geographic but on the magnetic pole.

Open University students should leave the text at this point and view Video 5 *Magnetic fields in space*. You should return to the text when you have finished viewing the video.

7 Closing items

7.1 Chapter summary

1 A particle of charge q and velocity $\boldsymbol{v}$ passing through a point specified by the position vector $\boldsymbol{r}$ experiences a magnetic force

$$\boldsymbol{F}_{\rm m} = q[\boldsymbol{v} \times \boldsymbol{B}(\boldsymbol{r})] \qquad (4.2)$$

where $\boldsymbol{B}(\boldsymbol{r})$ is the magnetic field at $\boldsymbol{r}$. Equation 4.2 shows that the direction of the magnetic force on the particle is perpendicular to both its velocity and the direction of the magnetic field at $\boldsymbol{r}$.

2 The magnitude of the magnetic force on the particle is

$$F_{\rm m} = |\,q\,|\,vB \sin\theta \qquad (4.3)$$

where θ is the angle between the particle's velocity $\boldsymbol{v}$ at $\boldsymbol{r}$ and $\boldsymbol{B}(\boldsymbol{r})$. The magnitude of the magnetic field at $\boldsymbol{r}$, $B(\boldsymbol{r})$ is called the magnetic field strength at that point, and is measured in SI units of teslas (T).

3 Magnetic fields throughout extended regions can be visualized by magnetic field lines.

4 Magnetic fields, like electric fields, are vector fields. The resultant magnetic field at a point can therefore be found by vector addition of the individual magnetic fields at that point.

5 Moving charges, that is, electric currents, produce magnetic fields.

6 The pattern of magnetic field lines due to a long straight wire carrying a steady current consists of a series of concentric circles centred on the wire and perpendicular to it. The sense in which the field lines circle the wire is related to the direction of conventional current flow along the wire by the right-hand grip rule. The magnitude of the field at a perpendicular distance r from a long straight current-carrying wire i is

$$B(r) = \frac{\mu_0 i}{2\pi r} \qquad (4.4)$$

where μ_0 is a constant with the defined value of $4\pi \times 10^{-7}\,\mathrm{T\,m\,A^{-1}}$.

7 The pattern of magnetic field lines around a circular loop of wire is known as a magnetic dipole field. The strength of the dipole is characterized by its magnetic dipole moment.

8 At the centre of a single coil with radius R, carrying a current i, the magnetic field is given by:

$$B_{\mathrm{centre}} = \frac{\mu_0 i}{2R}. \qquad (4.5)$$

9 Along the axis of an infinitely long cylindrical solenoid with N/l turns per unit length, the field is given by:

$$B = \frac{\mu_0 N i}{l}. \qquad (4.6)$$

10 The Earth's magnetic field is essentially dipolar and is generated by currents flowing in the Earth's outer core.

11 The nature of magnetic fields around permanent magnets is the same as that around current-carrying conductors.

12 All materials become magnetized to at least a small extent when placed in a magnetic field.

13 Ferromagnets can become strongly magnetized in even a weak applied magnetic field, and can retain this magnetization in the absence of an applied magnetic field.

14 Ferromagnetism arises because of the electrostatic interactions between electrons within atoms and between electrons in neighbouring atoms.

15 The domain structure of ferromagnets arises from a competition between the short range electrical forces that cause ferromagnetism and weaker, but longer range, magnetic forces.

16 The practical properties of ferromagnets are determined by the ease of motion of the walls of the domains.

17 The Lorentz force law describes the force on a charged particle moving in both electric and magnetic fields:

$$\boldsymbol{F}(\boldsymbol{r}) = q[\boldsymbol{\mathcal{E}}(\boldsymbol{r}) + \boldsymbol{v} \times \boldsymbol{B}(\boldsymbol{r})]. \qquad (4.10a)$$

In the case of uniform fields, this may be written as:

$$\boldsymbol{F} = q[\boldsymbol{\mathcal{E}} + (\boldsymbol{v} \times \boldsymbol{B})]. \qquad (4.10)$$

This form of the law was used to explain the Hall effect, and the operation of a mass spectrometer and a cyclotron.

18 The magnitude of the force on a current-carrying wire lying perpendicular to a uniform magnetic field is given by

$$F_{\rm m} = Bil. \tag{4.12}$$

19 Equation 4.12 was used to analyse: (i) the force between current-carrying conductors; (ii) the definition of the ampere (and hence the permeability of free space, μ_0); (iii) the torque on a current-carrying loop of wire in a uniform magnetic field.

20 A non-uniform magnetic field configuration known as a magnetic mirror has the property of 'reflecting' charged particles that are fired into it. A magnetic bottle has a magnetic mirror at each end and can therefore form a trap for charged particles. Such traps in the Earth's magnetic field give rise to the Van Allen belts.

21 The solar wind continually replenishes the supply of charged particles in the upper atmosphere. The particles are guided towards the poles by the magnetic field, and their interactions with oxygen and nitrogen in the atmosphere result in the auroral displays often seen at high latitudes.

7.2 Achievements

After studying this chapter, you should be able to:

A1 Explain the meaning of all the newly defined (emboldened) terms introduced in this chapter.

A2 Explain the concept of a magnetic field and define the magnetic field at a point in terms of the magnetic force on a moving charged particle.

A3 Sketch the pattern of field lines corresponding to a bar magnet, a long straight current-carrying wire, a current-carrying loop and a current-carrying solenoid.

A4 Explain how the phenomenon of ferromagnetism arises and describe some devices in which ferromagnets are used.

A5 Recall and apply the formulae for the magnetic field strength at a distance r from a long straight current-carrying conductor, and, at the centre of a circular current-carrying loop of radius R.

A6 Recall and use the right-hand grip rule.

A7 Recall and use the Lorentz force law.

A8 Describe the motion of charged particles in electric and magnetic fields, both qualitatively and quantitatively.

A9 Explain how forces on current-carrying conductors are used to define electromagnetic units.

A10 Recall and use the formula for the force on a long, straight, current-carrying conductor in a uniform magnetic field.

A11 Describe the main features of the motion of charged particles in the Earth's magnetic field.

7.3 End-of-chapter questions

Question 4.28 In London in 1990, the vertical component of the Earth's magnetic field was 44 μT vertically downward and the horizontal component was 19 μT northward. At Bangui in the Central African Republic, just 4.4° from the Equator, the vertical component of the Earth's magnetic field was 9.3 μT vertically upward and the horizontal component was 32 μT northward (Figure 4.76). What is the magnitude of the field at each location? At what angle to the horizontal does the field point at each location?

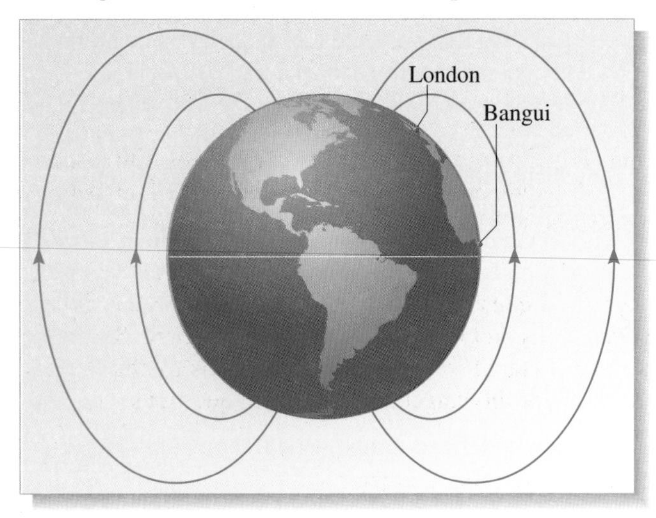

Figure 4.76 For use with Question 4.28.

Question 4.29 Suppose at some particular instant a proton moves with a speed of $2.0 \times 10^7\,\mathrm{m\,s^{-1}}$ in a direction parallel to a current-carrying wire, and at a perpendicular distance of 0.1 m from the wire. If the magnetic force experienced by the proton is of magnitude $6.4 \times 10^{-17}\,\mathrm{N}$, what is the current flowing along the wire?

Question 4.30 Approximately how many turns per metre of wire are required to make a solenoid which can generate a field of 10 T with a current of 100 A? If the wire is 1 mm in diameter, roughly how many layers of wire are required?

Question 4.31 A student comes across the statement that 'magnetic field lines can never cross one another' and asks you to explain why this is the case. Write down a brief justification of the statement (just a few sentences).

Question 4.32 A long straight wire lies along the z-axis in a Cartesian coordinate system, and carries a steady current in the $+z$-direction. A positively charged particle is observed on the $+x$-axis. In each of the following cases, what is the direction of the magnetic force (if any) on the charge when:

(a) the particle is stationary?

(b) the particle is moving in the $+x$-direction?

(c) the particle is moving parallel to the $+z$-axis?

Question 4.33 At a point near the Equator, the Earth's magnetic field is found to point due north, to be horizontal and of magnitude $5 \times 10^{-5}\,\mathrm{T}$. A current-carrying wire of length 2 m and mass 0.1 kg is lying in the east–west direction. If the wire's weight is exactly balanced by the field, what is the direction and magnitude of the current? (No calculator required!) Take g, the magnitude of the acceleration due to gravity, to be $10\,\mathrm{m\,s^{-2}}$. ■

Chapter 5 Consolidation and skills development

1 Introduction

This final chapter has two main aims: first to help you to consolidate what you have learned from this book, and second to help you to develop the skill of reading critically (and hence that of writing accurately and unambiguously).

Section 2 sets the scene by providing a brief overview of Chapters 1 to 4. It emphasizes the fundamental physical concepts of static fields and potentials, and highlights their relevance to many aspects of the physical world. Section 3 follows, with the new material on critical skills, including a set of exercises (based on the content of this book) that are designed to help you to read and write more effectively.

The remaining sections are devoted to question-based activities. Section 4 contains printed questions that will test the skills and knowledge you have acquired. Section 5 provides you with the opportunity to tackle the interactive questions for this book, and Section 6 invites you to try some of the longer skills-oriented questions contained in the *Physica* package.

You should aim to spend about 5 hours on this chapter.

2 Overview of Chapters 1 to 4

Chapter 1 introduced the concept of a *field,* a physical quantity to which some definite value can be ascribed at every point throughout a region of space. Examples of both *scalar* and *vector* fields were discussed, concentrating in particular on *gravitational fields* and *electric fields.* In these two cases, the value of the field at any point determines the gravitational or electrical force that would be experienced by a *test particle* at that point; hence these are both vector fields.

Newton's law of universal gravitation was stated and linked to the concept of a field. Beginning with the equation that describes the gravitational force on one particle due to another, we derived an expression for the gravitational field $\boldsymbol{g}$ at any point $\boldsymbol{r}$ due to a particle of mass M at the origin:

$$\boldsymbol{g}(\boldsymbol{r}) = \frac{-GM}{r^2}\hat{\boldsymbol{r}} \tag{5.1}$$

where $\hat{\boldsymbol{r}}$ is a unit vector that points in the $\boldsymbol{r}$ direction, radially away from the mass M.

A similar expression for the electric field $\boldsymbol{\mathscr{E}}(\boldsymbol{r})$ arises in electrostatics, where *Coulomb's law* plays a role analogous to Newton's law of universal gravitation. Table 5.1 summarizes the basic equations for both gravitation and electrostatics including Newton's law and Coulomb's law. Each of these force laws is an *inverse square law*: in each case the force diminishes as one over the square of the distance from the source. Despite their similarities, there is an important distinction between these two fundamental interactions. In the case of charge, both positive and negative charges exist, and unlike charges attract whereas like charges repel: i.e. both attractive and repulsive electrostatic forces are possible. In contrast, mass is always positive, and the gravitational force is always attractive.

Table 5.1 Gravitational and electric fields and forces.

	Gravitation	Electrostatics
Force laws for point particles	$\boldsymbol{F}_{\text{grav}} = \dfrac{-Gm_1m_2}{r^2}\hat{\boldsymbol{r}}$	$\boldsymbol{F}_{\text{el}} = \dfrac{1}{4\pi\varepsilon_0}\dfrac{q_1q_2}{r^2}\hat{\boldsymbol{r}}$
Force law in field	$\boldsymbol{F}_{\text{grav}} = m\boldsymbol{g}$	$\boldsymbol{F}_{\text{el}} = q\boldsymbol{\mathscr{E}}$
Field due to a particle at $\boldsymbol{r} = 0$ of mass M or charge Q	$\boldsymbol{g}(\boldsymbol{r}) = \dfrac{-GM}{r^2}\hat{\boldsymbol{r}}$	$\boldsymbol{\mathscr{E}}(\boldsymbol{r}) = \dfrac{1}{4\pi\varepsilon_0}\dfrac{Q}{r^2}\hat{\boldsymbol{r}}$

The *principle of superposition* can be used to work out the gravitational or electric field due to a distribution of mass or charge. For a simple situation in which there are several point masses, the gravitational field at any point can be calculated by taking the *vector sum* of the gravitational fields due to each of the point masses individually. Similarly, the electric field due to several point charges can be calculated from the vector sum of the fields due to each of the charges individually.

Extending this idea, the gravitational or electric field due to a continuous distribution of matter (e.g. an extended body) or charge can similarly be computed by treating it as a vector sum of the fields due to each of a large number of small elements making up the distribution. Usually this summation process is laborious, and needs to be done computationally, but often we can make use of symmetry to simplify the calculation. For example, for a *spherically symmetrical* distribution of mass or charge, the field *outside* the distribution is identical with that produced when the entire distribution is replaced by a single particle, with the same total mass or charge as the distribution, located at the point that was formerly the centre of the distribution. For example, the gravitational field of the Earth, beyond the Earth's own surface, can be calculated by ignoring the Earth itself and simply considering a point particle of mass M_{E} placed at the point that was the centre of the Earth.

Scalar fields, such as temperature or height, can be represented by contours, such as those indicating altitude on geographical maps or pressure isobars on weather charts. Vector fields require the additional directional information to be rendered. This can be done by an array of arrows indicating direction, with the length of the arrow proportional to the magnitude of the vector. Wind speed information on TV weather broadcasts is usually represented in this way. Electric and gravitational fields can also be elegantly represented by *field lines*. In this representation, the field at each point along the line is tangential to the line; the direction being indicated by arrowheads along the field line. The density of field lines in any region indicates the magnitude of the field there. The number of gravitational field lines terminating on a given mass is proportional to the value of the mass. For a single isolated mass, the field lines are *radial* straight lines originating at infinity and converging on the mass. For more complicated mass distributions, the resultant pattern of field lines is determined by the principle of superposition. Similar rules apply for electric field lines. With a little thought, these patterns can be sketched and interpreted for simple arrays of a few masses or charges. Strictly, the field line representation works in *three* dimensions, so a two-dimensional sketch on paper may result in apparent anomalies.

Chapter 2 concentrated on two examples of *scalar* fields: the electrostatic and gravitational potential. Both these potential fields are defined in relation to a corresponding potential energy. The electrostatic potential at any point is the electrostatic potential energy per unit charge at that point, and the gravitational potential is the gravitational potential energy per unit mass. So,

$$V_{\text{grav}}(\boldsymbol{r}) = \frac{1}{m} E_{\text{grav}} \text{ (of } m \text{ at } \boldsymbol{r}\text{)}$$

and $$V(\boldsymbol{r}) = \frac{1}{q} E_{\text{el}} \text{ (of } q \text{ at } \boldsymbol{r}\text{)}.$$

These general definitions lead to the following expressions for the potential due to a particle of mass M or charge Q at the origin:

$$V_{\text{grav}}(\boldsymbol{r}) = \frac{-GM}{r}$$

and $$V(\boldsymbol{r}) = \frac{Q}{4\pi\varepsilon_0 r}.$$

The relationship between potentials and potential energies is very similar to that between fields and forces. Consequently, fields are related to potentials in much the same way that (conservative) forces are related to the corresponding potential energies. In fact, if we confine our attention to one dimension for simplicity, we can draw the following illuminating diagram:

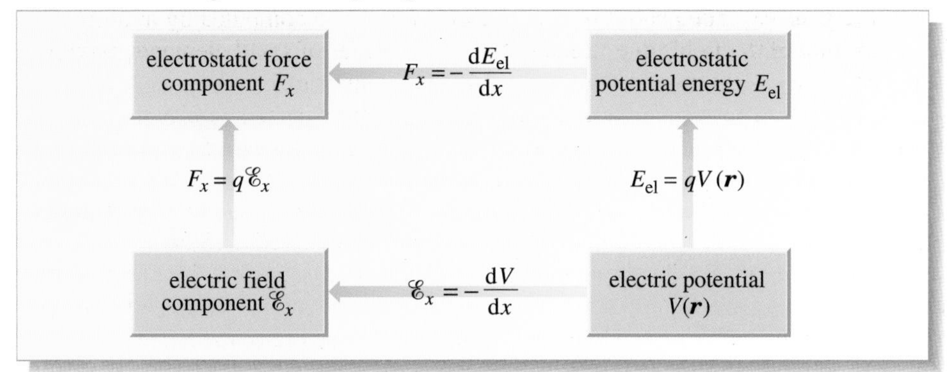

Notice particularly, in the top line, the application of the general relationship between a (conservative) force and the associated potential energy:

$$F_x = -\frac{dE_{\text{pot}}}{dx}. \quad (5.2)$$

Although only stated in a one-dimensional form, this powerful relationship can be generalized to three dimensions. With this generalization in mind, we can say that an electric field always points in the direction in which the electrostatic potential decreases most rapidly, and is proportional to the gradient of the electrostatic potential in that direction. Similarly, a gravitational field always points in the direction in which the gravitational potential decreases most rapidly, and is proportional to the gradient of the gravitational potential in that direction. It follows that, at every point, these fields are always *perpendicular* to the associated *equipotential* surfaces.

Why are potentials important? Because vector fields are difficult to deal with, especially when they have to be added together. Working with potentials and then expressing the answers in terms of fields is generally much simpler, and therefore expands the range of tractable problems. When solving a problem that involves electrostatic or gravitational fields, it is always worth considering whether the alternative formulation in terms of the associated potential might be simpler.

Gravitational potential energy has obvious manifestations, as anyone who ever dropped a hammer on their foot can testify. Harnessing potential energy on a large scale occurs, for example, in hydroelectric power stations, and is fundamental to the process of star formation.

Wherever an electrostatic potential gradient exists, charged particles will experience an electrostatic force. If the charged particles are free to move, like the conducting electrons in a copper wire, then a *current* will flow. Chapter 3 explored the properties and some of the consequences of current flow, and analysed simple circuits. The basic components of simple electrical circuits are:

- *batteries*, which are sources of the *electromotive force* that drives the current around the circuit;
- *resistors*, which dissipate energy as a result of current flow;
- *capacitors*, which store electrostatic potential energy.

Ohm's law ($V = iR$) relates the current through a resistor to the voltage drop across it; and Kirchhoff's laws allow us to analyse simple circuits. These laws are not fundamental in the same sense as, for example, Newton's laws. Ohm's law expresses his empirical finding that, for many conducting materials, the voltage drop across a sample is proportional to the current through it. Kirchhoff's laws arise from the more fundamental law of charge conservation, and the definition of the electrostatic potential. Applying these rules allows us to predict the current that will flow in a circuit of known properties, or to deduce the values of unknown components, given sufficient information about the remainder of the circuit. Specific applications of Kirchhoff's laws include the derivation of formulae for the *effective resistances* for networks of multiple resistors in series and in parallel.

The *differential equation* describing the charge $q(t)$ that remains on a capacitor as it discharges through a resistor is

$$\frac{\mathrm{d}q}{\mathrm{d}t} = -\frac{q}{RC}. \tag{5.3}$$

This has a solution of the form $q = q_0 \mathrm{e}^{-t/RC}$, which describes an exponential decay with time constant RC. Because the current i through the resistor is determined by the rate at which charge leaves the capacitor, it follows that i will also decay exponentially, with time constant RC.

Chapter 4 was mainly concerned with *magnetic fields*, their origin and their effects. Magnetic fields are produced by moving charges: whenever there is a current flowing, a magnetic field will be generated. Magnetism is associated with phenomena as diverse as the orientation of atoms and subatomic particles, and the generation of the large-scale magnetic fields in our Solar System and beyond.

When we consider magnetism, vector notation becomes indispensable, because the direction of the magnetic field generated by a current is perpendicular to the direction of the current. The phenomenon is therefore inherently three-dimensional. For example, the magnetic field lines due to a current flowing in an infinitely long straight wire are circles centred on the wire. Because, in this simplest of cases, there is *cylindrical symmetry*, we can specify the (vector) magnetic field at any point $\boldsymbol{r}$ by combining an expression for the (scalar) magnetic field strength at a distance r from the wire, $B(r) = \mu_0 i/2\pi r$, with the statement that the direction of the field, relative to the current, is given by the *right-hand grip rule.* Other examples of magnetic fields discussed in the chapter include those due to a circular current-carrying coil and an infinitely long cylindrical solenoid.

Magnetic fields give rise to magnetic forces on moving charged particles. A particle of charge q moving with velocity $\boldsymbol{v}$ at a point $\boldsymbol{r}$, in a magnetic field $\boldsymbol{B}(\boldsymbol{r})$, is subject to a magnetic force $\boldsymbol{F}_\mathrm{m} = q[\boldsymbol{v} \times \boldsymbol{B}(\boldsymbol{r})]$, where the vector cross product indicates that the direction of $\boldsymbol{F}_\mathrm{m}$ is perpendicular to $\boldsymbol{v}$ and $\boldsymbol{B}$ in the sense indicated by the *right-hand rule*. (Incidentally, note the distinction between the right-hand rule and the right-hand *grip* rule that was mentioned in the last paragraph.)

The Lorentz force law,

$$\boldsymbol{F}(\boldsymbol{r}) = q[\boldsymbol{\mathscr{E}}(\boldsymbol{r}) + \boldsymbol{v} \times \boldsymbol{B}(\boldsymbol{r})] \tag{5.4}$$

describes the combined effect of electric and magnetic fields on a charged particle, and provides a basis for the logical definition and experimental determination of the electric and magnetic fields at any point. It can be used to calculate the magnetic force on a current-carrying wire, to explain the Hall effect, to account for the trapping of particles in the Earth's magnetic field, and to explain the operation of devices such as cyclotrons, mass spectrometers, and electric motors.

For most of us, our first experience of magnetism comes from permanent magnets, which are samples of magnetized *ferromagnetic* material. In this case the currents causing the magnetic behaviour are on the atomic scale, but can be understood in the same framework as the macroscopic examples listed above.

Gravitational, electric, and magnetic interactions are of fundamental importance. Using the field concept we are able to understand, succinctly describe, and simply calculate the behaviour of many disparate phenomena on a wide range of scales in physics. Once you have mastered the material in this book you are ready to approach Maxwell's equations, and to appreciate the reasoning that led Einstein to develop his revolutionary special theory of relativity.

3 Critical skills

Here we begin development of a new set of skills. While studying the material in this course, you have been immersing yourself in the language of science. In the literary world, writers sometimes intentionally introduce ambiguities to produce a wealth of possible interpretation. In scientific writing, however, the aim is unambiguous clarity, generally with the goal of enabling others to reproduce (and thereby check) an experimental result or a theoretical conclusion. Because the subject matter of such scientific writing is often complex and abstract, it is easy to fall into the trap of writing something which is open to misinterpretation. As you read you should always be on the lookout for ambiguity or imprecision, and when you meet it you should try to resolve it as quickly as possible, perhaps by referring to a glossary or a dictionary of scientific terms. Without a clear grasp of basic definitions you will soon find yourself lost in a welter of half-understood terms and vaguely grasped concepts. Similarly, as you write, you should ask yourself if your prose conveys your intended meaning, without allowing any other interpretation. In general, provided this is achieved, the shortest piece of prose is to be preferred.

There are no hard and fast rules that will inevitably lead to clear communication, though there are plenty of well-founded conventions (such as the use of SI units) and many 'rules of thumb' that experienced communicators can easily recognize. Here are a few points that you may find helpful.

Often, one of the greatest barriers to clear communication is the different frame of mind of the writer (or speaker) and the audience. One way of trying to ensure that there is an appropriate meeting of minds is to use a diagram to establish common ground. When reading, always try to picture what is being described, even if no illustration is provided (this applies even to cases where the description concerns an abstract relationship, such as a mathematical equation). When writing, it is generally a good idea to introduce any relevant illustration as soon as possible. The details can be added later, once the 'big picture' has been established. In fact, it is a useful habit to sketch a diagram whenever you are attempting to understand something, or explain it to others. A diagram will often help you to organize your thoughts and pinpoint ambiguities.

On the whole, new technical terms should be introduced and grasped one at a time. When reading, it usually takes a while to digest technical concepts. As far as possible give yourself time to think about each new term or concept before you move on to the next. When writing, try to give the reader the opportunity to come to terms with one concept before you introduce another. You cannot have failed to notice that in this book the introduction of a new term, particularly an important one, is often followed by a question about that term. In the absence of such questions you should try to get into the habit of asking yourself questions about newly introduced terms. For example, when you meet some new physical quantity, ask yourself if it is extended, like a field, or localized, like the position of a particle, ask yourself what kind of quantity it is, what its unit is, and whether you would know how to go about measuring it. (What is the unit of magnetic field? How, in principle at least, would you set about determining the magnitude and direction of the magnetic field at a given point?)

When reading or writing be alert to (and aware of) the use of visual clues and highlights. When reading a printed text this certainly involves paying due attention to punctuation and italicization, but also be watchful for bulleted or numbered lists or for indented paragraphs and subheadings. It is very easy when reading pages of prose to simply pass over headings without really noticing them. It is not really practical to use italicization in your handwritten work, but you can use underlining to achieve the same effect.

You may be tempted to dismiss this section as peripheral. Critical and communication skills are, however, crucial to success as a scientist. Often one discovers something hitherto elusive during the process of explaining the difficulty; this is partly why so many scientific breakthroughs occur as a result of collaborative work.

To end this section, here are some statements for you to read and consider. Identify any ambiguities, and suggest an improved formulation for each statement.

1. We connected the three resistors in a circuit with the battery and measured a current of three amps.
2. While the capacitor discharged the voltage was 10 volts.
3. The contour lines on the map run directly from north to south. In which direction would a ball roll?
4. The particle experiences an electrostatic force of 10 N, and a gravitational force of 5 N.
5. The gravitational potential energy of a body near the Earth depends on one over the distance.
6. An electron moving in a uniform electric field reaches a speed of $100\,\mathrm{m\,s^{-1}}$ after 10 seconds. How strong is the electric field?
7. To measure the unknown resistor we put it in the Wheatstone bridge. We used the variable resistor to make changes. When it was right we knew the potential difference had vanished and so we could know the ratios and then we got its value. The others were $10\,\Omega$, $20\,\Omega$ and $3\,\Omega$, so the value we got was $6\,\Omega$.
8. The Lorentz motion is like a spiral, but if the field strengthens it can be a mirror. This is the cause of the light in the sky, oxygen is needed too.

Comments

1 We connected the three resistors in a circuit with the battery and measured a current of three amps.

How were the resistors connected? Were they in series, all three in parallel, or in a combination of series and parallel connections? Where was the current measured? The wording suggests, but does not specify, that the circuit was a simple series circuit, in which case the current is the same at all points around the circuit. An improved wording might be:

> We connected the three resistors in series with the battery, and measured the current through the circuit to be 3 A.

2 While the capacitor discharged the voltage was 10 volts.

Where was the voltage measured? Was it the voltage across the capacitor, or was it measured across some other component connected in a circuit with the capacitor? The discharge of a capacitor generally produces a *time-dependent* current and voltage, so the sentence should also specify at what time the voltage was measured. For example:

> The voltage across the capacitor when the discharge began was 10 V. As the discharge proceeded the voltage dropped.

Ideally, the diminishing voltage could be described by a table of measurements of $V(t)$, or by a graph depicting these measurements. Either of these methods will convey the precise information more succinctly and digestibly than prose.

3 The contour lines on the map run directly from north to south. In which direction would a ball roll?

If the contour lines run from north to south then the slope is perpendicular to this direction. We are not, however, told in which direction the contour values are decreasing, therefore we cannot say whether a ball will roll due east or due west. A possible improvement might be:

> The contour lines on the map run directly from north to south, with altitude increasing towards the west. In which direction would a ball roll?

4 The particle experiences an electrostatic force of 10 N, and a gravitational force of 5 N.

Because force is a *vector* quantity, it is not defined until it is given a direction, hence the possible rewording:

> The particle experiences an electrostatic force of 10 N directed towards the north, and a gravitational force of 5 N directed vertically downwards.

5 The gravitational potential energy of a body near the Earth depends on one over the distance.

Here the distance is not unambiguously defined. Is the statement implicitly referring to an expression such as $E_{grav} = mgh$, where h is the distance from the Earth's surface (in which case it is wrong), or is it referring to $E_{grav} = -GM_E/r$, where r is the distance from the Earth's centre (and the statement is correct)? Assuming that the intended statement is correct, here is a reformulation that removes the possible ambiguity:

> The gravitational potential energy of a body near the Earth is inversely proportional to the distance of the body from the centre of the Earth.

6 An electron moving in a uniform electric field reaches a speed of 100 m s^{-1} after 10 seconds. How strong is the electric field?

The question tempts us to apply the equations of motion (i.e. $v = u + at$) with the acceleration magnitude, a, set equal to eE/m. However, the initial speed of the particle is not specified, so we cannot deduce the value of the acceleration. An unambiguous rewording follows:

> An electron is subjected to a uniform electric field, which accelerates it from rest to a speed of 100 m s^{-1} in 10 s. Assuming that no other forces act on the particle, calculate the strength of the electric field.

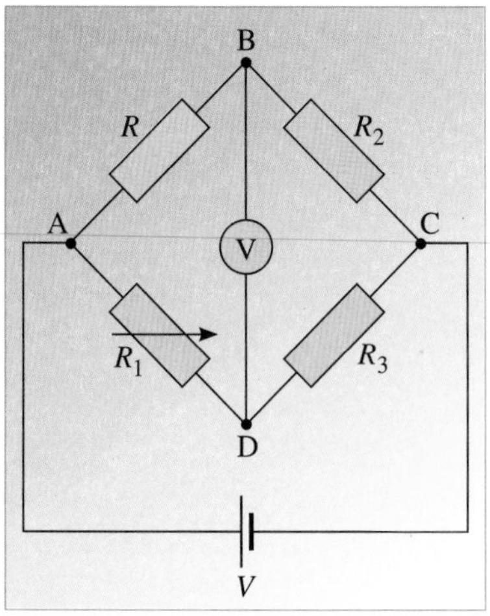

Figure 5.1 A Wheatstone bridge circuit containing a variable resistor R_1, an unknown resistor, R, and two fixed resistors R_2 and R_3.

7 To measure the unknown resistor we put it in the Wheatstone bridge. We used the variable resistor to make changes. When it was right we knew the potential difference had vanished and so we could know the ratios and then we got its value. The others were 10 Ω, 20 Ω and 3 Ω, so the value we got was 6 Ω.

This description of an experimental procedure refers to the use of an electrical circuit called a Wheatstone bridge to measure the resistance of an unknown resistor. Generally when describing an experiment, or a technical concept, a diagram is invaluable. Figure 5.1 is a diagram of the circuit in question. An improved description, including this diagram, might read:

> We used the Wheatstone bridge circuit shown in Figure 5.1 to determine the resistance of the unknown resistor, R. The circuit contains a variable resistor, R_1, which is connected in series with R_3. The resistors R_1 and R_3 are connected in parallel with the unknown resistor, R, and with R_2, as shown in the figure. A battery supplies a potential difference between points A and C, and a voltmeter measures the potential difference between points B and D. The fixed resistors R_2 and R_3 had values of 20 Ω and 10 Ω, respectively. By altering the value of the variable resistor, we found the value of R_1 for which the voltmeter read zero. This meant that the potential difference between A and B, V_{AB}, was equal to the potential difference between A and D, V_{AD}, i.e. $V_{AB} = V_{AD}$. Similarly, it means that $V_{BC} = V_{DC}$, and hence $R/R_2 = R_1/R_3$. The value of the variable resistor when the voltmeter read zero was 3 Ω, so we obtained the result:
>
> $$R = R_1R_2/R_3 = [3 \times 20/10]\,\Omega = 6\,\Omega.$$

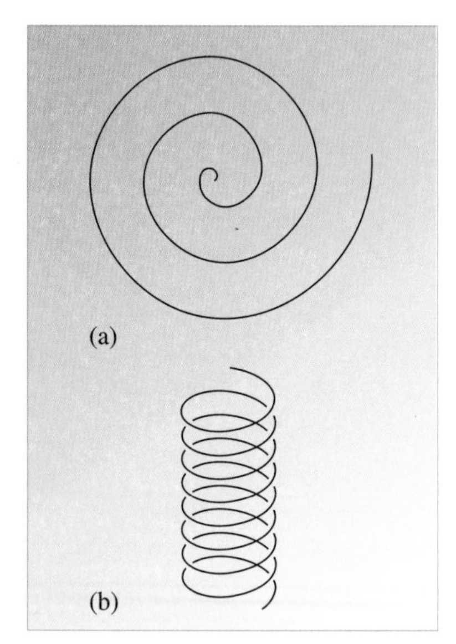

Figure 5.2 (a) A spiral is a two-dimensional curve. (b) A helix is a three-dimensional curve that could be drawn on the surface of a cylinder.

8 The Lorentz motion is like a spiral, but if the field strengthens it can be a mirror. This is the cause of the light in the sky, oxygen is needed too.

Here the difference between scientific and everyday meanings becomes important: mathematically, a spiral is a two-dimensional curve that starts from a point and diverges outwards, as shown in Figure 5.2a. The motion of a charged particle in a magnetic field is *helical*: a three-dimensional path like that of the bannister in a spiral staircase (Figure 5.2b). This is only one shortcoming of many in this inadequate summary of the cause of the aurora. An improved version might be:

> The motion of a charged particle in a constant magnetic field consists of circular motion perpendicular to the field and constant velocity parallel to the field, so that the path followed is a helix. If the field is converging, then the radius of the circular motion becomes smaller, and the helix becomes more tightly wrapped. The particle experiences a Lorentz force which slows its motion along the field until ultimately it 'reflects', beginning to move in a helical path in the opposite direction. In a dipolar field like the Earth's, particles can be trapped gyrating along the field lines and reflecting near the poles where the field converges. Leakage of these particles into the upper atmosphere occurs near the poles, and "excites" atoms or molecules, causing them to glow like the gases in fluorescent lights. This glow is observed in the aurora.

4 Basic skills and knowledge test

You should aim to answer these questions without referring to earlier chapters. Leave your answers in terms of π, $\sqrt{2}$, etc. where appropriate.

Question 5.1 The hydrogen atom may be crudely modelled as an electron orbiting a proton in a circular orbit of radius 5×10^{-11} m. Treating the proton and the electron as positive and negative point charges, respectively, write down an expression for the ratio of the electrostatic force to the gravitational force between the two particles. You may leave your answer in terms of symbols, but you should define all the symbols you use. If the electron's orbit had radius 5×10^{-10} m, how different would your answer be?

Question 5.2 Figure 5.3 is a contour map showing two hills with summits located at coordinates (2.65, 1.95) and (2.50, 4.03). The contours on the diagram are drawn at 10 m altitude intervals. (a) Which summit is higher? (b) Give the approximate coordinates of the steepest slope. (c) In what direction should walkers descend from the summit at (2.65, 1.95) if they wish to travel in a straight line along the gentlest available gradient? (d) In which direction does the gravitational potential energy per unit mass decrease most rapidly from position (2.50, 2.25)? (e) In which direction would a ball roll if released at (2.50, 2.25)? (f) In which direction would a ball roll if released from (2.25, 4.00)?

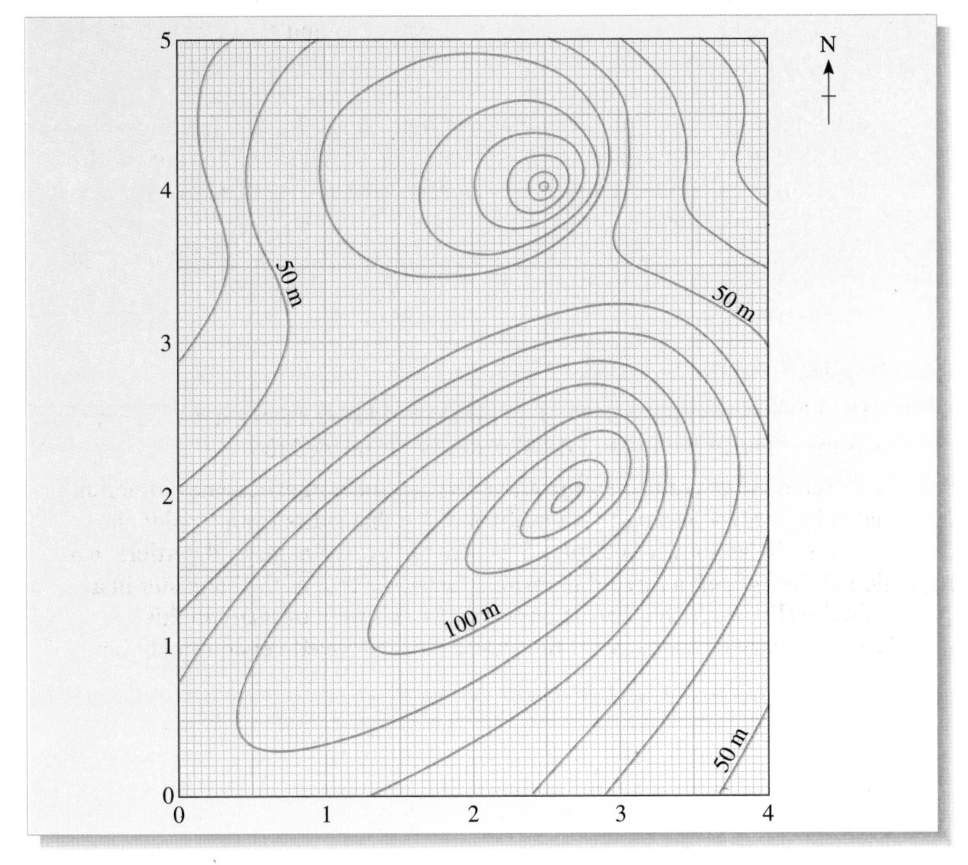

Figure 5.3 A contour map for use with Question 5.2. Altitude contours are drawn at 10 m height intervals. North is indicated, and grid reference coordinates are given along the bottom and the left-hand side.

Question 5.3 Two small objects of masses 1.00 g and 1.00 kg are placed 1.00 m apart. Given that $G = 6.67 \times 10^{-11}\,\mathrm{N\,m^2\,kg^{-2}}$, calculate the magnitude of the gravitational force that each of the objects exerts on the other.

Question 5.4 Figure 5.4 shows an array of fixed charges, each square on the graph representing 1 square metre. Calculate the simplest possible expression for the net force on the charge q at the origin.

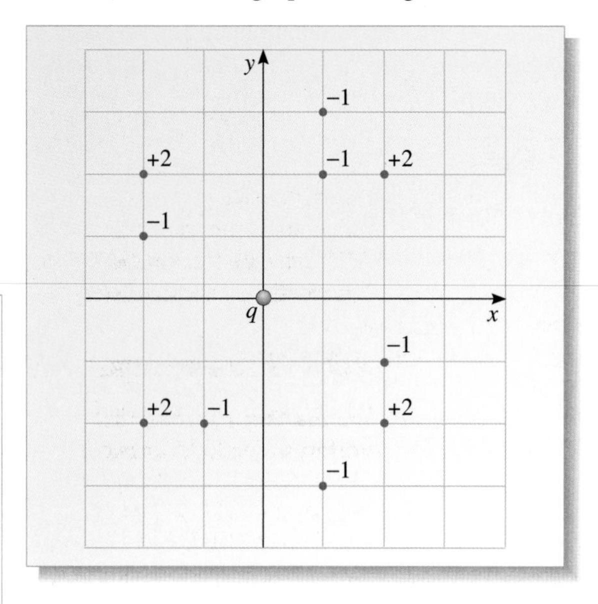

Figure 5.4 An array of charges for use with Question 5.4.

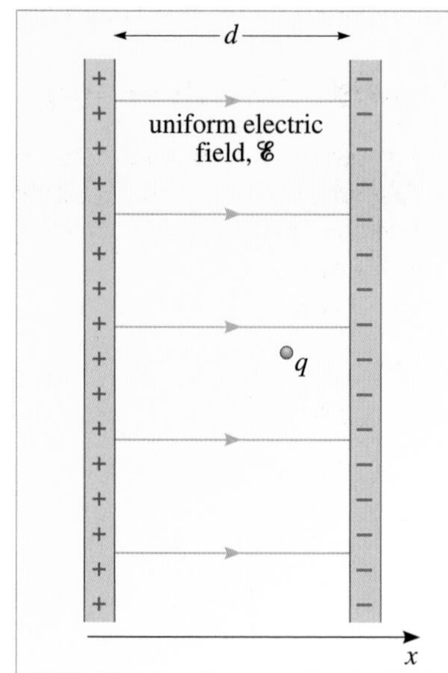

Figure 5.5 A test charge q in a uniform electric field.

Question 5.5 Calculate the change in gravitational potential energy when a satellite of mass $m_s = 4.00 \times 10^3$ kg is put into an orbit of radius $R_s = 4.20 \times 10^7$ m about the Earth. Treat the Earth as a point mass situated at its centre, and take its radius to be $R_E = 6.40 \times 10^6$ m, its mass $M_E = 6.00 \times 10^{24}$ kg and $G = 6.67 \times 10^{-11}\,\mathrm{N\,m^2\,kg^{-2}}$.

Question 5.6 A positive charge q is situated in a region where there is a uniform electric field that points in the positive x-direction (Figure 5.5). Sketch graphs to show how the electrostatic force, F_x, and the electrostatic potential energy, E_{el}, of the charge vary with the x-coordinate.

Question 5.7 A small moon of Saturn has a mass of 3.8×10^{19} kg and a diameter of 500 km. Calculate the maximum angular speed with which it can rotate on its axis if loose rocks on its surface at the equator are not to fly off.

Question 5.8 Figure 5.6 shows a simple circuit. (a) What is the effective resistance between points A and D? (b) What is the current at point A? (c) What is the current at points B, C, and D?

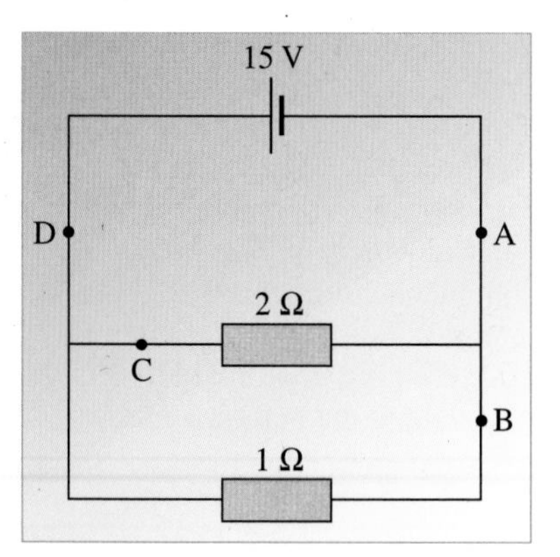

Figure 5.6 A simple circuit containing a battery and two resistors for use in Question 5.8.

Question 5.9 Figure 5.7 shows a Wheatstone bridge circuit, containing a resistor of unknown resistance R. (a) If the voltmeter reads 0.0 V, what is the potential difference between B and D, and what is the current in this part of the circuit? (b) What is the current through the 4.0 Ω resistor? (c) What is the potential drop across the 20 Ω resistor? (d) Hence what is the current through point B? (e) What is the value of the unknown resistance R? (f) What would the voltmeter read if the 20 Ω resistor were replaced by a 10 Ω resistor? (Assume that the voltmeter has effectively infinite resistance.)

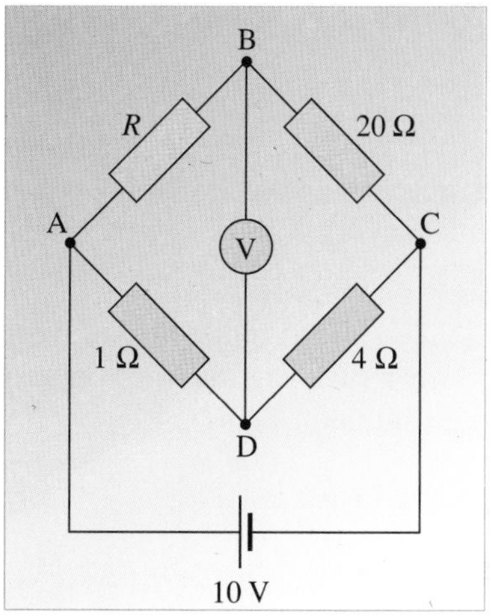

Figure 5.7 A Wheatstone bridge circuit for use in Question 5.9.

Question 5.10 (a) Referring to Chapter 3 Table 3.1 (the resistivities of some substances), which substance would you choose for the connecting wires in a circuit if you wanted to minimize the resistances introduced by these wires? (b) If you were forced to use copper instead, how much thicker in diameter d would your wires need to be to have the same resistance?

Question 5.11 Figure 3.21a in Chapter 3 shows a circuit containing a charged capacitor, a switch and a resistor. The switch is closed at time $t = 0$. (a) Write down the power, P, dissipated in the resistor in terms of $i(t)$ and $V(t)$, the current through and voltage across the resistor at time t, respectively. (b) By substituting expressions for $i(t)$ and $V(t)$ in terms of R, C, and q_0, the initial charge on the capacitor, derive an expression for $P(t)$ in terms of R, C, and q_0. (c) Sketch a graph of $P(t)$, and describe the form it takes with a simple sentence.

Question 5.12 Refer to Figure 5.8, which shows three points close to a wire carrying a current of 10 A. For each of the points A, B and C, calculate the magnetic field strength, and state whether the magnetic field points into or out of the page.

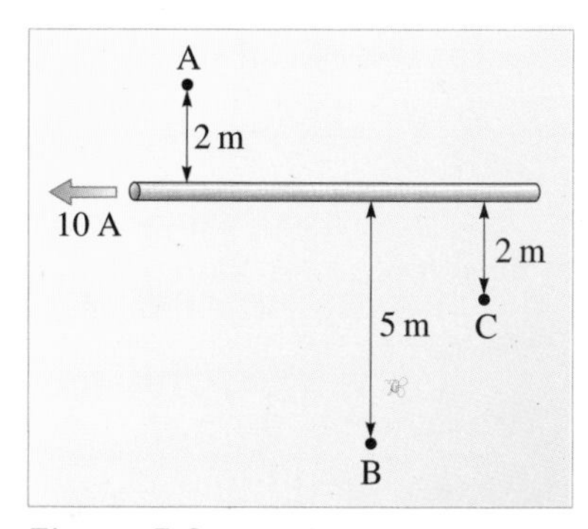

Figure 5.8 A wire carrying a current of 10 A and three points A, B and C referred to in Question 5.12.

Question 5.13 The free charge carriers in a conductor at room temperature are in random thermal motion, even though there is no current in the conductor. If the conductor is in a magnetic field, will there be a magnetic force on each of the charge carriers? Will the conductor experience a force?

Question 5.14 What is the cyclotron radius of a proton moving with speed $2.0 \times 10^3\,\mathrm{km\,s^{-1}}$ perpendicular to a uniform magnetic field of strength 2.1 T? ■

5 Interactive questions

Open University students should leave the text at this point and use the interactive question package for *Static fields and potentials*. When you have completed the questions you should return to this text.

The interactive questions package includes a random number feature that alters the values used in many of the questions each time those questions are accessed. This means that if you try the questions again, as part of your end-of-course revision for instance, you will find that many of them will have changed, at least in their numerical content.

6 *Physica* problems

Open University students should leave the text at this point and tackle the *Physica* problems that relate to *Static fields and potentials*.

Answers and comments

Q1.1 The unit of force, the newton, can also be written as mass times acceleration, i.e. $kg \times m\,s^{-2}$. Substituting this into $N\,m^2\,kg^{-2}$ gives $kg\,m\,s^{-2}\,m^2\,kg^{-2}$, which simplifies to $m^3\,s^{-2}\,kg^{-1}$.

Q1.2 (a) The magnitude of the gravitational force between two 1 kg masses 1 cm apart is found by substitution in Equation 1.1:

$$F_{grav} = \frac{G m_1 m_2}{r^2}$$

$$= 6.67 \times 10^{-11}\,N\,m^2\,kg^{-2} \times \frac{(1\,kg)^2}{(10^{-2}\,m)^2}$$

$$= 6.67 \times 10^{-7}\,N.$$

(b) The weight of either mass is given by:

$$W = mg = 1\,kg \times 9.81\,m\,s^{-2} = 9.81\,N.$$

The ratio $\frac{F_{grav}}{W} = 6.8 \times 10^{-8}$.

Q1.3 The mass m_2 will be in equilibrium if there is no resultant force on it. Since it is between m_1 and m_3, and it is attracted to each, this is possible. So we must have

$$\frac{G m_1 m_2}{(0.1\,m)^2} = \frac{G m_2 m_3}{(0.4\,m)^2}$$

so $$m_1 = \left(\frac{0.1}{0.4}\right)^2 m_3 = \frac{8 \times 10^{-2}\,kg}{16} = 5 \times 10^{-3}\,kg.$$

Comment *Note that the values of G and m_2 are not required for this calculation. Also, note that the presence of m_2 on the line between m_1 and m_3 does not affect the gravitational force that m_1 and m_3 exert upon each other. This is usually known as the principle of superposition.*

Q1.4 The force on m_A due to m_B acts along the line joining m_A to m_B and is towards m_B. Similarly, the force on m_A due to m_C acts along the line joining m_A to m_C and is toward m_C. The resultant force may be found by drawing the triangle (or parallelogram) of forces (Figure 1.58).

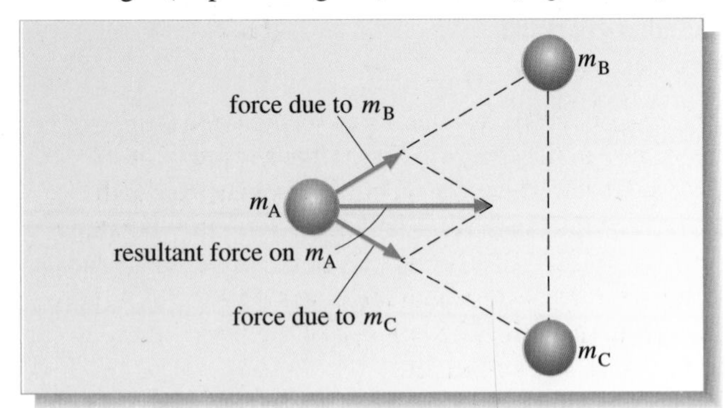

Figure 1.58 Solution to Q1.4.

Q1.5 (a) The gravitational force on the mass at P due to the mass M at the origin is given by Equation 1.6 (note that it is proportional to m_P):

$$\boldsymbol{F} = -\frac{G M m_P}{r^2}\hat{\boldsymbol{r}}$$

$$= -\left(6.67 \times 10^{-11}\,N\,m^2\,kg^{-2} \times 10\,kg \times \frac{1\,kg}{(1\,m)^2}\right)\hat{\boldsymbol{r}}$$

$$= -(6.67 \times 10^{-10}\,N)\hat{\boldsymbol{r}}.$$

(b) As we noted, the force is proportional to m_P, so when we double m_P from 1 kg to 2 kg, with all of the other values remaining unchanged, the force is also doubled, to $-(1.33 \times 10^{-9}\,N)\,\hat{\boldsymbol{r}}$.

Q1.6 Because the field is in the x-direction, so is the force and we can use the component form of Equation 1.9: $(F_{grav})_x = mg_x = (0.2\,kg) \times (10\,N\,kg^{-1}) = 2\,N$.

Q1.7 In the brushing process, charge will be transferred from the hairs to the brushes. This will leave the hairs carrying one type of charge and the brushes with the opposite charge.

(a) The hairs carry like charges and so repel one another.

(b) The hairs carry the opposite type of charge to the brush and will be attracted to the brush.

(c) The brushes carry like charges and so repel each other.

Q1.8 Figure 1.29c: When the spheres are separated, the one closer to the rod carries an excess positive charge, which is concentrated towards the rod. The other sphere is left with an excess negative charge, which is repelled towards the right by the charge on the rod.

Figure 1.29d: When the rod is removed and the spheres are still sufficiently close to interact, the distribution of the excess charge changes since unlike charges attract each other. The total excess charge on each of the spheres is unaffected.

Figure 1.29e: When the spheres are a large distance apart, they carry equal and opposite charges distributed uniformly over their surfaces.

Q1.9 Equation 1.12 can be rearranged to give

$$\frac{1}{4\pi\varepsilon_0} = \frac{F_{el}\,r^2}{(|q_1|\,|q_2|)}.$$

The unit of $\frac{1}{4\pi\varepsilon_0}$ is therefore those of force $\times \frac{\text{distance}^2}{\text{charge}^2}$, which is $N\,m^2\,C^{-2}$.

Q1.10 Coulomb's law states that the magnitude F_{el} of the electrostatic force in a vacuum is given by:

$$F_{el} = \frac{1}{4\pi\varepsilon_0}\frac{|q_1||q_2|}{r^2}$$

$$= (9\times10^9\,\mathrm{N\,m^2\,C^{-2}})\times\frac{(1\,\mathrm{C})^2}{(1\,\mathrm{m})^2} = 9\times10^9\,\mathrm{N}.$$

The magnitude of the gravitational force on a mass m at the surface of the Earth is given by $F_{grav} = mg$. The mass of iron on which the gravitational force is of magnitude 9×10^9 N is therefore

$$m = \frac{F_{grav}}{g} = \frac{(9\times10^9\,\mathrm{N})}{(9.8\,\mathrm{m\,s^{-2}})} = 0.92\times10^9\,\mathrm{kg}.$$

Comment *Because 1 metric ton (tonne) equals 10^3 kg, this is almost one million tonnes!*

Q1.11 The electrostatic force of attraction has magnitude

$$F_{el} = \left(\frac{1}{4\pi\varepsilon_0}\right)\left(\frac{|e||-e|}{r^2}\right)$$

and the gravitational force has magnitude

$$F_{grav} = \frac{Gm_pm_e}{r^2}.$$

The ratio of these is

$$\frac{F_{el}}{F_{grav}} = \frac{e^2}{(4\pi\varepsilon_0 Gm_pm_e)}.$$

$$= (9\times10^9\,\mathrm{N\,m^2\,C^{-2}})\times$$

$$\frac{(1.6\times10^{-19}\mathrm{C})^2}{(6.7\times10^{-11}\mathrm{N\,m^2\,kg^{-2}}\times1.6\times10^{-27}\,\mathrm{kg}\times9.1\times10^{-31}\mathrm{kg})}$$

$$= 2.4\times10^{39}$$

In this case, the electric force is over 10^{39} times stronger than the gravitational force!

Q1.12 (a) To determine the direction in which q_3 will move we need to look at the signs of the charges and the relative distances of q_1 and q_2 from q_3. Now, q_2 has the same sign as q_3 so they will repel each other and q_3 will tend to move off to the right. On the other hand, q_1 has the opposite sign to q_3 so will attract it and tend to pull it to the left. However, because q_1 and q_2 are of the same magnitude but q_1 is twice the distance from q_3 as q_2, the attractive force due to q_3 will not be sufficient to cancel the repulsive force from q_2. Hence q_3 will move to the right.

(b) The magnitude of the resultant force F can be calculated using Equation 1.12:

$$F = \frac{1}{4\pi\varepsilon_0}\left[\frac{(|q_2||q_3|)}{(0.1\,\mathrm{m})^2} - \frac{(|q_1||q_3|)}{(0.2\,\mathrm{m})^2}\right]$$

$$= (9\times10^9\,\mathrm{N\,m^2\,C^{-2}})\times$$

$$\left[\frac{(10^{-7}\mathrm{C})\times(10^{-7}\mathrm{C})}{(0.1\,\mathrm{m})^2} - \frac{(10^{-7}\mathrm{C})\times(10^{-7}\mathrm{C})}{(0.2\,\mathrm{m})^2}\right]$$

$$= 6.75\times10^{-3}\,\mathrm{N}.$$

Q1.13 The force on q_A due to charge q_B will act along the line joining the charges. Because q_A and q_B are like charges (both positive), the force will act away from q_B. Similarly, the force on q_A due to charge q_C will act along the line between q_A and q_C and will be directed towards q_C. Because charges q_B and q_C are of the same magnitude and at equal distances from q_A, the forces they exert on q_A will also be equal in magnitude. The resultant force may be found by drawing a triangle or parallelogram of forces (Figure 1.59).

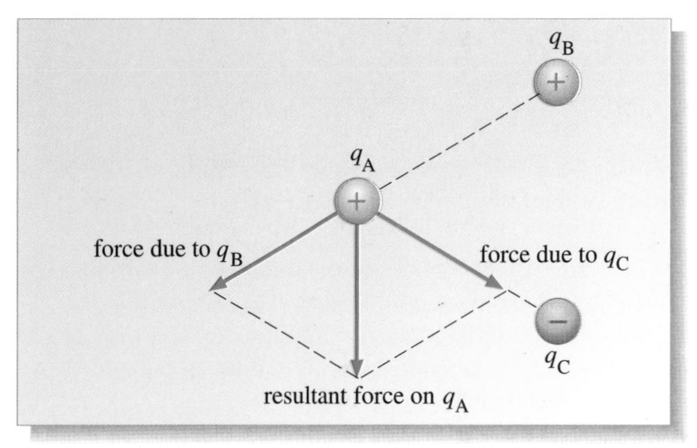

Figure 1.59 Solution to Q1.13.

Q1.14 (a) Because the two charges have the same sign, the force will be repulsive, and the force on the charge at point P will be in the direction of the position vector $\boldsymbol{r}$. The magnitude of the force is

$$F_{el} = \left(\frac{1}{4\pi\varepsilon_0}\right)\frac{Qq}{r^2}$$

$$= 9\times10^9\,\mathrm{N\,m^2\,C^{-2}}\times\frac{(10^{-6}\,\mathrm{C}\times10^{-7}\mathrm{C})}{(1\,\mathrm{m})^2}$$

$$= 9\times10^{-4}\,\mathrm{N}.$$

(b) From Coulomb's law, the force on a charge q is proportional to the charge q. In (a), we showed that $F_{el} = 9\times10^{-4}$ N. Thus if q is doubled in magnitude, the force will be doubled, but it will act in the same direction. Hence, $F_{el} = 1.8\times10^{-3}$ N and acts in the direction of the position vector $\boldsymbol{r}$. Notice that there was no need to refer to Q in finding this force; knowledge of the force on the 10^{-7} C charge was sufficient.

Q1.15 According to Equation 1.14, the unit for $|\mathscr{E}|$ must be the same as that for $\dfrac{|\boldsymbol{F}|}{|q|}$, i.e. newtons per coulomb (N C^{-1}).

Q1.16 The contributions to the field at P due to the positive and negative charges are shown in Figure 1.60. Their magnitudes are the same, because the magnitudes of the two charges are the same, as are their distances from P. The directions are along the line joining P to the charges and away from the positive charge but towards the negative charge. Thus, the two contributions make equal angles above and below the positive x-direction. The resultant field at P is therefore in the positive x-direction.

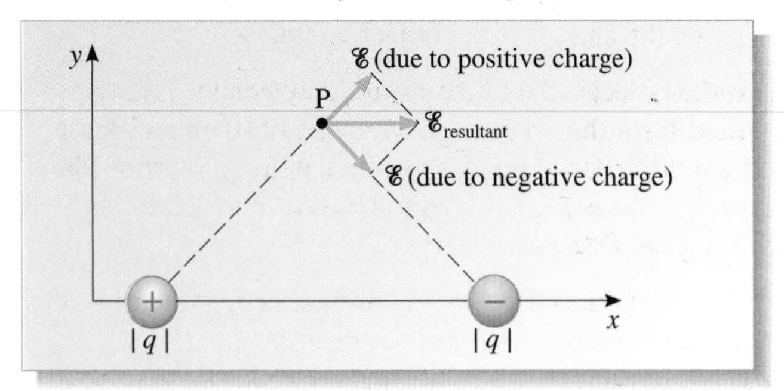

Figure 1.60 Solution to Q1.16.

Q1.17 The directions of the two contributions to the total field at P are shown on Figure 1.61. (Note that the magnitudes are not necessarily drawn to scale – they are to be calculated.) These contributions can be calculated separately, and then summed vectorially.

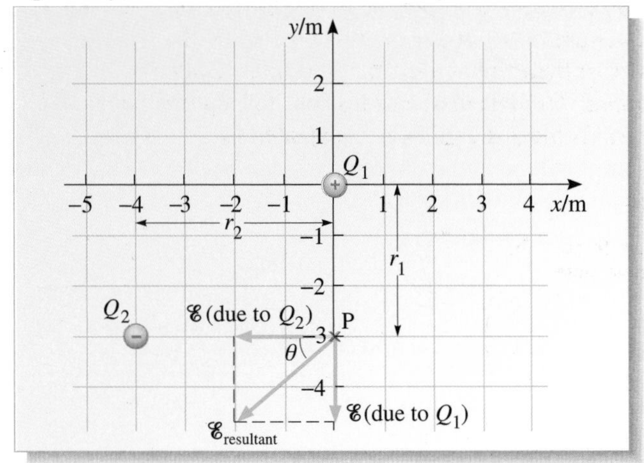

Figure 1.61 Solution to Q1.17.

The field contribution at P can be obtained either by using Equation 1.16 or by using Equation 1.16a together with the 'like charges repel, unlike charges attract' rule. By the latter method, the field at P due to Q_2 has components $\mathscr{E}_y = 0$ and

$$\mathscr{E}_x = -\frac{|Q_2|}{4\pi\varepsilon_0 r_2^{\,2}}$$

$$= -9\times10^9\,\text{N}\,\text{m}^2\,\text{C}^{-2}\times\frac{3\times10^{-6}\,\text{C}}{(4\,\text{m})^2}$$

$$= -1.69\times10^3\,\text{N}\,\text{C}^{-1}.$$

The field at P due to Q_1 has components $\mathscr{E}_x = 0$ and

$$\mathscr{E}_y = -\frac{|Q_1|}{4\pi\varepsilon_0 r_1^{\,2}}$$

$$= -9\times10^9\,\text{N}\,\text{m}^2\,\text{C}^{-2}\times\frac{5\times10^{-6}\,\text{C}}{(3\,\text{m})^2}$$

$$= -5\times10^3\,\text{N}\,\text{C}^{-1}.$$

Comment *In using the alternative vector equation*

$$\mathscr{E}(\boldsymbol{r}) = \frac{Q}{4\pi\varepsilon_0 r^2}\hat{\boldsymbol{r}}$$

it is important to remember that $\boldsymbol{r}$ is the position vector of P from the source charge. Thus, in working out the field at P due to Q_2, $\hat{\boldsymbol{r}}$ would point along the +x-direction and because Q_2 is negative, $\mathscr{E}$ would come out with a minus sign, i.e. pointing along the –x-direction. In working out the field at P due to Q_1, $\hat{\boldsymbol{r}}$ would point along the –y-direction, and, since Q_1 is positive, $\mathscr{E}$ would point in the same direction as $\hat{\boldsymbol{r}}$, i.e. along the –y-axis.

The magnitude of the resultant field at P is

$$|\mathscr{E}_{\text{resultant}}| = (\mathscr{E}_x^2 + \mathscr{E}_y^2)^{1/2} = 5.3\times10^3\,\text{N}\,\text{C}^{-1}.$$

The angle between the component fields is given by

$$\tan\theta = \frac{\mathscr{E}_y}{\mathscr{E}_x} = \frac{5}{1.69}.$$

Therefore $\theta = 71°$. This means that the electric field at P is of magnitude 5.3×10^3 N C^{-1} and is directed at an angle of 71° anticlockwise from the negative x-axis.

Q1.18 The field patterns are shown in Figure 1.62. **Note:** *The number of lines emerging from the double charge is twice the number ending on each single charge. This would still be so if more field lines were added.*

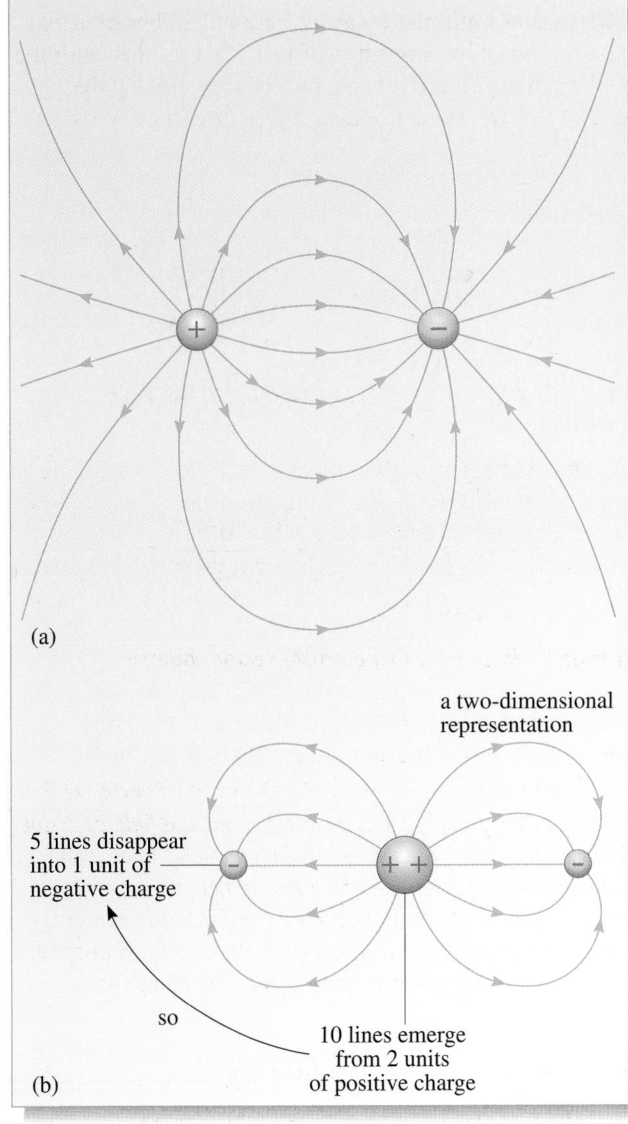

Figure 1.62 Solution to Q1.18.

Q1.19 No. It is true that the force will be parallel to the field if the charge is positive (anti-parallel if it is negative), but particles do not always move parallel to the force applied. For example, in the case of circular motion the particle moves at right angles to the direction of the force. It is, however, possible to make the approximation that particles do move along the electric field lines provided that (i) their initial velocity is negligible, (ii) the field is uniform, and (iii) there are no other forces acting on the particle.

Q1.20 (a) Figure 1.46 shows field lines that could represent either the gravitational field or the electric field, provided that the objects are both negatively charged. Notice that the object on the left is either twice the mass or has twice the charge of the object on the right.

(b) If the objects carry charges of the opposite sign, then the field lines should emerge from one of them and some of the field lines would be of finite length going from the positive to the negative charge. Thus, it must be the gravitational field that is shown.

Q1.21 Let the charge on the drop be $-ne$, where n is an integer. Equating the magnitudes of the electric and gravitational forces, we have $m_{\text{drop}}g = ne\mathscr{E}$, and therefore

$$n = \frac{m_{\text{drop}}g}{e\mathscr{E}}$$

$$= \frac{(4.31\times10^{-14}\,\text{kg})(9.81\,\text{m}\,\text{s}^{-2})}{(1.60\times10^{-19}\,\text{C})(1.15\times10^{5}\,\text{N}\,\text{C}^{-1})} = 23.$$

Q1.22 Near to the hole, the electric field will be approximately as in Figure 1.63. The hole is small and the plates are close together. Above the plate, the electric field is zero: there are no electric field lines. Only after falling through the hole do the drops start to be attracted back towards the positively charged plate.

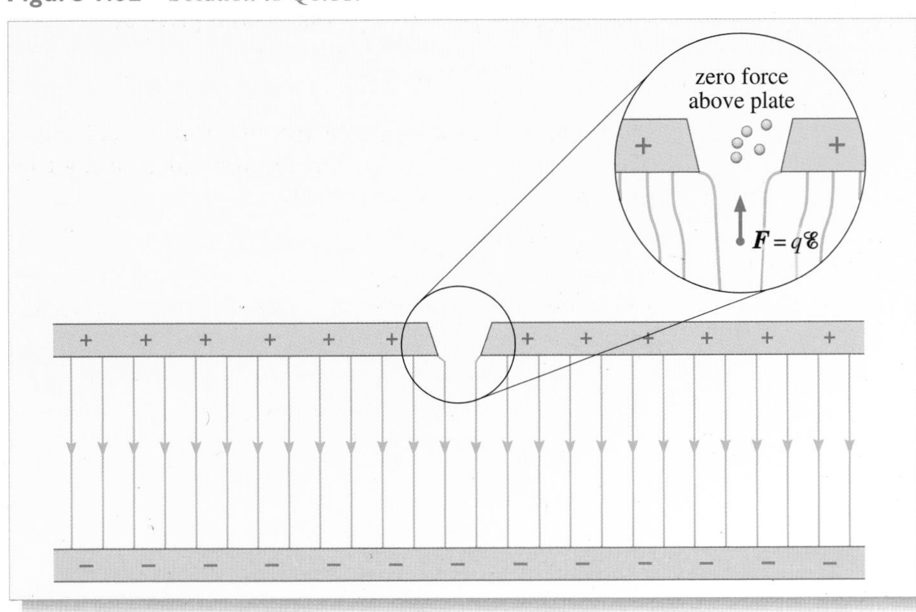

Figure 1.63 Solution to Q1.22.

Q1.23 (a) Newton's law of gravitation states that there is an attractive force between the two masses. If the masses are separated by a distance r, the force has magnitude $F_{\text{grav}} = \dfrac{Gm_1m_2}{r^2}$. The force is directed along the line joining the two masses.

(b) Figure 1.64a shows the direction of the forces on 1 kg at C. The direction of the gravitational field due to the two masses is in the direction of the resultant of these two forces.

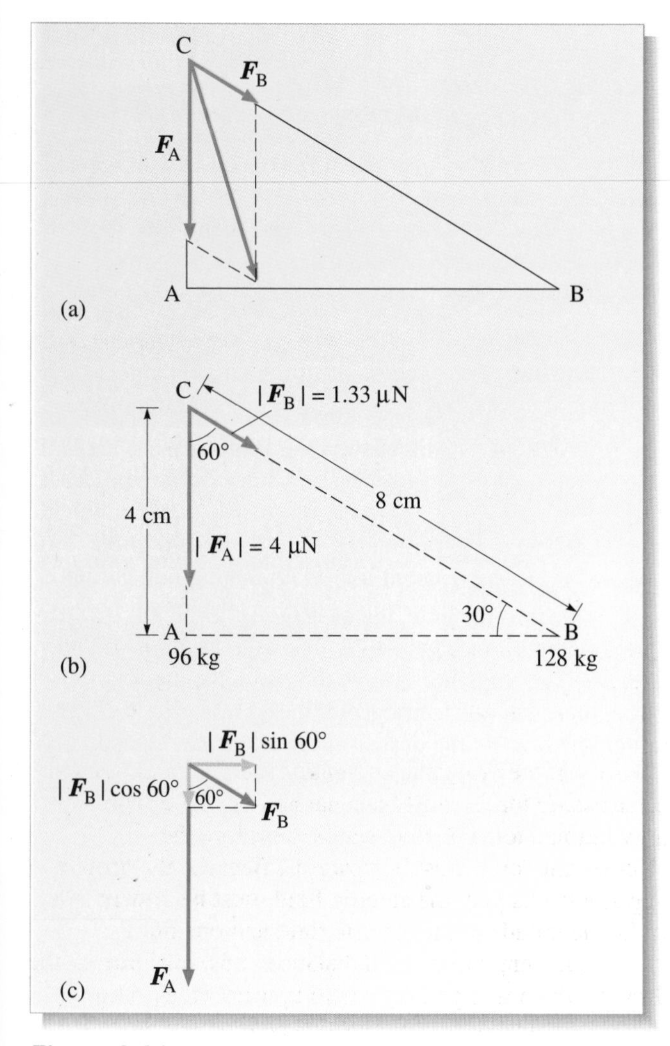

Figure 1.64 For the answer to Question 1.23. (a) The directions of the forces $\boldsymbol{F}_A$ and $\boldsymbol{F}_B$ and direction of the resultant force (field). (b) The directions and magnitudes of the forces. (c) The components of the two forces in perpendicular directions.

(c) The magnitude of the force on 1 kg at C due to 96 kg at A is

$$F_A = \frac{6.67\times10^{-11}\,\text{N m}^2\,\text{kg}^{-2}\times1\,\text{kg}\times96\,\text{kg}}{(4\times10^{-2}\,\text{m})^2}$$

$$= 4.00\times10^{-6}\,\text{N}.$$

The magnitude of the force on 1 kg at C due to 128 kg at B is

$$F_B = \frac{6.67\times10^{-11}\,\text{N m}^2\,\text{kg}^{-2}\times1\,\text{kg}\times128\,\text{kg}}{(8\times10^{-2}\,\text{m})^2}$$

$$= 1.33\times10^{-6}\,\text{N}.$$

Figure 1.64b shows the two forces. Note that the angle between them is 60°.

Figure 1.64c shows the resolution of the two forces into perpendicular components. The components are
(i) $F_A + F_B\cos 60° = 4.67\times10^{-6}$ N and
(ii) $F_B\sin 60° = 1.16\times10^{-6}$ N. Using Pythagoras' theorem, $F_{\text{resultant}} = ((4.67)^2 + (1.16)^2)^{1/2}\times10^{-6}\,\text{N} = 4.81\times10^{-6}$ N, or 4.81 μN.

Q1.24 Above the surface of the Earth, we can treat the Earth as a point mass located at the centre of the Earth. The magnitude of the gravitational field is given by

$$a_{\text{grav}} = \frac{GM_E}{R_E^2}.$$

The ratio $a_{\text{grav}}(h) : a_{\text{grav}}(0)$ of the gravitational field at height h above sea-level to that at sea-level is given by

$$\frac{a_{\text{grav}}(h)}{a_{\text{grav}}(0)} = \left(\frac{GM_E}{(R_E+h)^2}\right)\left(\frac{R_E^2}{GM_E}\right) = \left(\frac{R_E}{R_E+h}\right)^2$$

where R_E is the radius of the Earth. When the gravitational field has fallen to 81% of its value at sea-level, this ratio will be equal to 0.81. Thus,

$$0.81 = \left(\frac{R_E}{R_E+h}\right)^2$$

$$R_E + h = \frac{R_E}{0.9} = 7089\,\text{km}.$$

So, because $R_E = 6380$ km, $h = 709$ km.

Q1.25 When a charged object is brought up to an insulated metal rod, the electrons in the metal move under the influence of the electrostatic force they experience from the charged object. Once a certain amount of excess negative charge has built up at one end of the rod, this will itself exert an electrostatic repulsion on the other electrons in the metal and be attracted back towards the positive charge. The flow of electrons ceases when the opposing electrostatic forces are equal.

Q1.26 (a) If the two spheres repel each other, then they must carry like charges.

(b) If the spheres attract each other, then there are two possibilities: either (i) the spheres carry opposite charges or (ii) one sphere is charged and the other, initially uncharged, has its charge redistributed by induction when the other sphere gets close to it. This second situation is analogous to the attraction between a charged plastic ruler and an uncharged piece of tissue paper, and we shall return to a more detailed description of the phenomenon in Section 4.2. The initially uncharged sphere may be either a conductor or an insulator.

Q1.27 The field in the y-direction is given by

$$\mathscr{E}_y = \frac{F_y}{q} = \frac{10^{-14}\,\text{N}}{(-1.6\times10^{-19}\,\text{C})} = -6.25\times10^4\,\text{N}\,\text{C}^{-1}.$$

The magnitude of the force on an α-particle is $F = \mathscr{E}\times(2e)$, and since the only component of the field is in the y-direction, the only component of the force will also be in the y-direction. Thus $F_y = \mathscr{E}_y\times(2e) = -6.25\times10^4\,\text{N}\,\text{C}^{-1}\times3.2\times10^{-19}\,\text{C} = -2\times10^{-14}\,\text{N}$. (This is, of course, -2 times the force on the electron, as it should be.)

Q1.28 (a) Figure 1.65a is a suitable sketch. Electrons in the object will be repelled from the wire, leaving a net positive charge on the side nearer the wire and a net negative charge on the side farther away.

(b) The negative charge on one side of the object is of the same magnitude as the positive charge on the other. If the field were completely uniform, the electrostatic forces acting on the opposite sides of the object would exactly balance each other, and it would not move.

(c) If the field were non-uniform e.g. if it were that due to the charged wire itself (Figure 1.65b), then the object would be attracted towards the wire, as shown in Figure 1.65b. In order to estimate the number n of electrons displaced, we will make the rather crude assumption that there is a positive charge ne on the side near the wire and a negative charge $-ne$ on the other side. If the magnitude of the electric field at the positive side of the object is $\mathscr{E}$ then on the other side it will be $\mathscr{E} + \Delta\mathscr{E}$ and the magnitude of the resultant force F on the object will be given by

$$F = ne\mathscr{E} - ne(\mathscr{E} + \Delta\mathscr{E}) = ne\,\Delta\mathscr{E}.$$

Thus $$n = \frac{F}{e\,\Delta\mathscr{E}} = \frac{10^{-9}\,\text{N}}{1.6\times10^{-19}\,\text{C}\times1\,\text{N}\,\text{C}^{-1}} \approx 6\times10^9.$$

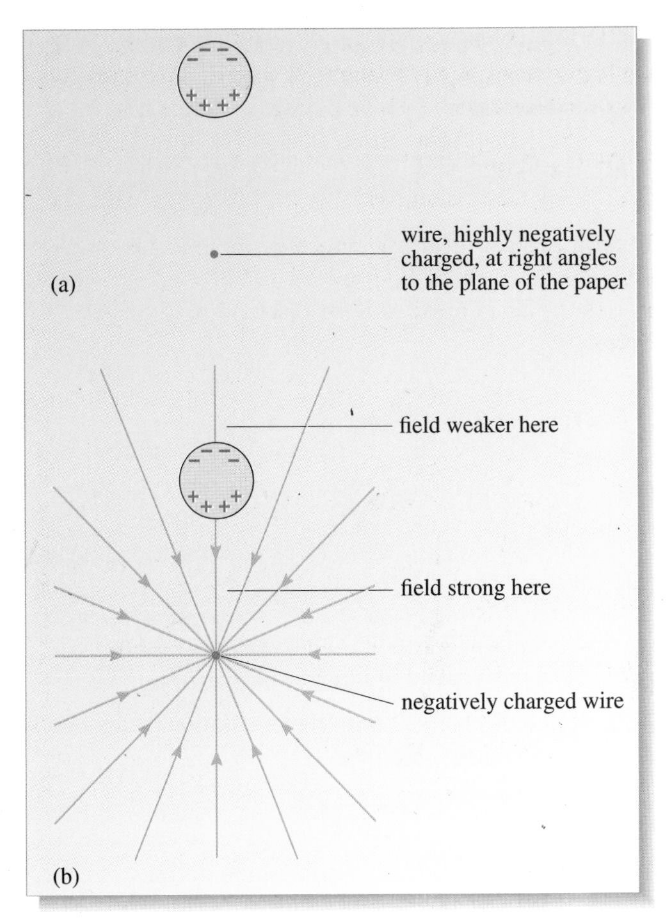

Figure 1.65 Solution to Q1.28.

Comment *This is only a tiny fraction of the electrons available even in an object made of a material that is not a very good conductor.*

Q1.29 If the droplet is stationary, the gravitational and electrostatic forces must be equal and opposite. The gravitational force is downwards, therefore the electrostatic force must be upwards. Because the droplet is negatively charged, the electric field must be downwards. If the magnitude of the electric field is momentarily increased, there will be an unbalanced upward force on the droplet which will give it a small upward acceleration. When the magnitude of the field is returned to the value at which the forces balance, the droplet will continue to move upwards with the velocity it acquired during the momentary field increase. The droplet will, of course, continue to move with a constant velocity when the total force acting on it is zero.

Q2.1 Remember the rule for differentiating positive or negative powers: if $y = x^n$, then $\frac{dy}{dx} = nx^{n-1}$. Applying this to the functions of r:

(a) $E_{pot} = kr + C;\ \frac{dE_{pot}}{dr} = k$

(b) $E_{pot} = \frac{k}{r} + C;\ \frac{dE_{pot}}{dr} = -\frac{k}{r^2}$

(c) $E_{pot} = \frac{k}{r} - C;\ \frac{dE_{pot}}{dr} = -\frac{k}{r^2}$

(d) $E_{pot} = -\frac{k}{r} + C;\ \frac{dE_{pot}}{dr} = \frac{k}{r^2}$

(e) $E_{pot} = \frac{k}{r^2} + C;\ \frac{dE_{pot}}{dr} = -\frac{2k}{r^3}$

(f) $E_{pot} = \frac{k}{r^3} - C;\ \frac{dE_{pot}}{dr} = -\frac{3k}{r^4}.$

The results of both (b) and (c) give expressions for dE_{pot}/dr that correspond to Equation 2.4.

Q2.2 Using Equation 2.6 for the gravitational potential energy of two masses, we have

$$E_{grav} = -\frac{GM_{Earth}M_{Moon}}{r}$$

where r is the average mean Earth–Moon distance, so

$$E_{grav} = -\frac{6.7\times10^{-11}\,\mathrm{N\,m^2\,kg^{-2}}\times6.0\times10^{24}\,\mathrm{kg}\times7.4\times10^{22}\,\mathrm{kg}}{3.8\times10^{8}\,\mathrm{m}}$$

$$= -7.8\times10^{28}\,\mathrm{J}.$$

Q2.3 (a) The mass at height h has gravitational potential energy E_{grav} given by Equation 2.8a:

$$E_{grav} = mgh = 5\,\mathrm{kg}\times9.8\,\mathrm{N\,kg^{-1}}\times20\,\mathrm{m} = 980\,\mathrm{J}.$$

(b) The mass is released from rest. Its initial kinetic energy is, therefore, zero. Its total energy is equal to its initial potential energy. When it hits the Earth's surface, all of its potential energy has now been converted into kinetic energy, i.e. $\frac{1}{2}mv^2 = mgh$. This gives an expression for the speed,

$$v = \sqrt{2gh}$$

$$= \sqrt{2\times9.8\,\mathrm{m\,s^{-2}}\times20\,\mathrm{m}}$$

$$= 19.8\,\mathrm{m\,s^{-1}}.$$

Q2.4 Using Equation 2.7:

$$E_{el} = \frac{q_1q_2}{4\pi\varepsilon_0 r}.$$

energy required = potential energy at 10 cm separation − potential energy at 10 m separation

$$= (9\times10^9\times\frac{10^{-6}\times10^{-6}}{0.1})\,\mathrm{J} - (9\times10^9\times\frac{10^{-6}\times10^{-6}}{10})\,\mathrm{J}$$

$$= 9\times10^9\times10^{-12}\left(\frac{1}{0.1}-\frac{1}{10}\right)\mathrm{J} = 0.089\,\mathrm{J}.$$

Q2.5 The electrostatic potential energy in this case is given by

$$E_{el} = \frac{q_1q_2}{4\pi\varepsilon_0 r} = \frac{(-e)e}{4\pi\varepsilon_0\,(10^{-10}\,\mathrm{m})}$$

$$= -(9\times10^9)\times\frac{(1.6\times10^{-19})^2}{10^{-10}}$$

$$= -2.3\times10^{-18}\,\mathrm{J}$$

Q2.6 **Preparation** The α-particle is small, and can be modelled as a point charge, with $q_\alpha = +2e$. The residual nucleus is large in comparison, with radius $r_{nuc} = 10^{-14}$ m. However, it will repel or attract external charges as though all of its own charge of $Q_{nuc} = +81e$ were concentrated at its centre. (This is the same principle that allows us to calculate the gravitational force on an object on or above the Earth's surface by taking the whole mass of the Earth to be concentrated at its centre; in other words, it is the electrostatic analogue of Newton's theorem.) Thus, when the α-particle has been expelled from, but is still 'just at the edge of', the nucleus, the magnitude of the force is

$$F_{el} = \frac{|q_\alpha||Q_{nuc}|}{4\pi\varepsilon_0 r_{nuc}^2}$$

and the electrostatic potential energy is

$$E_{el} = \frac{q_\alpha Q_{nuc}}{4\pi\varepsilon_0 r_{nuc}}.$$

When the α-particle escapes, the potential energy, E_{el}, it had 'at the edge of the nucleus' is converted into kinetic energy:

$$E_{trans} = \tfrac{1}{2}m_\alpha v^2.$$

Working Substituting $Q_{nuc} = +81e$ and $q_\alpha = +2e$, the magnitude of the force with which the α-particle is expelled is

$$F_{el} = \frac{(2e)(81e)}{4\pi\varepsilon_0 r_{nuc}^2}$$

$$= 9 \times 10^9\,\mathrm{N\,m^2\,C^{-2}} \times 162 \times \frac{(1.6 \times 10^{-19}\,\mathrm{C})^2}{(10^{-14}\,\mathrm{m})^2}$$

$$\approx 373\,\mathrm{N}.$$

(This is a very large force, especially since it is acting on such a tiny object.)

The potential energy when the α-particle is poised 'at the edge of the nucleus' is

$$E_{el} = \frac{(2e)(81e)}{4\pi\varepsilon_0 r_{nuc}}.$$

This expression is just equal to the value of the force we calculated above, multiplied by the radius of the nucleus, that is

$$E_{el} = 373\,\mathrm{N} \times 10^{-14}\,\mathrm{m}$$

$$\approx 3.7 \times 10^{-12}\,\mathrm{J}.$$

Assuming the nucleus remains stationary during the decay, all this potential energy will be converted into the kinetic energy of the α-particle.

Note: *This is a reasonable assumption, given the disparity in mass between the two particles. A nucleus with 81 protons will have a mass number in the region of 200 and the α-particle has a mass number of 4. Thus the mass ratio is about 50 to 1.*

So $\frac{1}{2} m_\alpha v^2 = 3.7 \times 10^{-12}\,\mathrm{J}$.

The α-particle consists of two protons and two neutrons, so

$$m_\alpha = 2m_p + 2m_n$$

$$= 4 \times 1.7 \times 10^{-27}\,\mathrm{kg}.$$

Therefore
$$v = \left(\frac{2 \times 3.7 \times 10^{-12}\,\mathrm{J}}{4 \times 1.7 \times 10^{-27}\,\mathrm{kg}}\right)^{1/2}$$

$$\approx 3.3 \times 10^7\,\mathrm{m\,s^{-1}}.$$

(The question only asked for an estimate, so one or two significant figures are enough.)

Checking The units have thus worked out correctly:

$$1\,\mathrm{J} = 1\,\mathrm{N\,m} = 1\,\mathrm{kg\,m\,s^{-2}} \text{ so } \left(\frac{1\,\mathrm{J}}{1\,\mathrm{kg}}\right)^{1/2} = 1\,\mathrm{m\,s^{-1}}$$

and v has come out in metres per second. The speed of the ejected α-particle is about $\frac{1}{10}$ the speed of light, which does not seem unreasonable.

Q2.7 The electric field goes from positive to negative. We can define the positive y-direction in Figure 2.7 to be the direction of the field. Then, both the electric field component, $\mathscr{E}_y$, and the change in position of the electron, Δy, are positive.

(It doesn't matter which direction you choose as positive so long as, once you have chosen, you take care to get the signs correct. Similarly, it doesn't matter whether you call it x, y, z, r or anything else, so long as you remain consistent.)

The force on the electron is $F_y = -e\mathscr{E}_y$ so the work done by the electric field in moving the charge a distance Δy is $W_{cons} = -e\mathscr{E}_y \Delta y$. The change in potential energy of the electron is the negative of this. Thus

$$\Delta E_{pot} = -(-1.6 \times 10^{-19}\,\mathrm{C} \times 10^4\,\mathrm{N\,C^{-1}} \times 10^{-3}\,\mathrm{m})$$

$$= +1.6 \times 10^{-18}\,\mathrm{J}.$$

Thus the potential energy of the electron increases by $1.6 \times 10^{-18}\,\mathrm{J}$.

Q2.8 From Equation 2.13, $\Delta E_{el} = q\,\Delta V$ where ΔV is the potential difference between the terminals. Thus, the potential energy change is $2\,\mathrm{C} \times (-9\,\mathrm{V}) = -18\,\mathrm{J}$.

Q2.9 Let the direction from one terminal to the other be the x-direction. Then the average magnitude of the electric field between the terminals is given by

$$|\mathscr{E}_x| = \left|\frac{\Delta V}{\Delta x}\right| = \frac{12}{0.015}\,\mathrm{V\,m^{-1}} = 800\,\mathrm{V\,m^{-1}}.$$

Q2.10 Since the electric field is uniform between the two points we can write (using Equation 2.18)

$$\mathscr{E}_x = -\left(\frac{\Delta V}{\Delta x}\right)$$

giving $-\Delta V = \mathscr{E}_x \Delta x$.

Thus, the magnitude of the maximum sustainable difference in electric potential between two points 1 cm apart is given by

$$|\Delta V|_{max} = |\mathscr{E}|_{max} \times 1\,\mathrm{cm} = 3 \times 10^6\,\mathrm{N\,C^{-1}} \times 0.01\,\mathrm{m}$$

$$= 3 \times 10^4\,\mathrm{V}.$$

Q2.11 (a) Since the field between the plates is uniform, the magnitude of the potential difference between the plates is given by

$$|\Delta V| = |\mathscr{E}|d = 10^4\,\mathrm{N\,C^{-1}} \times 10^{-3}\,\mathrm{m} = 10\,\mathrm{V}$$

where d is the distance between the plates.

(b) The magnitude of the change in potential energy of the electron is

$$|\Delta E_{el}| = |q\Delta V| = 1.6 \times 10^{-19}\,\mathrm{C} \times 10\,\mathrm{V} = 1.6 \times 10^{-18}\,\mathrm{J}$$

which, of course, is the same answer as before.

Q2.12 The unit of the gravitational field, g, is that of force/mass. The unit of gravitational potential, V_{grav}, is the unit of gravitational field times the unit of distance, from Equation 2.21. The unit of V_{grav} is therefore (force × distance)/(mass) or (energy/mass), i.e. $J\,kg^{-1}$. (Alternatively, the unit of V_{grav} is (acceleration × distance), i.e. $(m\,s^{-2} \times m) = m^2\,s^{-2}$.)

Q2.13 From Figure 2.13, it is clear that the spacing between equipotentials *increases* as the distance from the charge increases. The magnitude of the electric field due to a point charge *decreases* as the square of the distance from the charge increases. The equipotentials are, therefore, farthest apart where the field is weakest, and closest together where the field is strongest. The field lines for a point charge are radial (see Chapter 1, Section 4.5), and so they cross the equipotentials at right angles.

Q2.14 See Figure 2.37. Radially outwards from the tree, $\Delta V = -\mathscr{E}_r \Delta r$. Cow A has her head on one equipotential and her tail on another. The potential difference will therefore be high since Δr is large. The cow is in grave danger! Cow B, however, is standing approximately on an equipotential, so the potential difference across her is much smaller — she may be lucky.

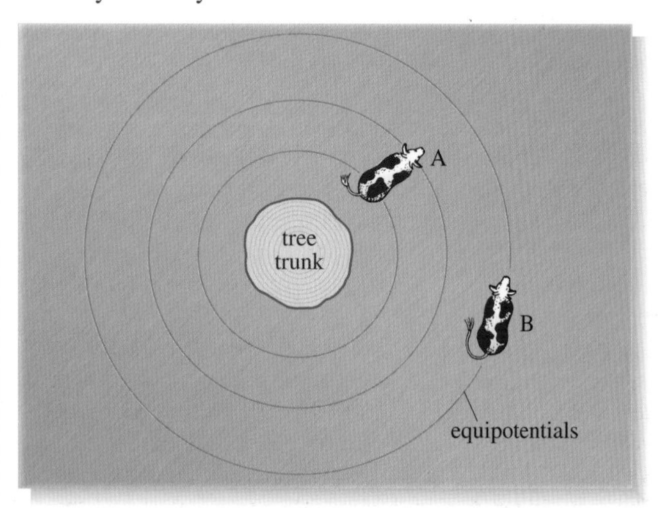

Figure 2.37 For use with answer to Q2.14.

Q2.15 Since the field lines must always be at right angles to the equipotential lines, they may be sketched in as shown in Figure 2.38. Since the direction of the field is that of the force on a *positive* charge, the field will be away from the positive plate, as shown by the arrows on the field lines. In the central region, the equipotentials are approximately equally spaced. The field strength will therefore be about the same at all points in this region (a uniform field). In the two peripheral regions, the field lines bulge outwards and the distance between the intersections with adjacent equipotential lines is larger here than in the central region. The strength of the field is therefore weaker here.

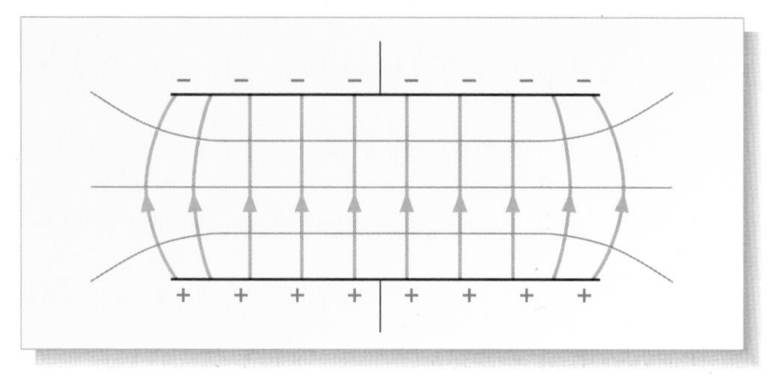

Figure 2.38 For use with answer to Q2.15.

Q2.16 Since a perfect conductor must be an equipotential and field lines always cross equipotentials at right angles, field lines must always leave a perfect conductor at right angles to the surface.

Q2.17 The aircraft fuselage is made of metal, which is a good conductor. Passengers are therefore surrounded by a conducting shell, inside which the electric field must be zero because of electrostatic screening. Conducting shells of this type are called *Faraday cages* (after Michael Faraday, who first demonstrated this effect). If it is necessary to protect people or instruments from intense electric fields, they are therefore enclosed in hollow conductors. Remember, when a cage is used to shield something *inside* it, it does not have to be earthed.

Q2.18 The capacitance of the aircraft's wings is given by

$$C = \varepsilon_{air}\varepsilon_0 \frac{A}{d}$$

$$= 1 \times 8.854 \times 10^{-12}\,C^2\,N^{-1}\,m^{-2} \times \frac{300\,m^2}{3\,m}$$

$$= 8.9 \times 10^{-10}\,F.$$

The potential difference between the wings and the ground is therefore

$$V = \frac{q}{C} = \frac{10^{-6}\,C}{8.9 \times 10^{-10}\,F} = 1100\,V.$$

Before a plane is refuelled, great care is taken to ensure that it is earthed. Otherwise, these high voltages may lead to sparks that could easily cause the fuel to ignite.

Q2.19 They both fall by a factor of five. The charge on each plate is not affected by the introduction of the mica since the capacitor is isolated. From the equation $q = CV$, we can write:

$$C_{\text{before}}V_{\text{before}} = q = C_{\text{after}}V_{\text{after}}$$

$$\frac{V_{\text{after}}}{V_{\text{before}}} = \frac{C_{\text{before}}}{C_{\text{after}}}.$$

The capacitance of a parallel plate capacitor is proportional to the relative permittivity of the dielectric. Therefore:

$$\frac{C_{\text{before}}}{C_{\text{after}}} = \frac{\varepsilon_0}{5\varepsilon_0} = \frac{1}{5}.$$

Thus, the potential difference falls by a factor of five. From the equation $E_{\text{el}} = \frac{1}{2}qV$, it immediately follows that the recoverable electric potential energy also falls by a factor of five.

Q2.20 If 10 MW of power is to be delivered and the conversion is 90% efficient, the power extracted must be $10 \times \frac{100}{90} = 11.1$ MW. This power comes from the rate of loss of potential energy $m_{\text{f}}gh$, where h is the height through which the water falls (1000 m), g is the acceleration due to gravity (9.81 m s^{-2}) and m_{f} is the rate of flow of water in kilograms per second. Therefore,

$$m_{\text{f}} = \frac{11.1 \times 10^6 \text{ W}}{(1000 \text{ m} \times 9.81 \text{ m s}^{-2})} = 1130 \text{ kg s}^{-1}.$$

Q2.21 The gravitational potential energy is given by $E_{\text{grav}} = mgh$. So in this case we have 3.5×10^5 J = $m \times 9.81$ m s$^{-2} \times 30$ m. So

$$m = \frac{3.5 \times 10^5 \text{ J}}{9.81 \text{ m s}^{-2} \times 30 \text{ m}} = 1200 \text{ kg}.$$

The person on the beach below should certainly be worried!

Q2.22 The potential difference of 1 kV (1000 V) is the potential energy change *per unit charge*. The change in the potential energy of the electron is therefore 1000 V × -1.6×10^{-19} C = -1.6×10^{-16} J. This loss in potential energy is equal to the gain in kinetic energy. So

$$\frac{mv_{\text{final}}^2}{2} = 1.6 \times 10^{-16} \text{ J}$$

$$v_{\text{final}} = \left(\frac{2 \times 1.6 \times 10^{-16} \text{ J}}{9.1 \times 10^{-31} \text{ kg}}\right)^{1/2} = 1.9 \times 10^7 \text{ m s}^{-1}.$$

One electronvolt is defined to be the energy acquired by an electron falling through a potential difference of one volt. Since, in the present example, the electron falls through 1000 volts, its final energy will be 1000 eV.

Q2.23 The potential energy of the Earth–Moon system is -7.8×10^{28} J.

If the Moon travels a distance $2\pi r_{\text{Moon}}$ in 28 days, its average speed is:

$$v_{\text{Moon}} = \frac{(2\pi \times 3.8 \times 10^8 \text{ m})}{(28 \times 24 \times 60 \times 60 \text{ s})} = 990 \text{ m s}^{-1}.$$

The mean kinetic energy of the Moon is

$$\frac{M_{\text{Moon}}v_{\text{Moon}}^2}{2} = \tfrac{1}{2} \times 7.4 \times 10^{22} \text{ kg} \times (990 \text{ m s}^{-1})^2$$

$$= 3.6 \times 10^{28} \text{ J}.$$

This is half the *magnitude* of the potential energy of the Earth–Moon system. The latter is negative, as are all gravitational potential energies due to the universally attractive nature of the gravitational force. The *total* energy of the Moon in its orbit around the Earth is thus negative, which is as it must be, otherwise the Moon would not be bound to the Earth: it would fly off into space!

Q2.24 The electrostatic energy stored is given by Equation 2.25:

$$E_{\text{el}} = \frac{q^2}{2C} = \frac{(2.5 \times 10^{-5} \text{ C})^2}{(2 \times 10 \times 10^{-6} \text{ C V}^{-1})} = 3.1 \times 10^{-5} \text{ J}.$$

Q2.25 We start from the baseline of the large gravitational potential energy of the isolated dust-cloud system from which the star is formed. The gravitational attraction causes the system to collapse towards its centre of mass, so continually reducing its potential energy (making it more negative). The kinetic energy of the constituents increases to conserve energy, there being no other source of energy at this stage. This amounts to an increase in temperature in the cloud. At a certain level of temperature and density, the kinetic energy of the hydrogen nuclei is enough to overcome electrostatic repulsion and nuclear fusion can take place, providing a new source of energy. Eventually, when all the available hydrogen has been exhausted by fusion, the core will collapse, liberating more gravitational potential energy. The resulting radiation expands the outer layers and the star becomes a red giant. The core eventually becomes a stable white dwarf when the outward Pauli pressure of the electrons counterbalances the inward gravitational force.

Q3.1 Apply Ohm's law:

$$R = \frac{|V_{\text{R}}|}{|i|} = \frac{1.5}{0.3}\,\Omega$$

$$= 5\,\Omega.$$

If $R = 2.5\,\Omega$, then

$$i = \frac{|V_{\text{R}}|}{R} = \frac{1.5}{2.5}\text{ A}$$

$$= 0.6 \text{ A}.$$

Q3.2 From Table 3.1 we find the resistivities of copper and aluminium:

$$\rho_{\text{copper}} = 1.7 \times 10^{-8}\,\Omega\,\text{m}$$

$$\rho_{\text{aluminium}} = 2.7 \times 10^{-8}\,\Omega\,\text{m}.$$

Using Equation 3.4

$$R = \frac{\rho L}{A}$$

we find

(a) $$R_{\text{copper}} = \frac{1.7 \times 10^{-8}\,\Omega\ \text{m} \times 10^{-5}\,\text{m}}{10^{-6}\,\text{m} \times 10^{-7}\,\text{m}} = 1.7\,\Omega.$$

(b) Similarly

$$R_{\text{aluminium}} = \frac{2.7 \times 10^{-8}\,\Omega\ \text{m} \times 10^{-5}\,\text{m}}{10^{-6}\,\text{m} \times 10^{-7}\,\text{m}} = 2.7\,\Omega.$$

Q3.3

(i) A single resistor gives $10\,\Omega$.

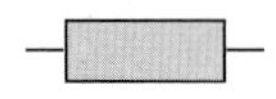

(ii) Two resistors can be combined in series giving

$$R_{\text{eff}} = 10\,\Omega + 10\,\Omega = 20\,\Omega$$

(iii) …or combined in parallel giving

$$\frac{1}{R_{\text{eff}}} = \left(\frac{1}{10} + \frac{1}{10}\right)\Omega^{-1} = \frac{2}{10}\Omega^{-1}$$

giving $R_{\text{eff}} = 5\,\Omega$.

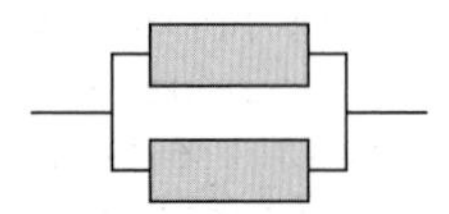

(iv) Three resistors can be combined in series ($R_{\text{eff}} = 30\,\Omega$)

(v) …or all in parallel

$$\frac{1}{R_{\text{eff}}} = \left(\frac{1}{10} + \frac{1}{10} + \frac{1}{10}\right)\Omega^{-1} = \frac{3}{10}\Omega^{-1}$$

i.e. $R_{\text{eff}} = 3.33\,\Omega$.

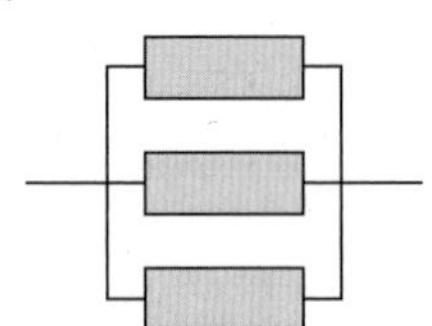

(vi) The three resistors can also be arranged so that one is in series with a parallel pair. From part (iii) we know that the effective resistance of a parallel pair is $5\,\Omega$ so:

$$R_{\text{eff}} = 10\,\Omega + 5\,\Omega = 15\,\Omega.$$

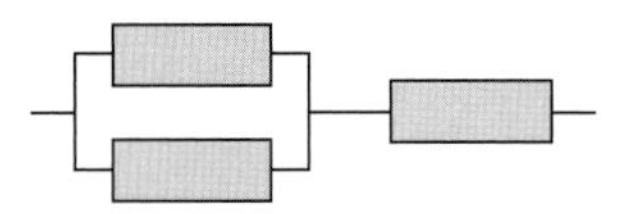

(vii) The three resistors can also be arranged so a single resistor is connected in parallel across a pair in series. From part (ii) we know that the effective resistance of a series pair is $20\,\Omega$ so:

$$\frac{1}{R_{\text{eff}}} = \left(\frac{1}{10} + \frac{1}{20}\right)\Omega^{-1} = \frac{3}{20}\Omega^{-1}$$

giving $R_{\text{eff}} = 6.67\,\Omega$.

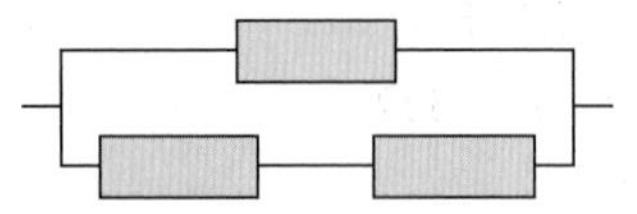

Q3.4

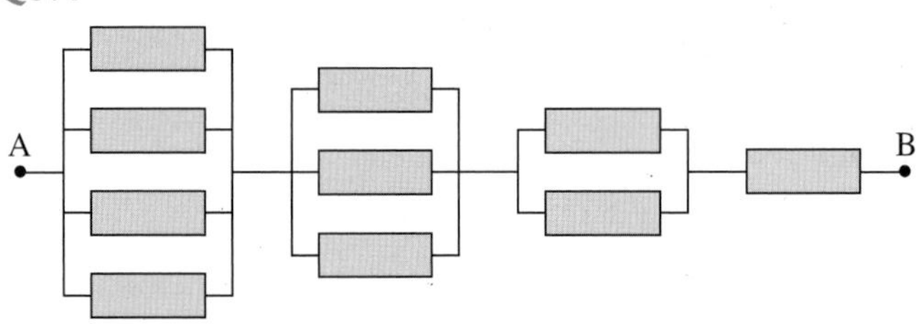

The effective resistance of the first set of four parallel resistors is calculated from

$$\frac{1}{R_{\text{eff}}} = \left(\frac{1}{R} + \frac{1}{R} + \frac{1}{R} + \frac{1}{R}\right) = \frac{4}{R}$$

i.e. $R_{\text{eff}} = R/4$.

Similarly, the effective resistance of the second set of three parallel resistors is $R_{\text{eff}} = R/3$, and the two resistors in parallel have an effective resistance of $R_{\text{eff}} = R/2$.

Thus the total effective resistance is

$$R_{\text{eff}} = \frac{R}{4} + \frac{R}{3} + \frac{R}{2} + \frac{R}{1} = R\left(\frac{25}{12}\right) = 2.08R.$$

Q3.5 The circuit network is drawn in Figure 3.39 with current directions assigned at A and C.

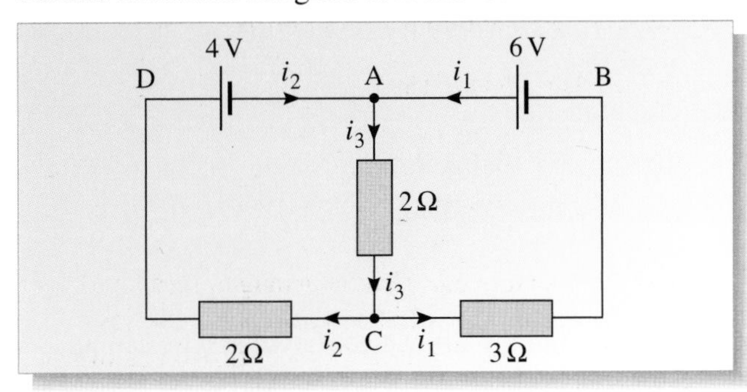

Figure 3.39 For use with Q3.5.

Kirchhoff's second law at point A gives

$$i_3 = i_1 + i_2. \qquad \text{(i)}$$

Applying Kirchhoff's first law to the circuit loop ADCBA:

$$4\,\text{V} + 2\,\Omega \times i_2 - 3\,\Omega \times i_1 + 6\,\text{V} = 0$$

or, omitting units,

$$2i_2 - 3i_1 + 10 = 0. \qquad \text{(ii)}$$

Now, applying Kirchhoff's first law to circuit loop ACBA:

$$-2\,\Omega \times i_3 - 3\,\Omega \times i_1 + 6\,\text{V} = 0$$

or, omitting units,

$$-2i_3 - 3i_1 + 6 = 0. \qquad \text{(iii)}$$

Substituting for i_3 into (iii) using (i):

$$-2(i_1 + i_2) - 3i_1 + 6 = 0$$

$$-5i_1 - 2i_2 + 6 = 0. \qquad \text{(iv)}$$

Now, adding (ii) and (iv)

$$-8i_1 + 16 = 0$$

and hence

$$i_1 = 2\,\text{A}.$$

Using this value for i_1 in (iv):

$$-10 - 2i_2 + 6 = 0$$

and hence

$$i_2 = -2\,\text{A}.$$

This tells us that the current i_2 flows in the opposite direction to that shown in Figure 3.39.

Finally, from (i),

$$i_3 = 2\,\text{A} - 2\,\text{A} = 0\,\text{A}$$

so points A and C are at the same potential.

Q3.6 Using the suggested substitutions, the unit of RC is found to be

$$\frac{\text{V}}{\text{A}} \times \frac{\text{C}}{\text{V}} = \frac{\text{C}}{\text{A}} = \frac{\text{C}}{\text{C}\,\text{s}^{-1}} = \text{s}$$

as expected.

Q3.7 Electrical power = iV, and so current drawn = 100 W/220 V = 0.45 A.

From Ohm's law, resistance = V/i = 220 V/0.45 A = 489 Ω.

The bulb is rated at 100 W or equivalently 0.1 kW. Therefore in 24 hours it uses 0.1 × 24 = 2.4 kW h of energy, which costs 2.4 × 7 p = 16.8 p.

Q3.8 (a) The initial charge q_0 stored on the capacitor is given by

$$q_0 = CV = 10^{-3}\,\text{F} \times 10^2\,\text{V} = 0.1\,\text{C}.$$

(b) The initial stored energy is given by

$$E_{\text{el}} = \tfrac{1}{2}CV^2 = \tfrac{1}{2} \times 10^{-3}\,\text{F} \times (10^2\,\text{V})^2 = 5\,\text{J}.$$

(c) The initial current i_0 is given by

$$i_0 = V/R = 100\,\text{V}/10^{-1}\,\Omega = 10^3\,\text{A}.$$

(d) The time constant $\tau = RC = 10^{-1}\,\Omega \times 10^{-3}\,\text{F} = 10^{-4}\,\text{s}$ = 100 μs.

(e) Initial power = initial current × initial potential difference = $10^3\,\text{A} \times 10^2\,\text{V}$ = 100 kW.

(f) 200 μs is the interval covered by two time constants. After each time constant, the current and the potential difference will *each* have decayed by a factor e, and the power will therefore have decayed by a factor e^2. So, after *two* time constants

$$\text{power after } 200\,\mu\text{s} = \frac{\text{initial power}}{e^2 \times e^2}$$

$$= \frac{10^5\,\text{W}}{54.6} = 1.8 \times 10^3\,\text{W} = 1.8\,\text{kW}.$$

Comment*: The assumption that the resistance remains constant at 0.1 Ω is unrealistic. The metal pieces are being heated to a very high temperature and are even melting, and under such circumstances, the resistance would increase markedly, perhaps by a factor of 10. This would tend to lengthen the pulse, but reduce the peak power.*

Q3.9 After the voltmeter is placed across the resistor, there are two resistances in parallel. The total effective resistance is given by

$$\frac{1}{R_{\text{eff}}} = \frac{1}{R_{\text{resistor}}} + \frac{1}{R_{\text{voltmeter}}}$$

$$= \left(\frac{1}{10^5} + \frac{1}{10^7}\right)\Omega^{-1}$$

so $\quad R_{\text{eff}} = 9.90 \times 10^4\,\Omega.$

(a) The potential difference across the meter *and* the resistor is therefore:

$$V = iR_{\text{eff}} = 10^{-4}\,\text{A} \times 9.9 \times 10^4\,\Omega = 9.90\,\text{V}.$$

(b) Compare this potential difference with the potential difference $V_{initial}$, which existed before the meter was connected. This is simply given by

$$V_{initial} = 10^{-4}\,\text{A} \times 10^5\,\Omega = 10\,\text{V}.$$

So, installing the meter has changed the potential difference that was being measured. The larger the meter's resistance, the smaller is the reduction of the potential difference when it is connected. This is why it is important that voltmeters should have very high resistances.

Comment: *Whenever measurements are made, on electrical or any other systems, the experimenter must attempt to minimize the effect that the measurement has on the system. In fact, it is impossible to eradicate completely the perturbing effect of making measurements. In this example, the effect can be reduced by increasing the resistance of the voltmeter.*

(c) An ammeter, on the other hand, is designed to measure the current *through* a circuit component, and must therefore be placed *in series* with that component. If the ammeter had a high resistance, Ohm's law shows that the overall current in the circuit would fall when the ammeter was connected.

Q3.10 The current i through the $4\,\Omega$ light-bulb, and the voltage, V, across it, are related by the equation $V = iR_{bulb}$. Equation 3.19 gives the value of V as

$$V = V_{EMF} - ir$$

Therefore

$$iR_{bulb} = V_{EMF} - ir$$

$$i \times 4\,\Omega = 2\,\text{V} - (i \times 10\,\Omega)$$

$$i = \tfrac{1}{7}\,\text{A}.$$

So the power dissipated is

$$i^2 R_{bulb} = \left(\tfrac{1}{7}\right)^2 \times 4\,\text{W}$$

$$= 0.08\,\text{W}.$$

When $r = 0$, as in a new battery,

$$V = V_{EMF} = 2\,\text{V}$$

$$\text{power dissipated} = \frac{V^2}{R_{bulb}}$$

$$= \frac{4\,\text{V}^2}{4\,\Omega}$$

$$= 1\,\text{W}.$$

This is more than ten times the rate at which energy is dissipated in the bulb when the battery is run down.

Q3.11 When current is passing through the $10\,\Omega$ resistor, the potential difference at the terminals is given by Ohm's law:

$$V = iR_{load} = 1\,\text{A} \times 10\,\Omega = 10\,\text{V}.$$

The voltmeter draws very little current and will, to a good approximation, read the actual electromotive force of the battery. Applying Equation 3.19,

$$V = V_{EMF} - ir.$$

Substituting, we find

$$10\,\text{V} = 11\,\text{V} - (1\,\text{A} \times r)$$

which is solved for r to yield:

$$r = 1\,\Omega.$$

Q3.12 The energy needed to excite an ion enough to cause it to emit light is a few electronvolts: let's call it 5 eV. The energy transferred to an electron or an ion carrying a single electron charge by an electric field of magnitude $\mathscr{E}$ over a distance d is

$$\Delta E_{el} = e\mathscr{E}d.$$

Thus to acquire an energy of 5 eV, the particle must travel a distance of at least

$$d = \frac{\Delta E_{el}}{e\mathscr{E}} = \frac{5\,\text{V} \times e}{e\mathscr{E}} = \frac{5}{100}\,\text{m} = 5\,\text{cm}.$$

Comment: *The value of 5 cm derived here is the mean distance travelled by the particles between inelastic collisions. That is, those in which kinetic energy is converted into electronic energy within the ion, thus leaving it in an excited state from which it can emit light. The particles suffer many elastic collisions in between the inelastic ones, but in these cases, by definition, they do not lose any of their kinetic energy.*

Q3.13 We can rearrange Equation 3.2 to obtain the drift speed

$$v = \frac{i}{neA}.$$

The cross-sectional area of the wire $A = \pi(d/2)^2$, where d is its diameter, so we have

$$v = \frac{4i}{ne\pi d^2}$$

$$= \frac{4 \times 0.1}{5.9 \times 10^{28} \times 1.6 \times 10^{-19} \times \pi \times (0.5 \times 10^{-3})^2}$$

$$= 5.4 \times 10^{-5}\,\text{m}\,\text{s}^{-1} = 0.054\,\text{mm}\,\text{s}^{-1}.$$

Q3.14 (a) The resistance of a piece of wire of length L, cross-sectional area A ($= \pi(d/2)^2$, where d is its diameter) and resistivity ρ is given by Equation 3.4:

$$R = \frac{\rho L}{A} = \frac{4\rho L}{\pi d^2} = \frac{4 \times 1.6 \times 10^{-8} \times 0.05}{\pi \times (0.5 \times 10^{-3})^2} = 4.1 \times 10^{-3}\,\Omega.$$

Using Ohm's law, we find

$$V = iR = (0.1 \times 4.1 \times 10^{-3})\,\text{V} = 4.1 \times 10^{-4}\,\text{V}.$$

(b) The power dissipated in the wire is given simply by

$$P = iV = (0.1 \times 4.1 \times 10^{-4})\,\text{W} = 4.1 \times 10^{-5}\,\text{W}.$$

Q3.15 (a) The energy transferred to an electron on 'falling' through a potential difference of 4.1×10^{-4} V is

$$E_{\mathrm{el}} = eV = 1.6 \times 10^{-19} \times 4.1 \times 10^{-4}\,\mathrm{J} = 6.6 \times 10^{-23}\,\mathrm{J}.$$

(b) The number density, n, of conduction electrons in silver was given in Question 3.13 as $5.9 \times 10^{28}\,\mathrm{m}^{-3}$. The total number, N, of conduction electrons in a piece of wire of length L and cross-sectional area A is

$$N = nLA = nL\pi(d/2)^2$$
$$= \frac{5.9 \times 10^{28} \times 0.05 \times \pi \times (0.5 \times 10^{-3})^2}{4} = 5.8 \times 10^{20}.$$

(c) If all the conduction electrons in the piece of wire were to travel from one end to the other, the total energy transferred to them by the applied voltage would be

$$NE_{\mathrm{el}} = 5.8 \times 10^{20} \times 6.6 \times 10^{-23}\,\mathrm{J} = 0.038\,\mathrm{J}.$$

The time, T, taken for all the electrons to travel the length of the wire is just L divided by the drift speed (calculated in Question 3.13), so

$$T = \frac{L}{v} = \frac{0.050}{0.054 \times 10^{-3}}\,\mathrm{s} = 9.3 \times 10^2\,\mathrm{s}.$$

The power is given by

$$\frac{\text{energy}}{\text{time}} = \frac{0.038}{9.3 \times 10^2}\,\mathrm{W} = 4.1 \times 10^{-5}\,\mathrm{W}.$$

It is satisfying that the two calculations of the power give the same result. The first is based on the empirically determined value of the resistivity, whereas the second requires a knowledge of the number density of conduction electrons in the metal. This quantity cannot be measured directly, but rather will be inferred from some other measured quantity, which may well be related to the resistivity! Nevertheless, approaching the problem from these two different angles should help you get to grips with how the various quantities are interconnected.

Q3.16 To minimize the current through point A, we must maximize the effective resistance from the combination of resistors. To do this, we will need the $100\,\Omega$ resistor in position 1. If it were in position 2 or 3 the current would flow predominantly through the other resistor connected in parallel. It makes no difference to the current at A which way round the $10\,\Omega$ and $1\,\Omega$ resistors are placed in positions 2 and 3.

Q4.1 To convince yourself that the assignment made in the figure is true, point the straightened fingers of your right hand in the direction of vector $\boldsymbol{v}$ in Figure 4.3. Then turn your wrist until you find that you can bend your fingers to align them with the vector $\boldsymbol{B}$. Your extended thumb now points in the direction of $(\boldsymbol{v} \times \boldsymbol{B})$. Since this is the same direction as $\boldsymbol{F}_{\mathrm{m}}$, it must be that q is positive. Had q been negative, the direction of $\boldsymbol{F}_{\mathrm{m}}$ would have been opposite to that of $(\boldsymbol{v} \times \boldsymbol{B})$.

Q4.2 (a) If the charged particle is moving parallel or antiparallel to the direction of the field, then $\theta = 0$, so $\sin\theta = 0$, and, according to Equation 4.3, F_{m} is also zero. Thus, a particle must be 'crossing' the field lines to experience a magnetic force.

(b) For a given speed v, Equation 4.3 shows that the maximum force will be experienced by the particle when $\theta = 90°$ (i.e. $\sin\theta = 1$), that is when the particle is moving in a direction perpendicular to the magnetic field. Notice that there are many possible directions that are perpendicular to the magnetic field, but all are confined to a single plane.

Q4.3 Since the question asks only about the *magnitude* of the magnetic force, Equation 4.3 is the appropriate one to use:

$$F_{\mathrm{m}} = |q|vB\sin\theta.$$

In this case, $\boldsymbol{B}$ and $\boldsymbol{v}$ are at right angles to each other. Therefore, $\sin\theta = 1$, and $F_{\mathrm{m}} = |q|vB$ and hence

$$B = \frac{F_{\mathrm{m}}}{|q|v}.$$

Q4.4 You now need to creatively apply the right-hand rule. Point the straightened fingers of your right hand in the x-direction of the vector $\boldsymbol{v}$, with your thumb extended 'sideways' from your hand. While keeping your fingers pointed in the direction of $\boldsymbol{v}$, rotate your wrist until your thumb points in the $-y$-direction. Bending your fingers will align them with the vector $\boldsymbol{B}$ which you should find is pointing in the $+z$-direction. The situation is illustrated in Figure 4.77.

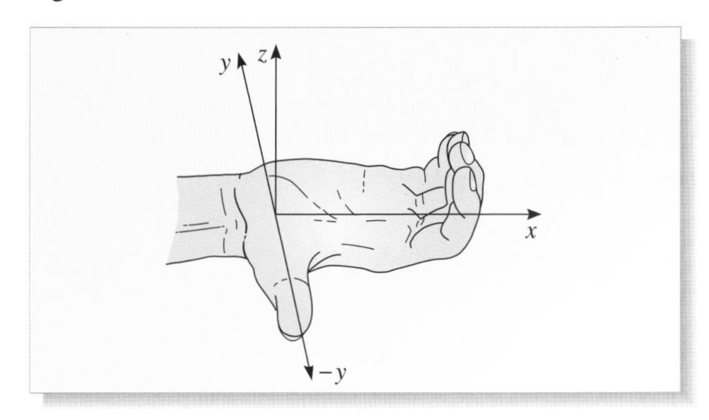

Figure 4.77 Answer to Q4.4.

Q4.5 Using the formula from Q4.3:

$$B = \frac{F_{\mathrm{m}}}{|q|v} = \frac{10^{-6}\,\mathrm{N}}{10^{-9}\,\mathrm{C} \times 100\,\mathrm{m\,s^{-1}}} = 10\,\mathrm{T}.$$

Q4.6 (a) 'Perpendicular to the plane of the page and out of the paper' is the answer. This can be checked easily by using the right-hand rule shown in Figure 4.8.

(b) To rotate your palm from $\boldsymbol{v}$ to $\boldsymbol{B}$ in this case, you must point your thumb into the page. So the answer is: 'perpendicular to the plane of the page and into the paper', which can be shown by a ×.

(c) $\boldsymbol{v} \times \boldsymbol{B}$ is into the page again, but this time q is negative, so $\boldsymbol{F}_{\rm m}$ is perpendicular to the plane of the page and outwards, illustrated by a •.

(d) $\boldsymbol{v} = 0$ means that $v = 0$, so $F_{\rm m} = 0$ and there is no force.

(e) $\boldsymbol{v} \times \boldsymbol{B}$ is in the plane of the page and points to the right. However, q is negative so $\boldsymbol{F}_{\rm m}$ points to the left.

(f) $\boldsymbol{F}_{\rm m}$ is in the plane of the page and points to the left.

Q4.7 In both cases, first substitute into Equation 4.3:

(a) $F_{\rm m} = |q|vB\sin\theta$

$F_{\rm m} = 2\,{\rm C} \times 10^4\,{\rm m\,s^{-1}} \times 2{\rm T}\sin 30°$

$F_{\rm m} = 2 \times 10^4\,{\rm N}.$

(Remember $\sin 30° = 0.5$.)

$\boldsymbol{v} \times \boldsymbol{B}$ points into the page and therefore so does $\boldsymbol{F}_{\rm m}$ (since q is positive).

(b) $F_{\rm m} = |q|vB\sin\theta$

$F_{\rm m} = 1.6 \times 10^{-19}\,{\rm C} \times 10^{-1}\,{\rm m\,s^{-1}} \times 10^{-4}\,{\rm T}\sin 90°$

$F_{\rm m} = 1.6 \times 10^{-24}\,{\rm N}.$

(Remember $\sin 90° = 1$.)

In this case, $\boldsymbol{v} \times \boldsymbol{B}$ points into the page, but q is negative so $\boldsymbol{F}_{\rm m}$ points out of the page.

Q4.8 At point A, the magnitude of the field is, according to Equation 4.4,

$$B = \frac{\mu_0 i}{2\pi r} = \frac{4\pi \times 10^{-7} \times 12}{2\pi \times 2}\,{\rm T} = 1.2 \times 10^{-6}\,{\rm T}.$$

Using the right-hand grip rule, you should find that the field at point A is directed *into* the page. At point B, the field strength is

$$B = \frac{4\pi \times 10^{-7} \times 12}{2\pi \times 5}\,{\rm T} = 4.8 \times 10^{-7}\,{\rm T}$$

and the field points *out* of the page. At point C, the field strength is the same as at A, i.e. $B = 1.2 \times 10^{-6}$ T and the field points *out* of the page.

Q4.9 (a) Using Equation 4.4:

$$B = \frac{\mu_0 i}{2\pi r} = \frac{4\pi \times 10^{-7} \times 1}{2\pi \times 0.1}\,{\rm T}$$
$$= 2 \times 10^{-6}\,{\rm T} = 2\,\mu{\rm T}.$$

(b) Using Equation 4.5:

$$B = \frac{\mu_0 i}{2R} = \frac{4\pi \times 10^{-7} \times 1}{2 \times 0.1}\,{\rm T}$$
$$= 2\pi \times 10^{-6}\,{\rm T} = 6.28\,\mu{\rm T}.$$

(c) Using Equation 4.5a:

$$B = \frac{\mu_0 N i}{2R} = \frac{4\pi \times 10^{-7} \times 10 \times 1}{2 \times 0.1}\,{\rm T}$$
$$= 2\pi \times 10^{-5}\,{\rm T} = 62.8\,\mu{\rm T}.$$

Q4.10 (a) Your completed version of Figure 4.32 should have looked roughly like Figure 4.78. Small differences between your drawing and this one are not important — you were only asked to produce a sketch. However, you should have got the direction of the magnetic field correct. Check again using the right-hand grip rule if necessary.

(b) The magnetic field of a cylindrical solenoid is shown in Figure 4.79a. The main thing you should have noticed is that the magnetic field of the cylindrical solenoid is reminiscent of that of a bar magnet. In fact, the field is nearly uniform close to the long axis at the solenoid but you are not expected to have brought this feature out in your sketch. Figure 4.79b illustrates a little trick for remembering how the field lines are directed: the end from which the lines emerge is the one (viewed end on) at which the current is flowing anticlockwise.

(c) The field lines diverge near one end of the solenoid and converge at the other. Therefore, if the solenoid is rather short, the field along it will only be uniform near the centre of its long axis, and Equation 4.6 will apply only to a rather small volume. However, if the solenoid is 'infinitely' long, the end-effects may be neglected and Equation 4.6 applies to the entire volume inside the solenoid. In practice, 'infinitely' long means 'long in relation to its diameter'. The quantity N/l may be interpreted as the number of turns per unit length. The form of Equation 4.6 does indeed seem reasonable: the more tightly wound the coil, or the higher the current for a given coil configuration, the greater the field strength.

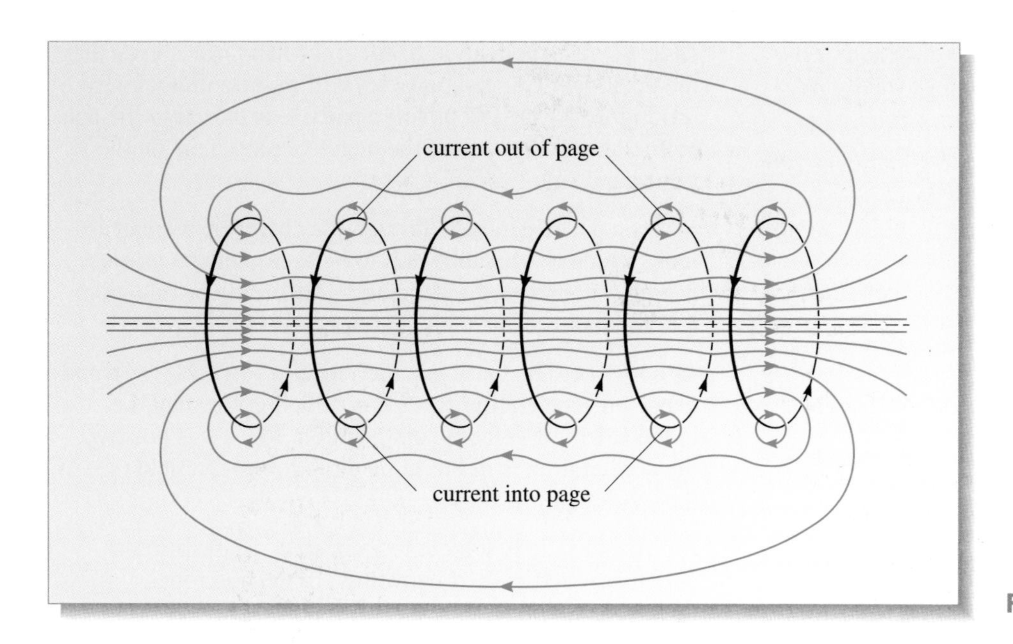

Figure 4.78 Answer to Q4.10a.

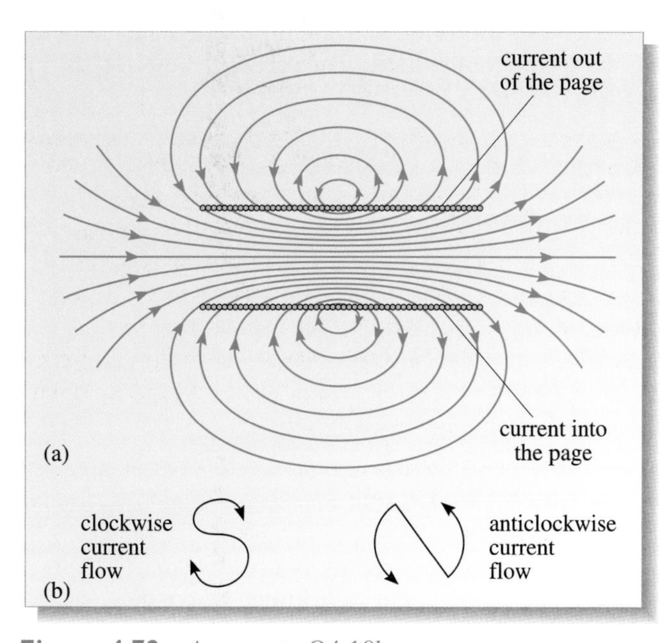

Figure 4.79 Answer to Q4.10b.

Q4.11 Inside an infinitely long cylindrical solenoid, the field strength is uniform. So, the strength at a point halfway between the axis and the winding is exactly the same as the field strength at any other point inside the solenoid. It is given by Equation 4.6 as

$$B = \frac{\mu_0 N i}{l} = \frac{4\pi \times 10^{-7}\,\mathrm{T\,m\,A^{-1}} \times 500 \times 4\,\mathrm{A}}{1\,\mathrm{m}}$$

$$= 2.5 \times 10^{-3}\,\mathrm{T}.$$

Q4.12 No. Magnetic fields are magnetic fields and the nature of their source cannot be determined by any experiment on the field itself.

Q4.13 The force on the particle is given by $\boldsymbol{F} = q(\boldsymbol{v} \times \boldsymbol{B})$. Since q is positive, $\boldsymbol{F}$ is in the same direction as $(\boldsymbol{v} \times \boldsymbol{B})$. It follows from Newton's second law that the acceleration is in the same direction as $\boldsymbol{F}$. Figure 4.80 shows what your answer should look like.

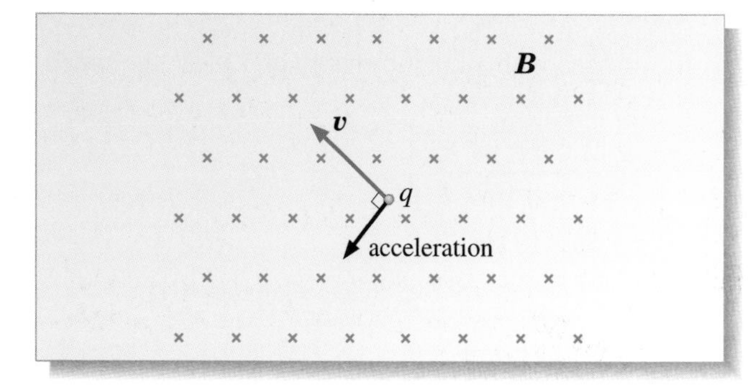

Figure 4.80 Answer to Q4.13.

Q4.14 If the sign of q is changed, the direction of the acceleration is reversed. Figure 4.81 shows what your answer should look like.

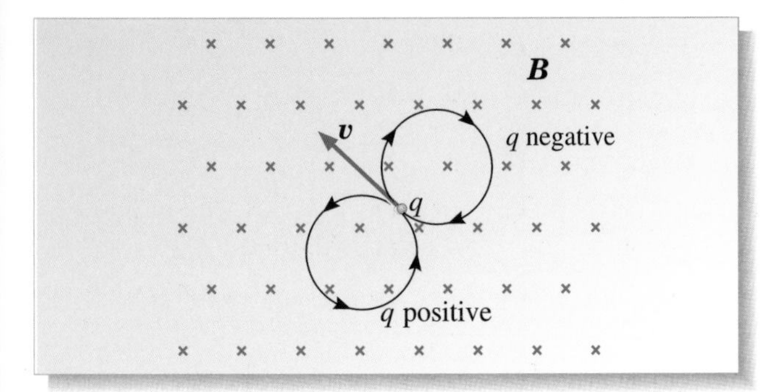

Figure 4.81 Answer to Q4.14.

Q4.15 The cyclotron radius and frequency are given by Equations 4.8 and 4.9 respectively:

$$R_\mathrm{C} = \frac{mv}{|q|B} \text{ and } f_\mathrm{C} = \frac{1}{2\pi}\frac{|q|B}{m}.$$

The proton has mass $m_\mathrm{p} = 1.67 \times 10^{-27}$ kg and its charge has a magnitude $|e| = 1.6 \times 10^{-19}$ C. With this data we can estimate the cyclotron frequency, because it does not depend on the particle's speed. Thus, for a magnetic field of 1 T:

$$f_\mathrm{C} = \frac{1}{2\pi}\frac{1.6\times10^{-19}\,\mathrm{C}\times1\,\mathrm{T}}{1.67\times10^{-27}\,\mathrm{kg}} = 15.2\,\mathrm{MHz}.$$

In order to work out the cyclotron radius, we need to work out the speed of the protons. Neglecting any relativistic effects, we equate the proton energy to the kinetic energy, and then solve for the speed. Recall that to convert electronvolts into joules we multiply by e, the charge on a proton.

$$(\mathrm{KE\,in\,eV})\times e = \tfrac{1}{2}mv^2$$

$$v = \sqrt{\frac{2\times(\mathrm{KE\,in\,eV})\times e}{m}}.$$

Thus for protons with a kinetic energy of 1 MeV, the speed is

$$v = \sqrt{\frac{2\times(1\times10^6)\times1.6\times10^{-19}}{1.67\times10^{-27}}}\,\mathrm{m\,s^{-1}}$$

$$= 1.38\times10^7\,\mathrm{m\,s^{-1}}$$

Their cyclotron radius in a magnetic field of 1 T is

$$R_\mathrm{C} = \frac{1.67\times10^{-27}\,\mathrm{kg}\times1.38\times10^7\,\mathrm{m\,s^{-1}}}{1.6\times10^{-19}\,\mathrm{C}\times1\,\mathrm{T}}$$

$$= 0.144\,\mathrm{m} = 14.4\,\mathrm{cm}.$$

Q4.16 (a) Application of the right-hand rule shows that the positively charged particles will move anticlockwise (cf. Figure 4.81). The positive particle is therefore the one on the left-hand side and the negative particle is on the right-hand side.

(b) As a particle moves through the chamber, it creates ions (which is what allows it to be seen), and so it loses energy, i.e. its speed v decreases. According to Equation 4.8, the radius of its path decreases as v falls.

Q4.17 (a) The particle experiences a force $\boldsymbol{F}_\mathrm{el} = q\boldsymbol{\mathcal{E}}$ and hence an acceleration $\boldsymbol{a} = q\boldsymbol{\mathcal{E}}/m$ which is constant, i.e. independent of velocity and position. Since q is positive, the acceleration is in the same direction as the field (i.e. in the $-y$-direction).

(b) The particle is subject to a constant acceleration $\boldsymbol{a}$ which acts vertically downwards. In the horizontal direction, the particle travels with constant speed. Hence its path in the field is parabolic — an exact analogy with the motion of a projectile in the uniform gravitational field near the surface of the Earth. The particle's trajectory is shown in Figure 4.82.

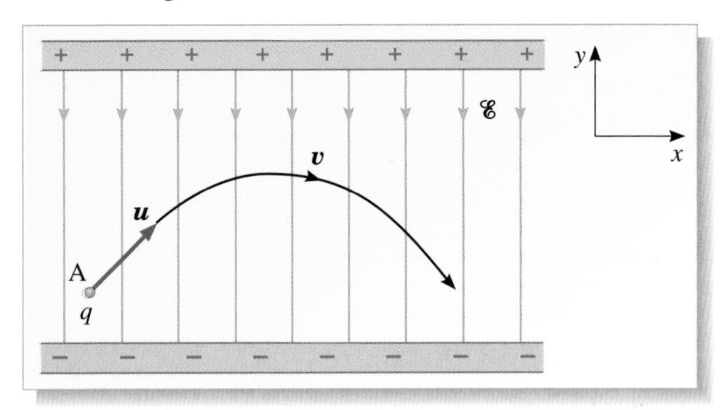

Figure 4.82 Answer to Q4.17.

(c) The x-component of velocity is a constant u_x, independent of time. In the y-direction, the constant acceleration equations apply. Here we have $a_y = -q\mathcal{E}/m$, where $\mathcal{E}$ is the *magnitude* of the electric field and therefore a positive number, so after time t, the y-component of the velocity, v_y, is given by

$$v_y = u_y - \left(\frac{q\mathcal{E}}{m}\right)t$$

where u_y is the y-component of the initial velocity $\boldsymbol{u}$.

Q4.18 (a) The expression for the Hall voltage is

$$V_H = \frac{iB}{nqt}$$

so $n = \dfrac{iB}{V_H qt}$.

$$i = 1\,\text{mA} = 10^{-3}\,\text{A},$$
$$B = 0.3\,\text{T},$$
$$|V_H| = 31.8\,\text{nV} = 3.18 \times 10^{-8}\,\text{V},$$
$$q = |e| = 1.6 \times 10^{-19}\,\text{C},$$
$$t = 1\,\mu\text{m} = 1 \times 10^{-6}\,\text{m}.$$

Substituting into the expression for n gives

$$n = \frac{10^{-3}\,\text{A} \times 0.3\,\text{T}}{3.18 \times 10^{-8}\,\text{V} \times 1.6 \times 10^{-19}\,\text{C} \times 10^{-6}\,\text{m}}$$
$$= 5.9 \times 10^{28}\,\text{m}^{-3}.$$

Notice that we have not taken account of the sign of the charge carriers because the sign of the measured voltage is not given.

(b) The number of volts/tesla generated across the sample is given by V_H/B:

$$\frac{V_H}{B} = \frac{i}{nqt}.$$

Substituting the values given

$$i = 1\,\text{mA} = 10^{-3}\,\text{A}$$
$$q = |e| = 1.6 \times 10^{-19}\,\text{C}$$
$$n = 10^{20}\,\text{m}^{-3}$$
$$t = 1\,\text{mm} = 1 \times 10^{-3}\,\text{m}$$

$$\frac{V_H}{B} = \frac{10^{-3}\,\text{A}}{10^{20}\,\text{m}^{-3} \times 1.6 \times 10^{-19}\,\text{C} \times 10^{-3}\,\text{m}}$$
$$= 0.0625\,\text{V}\,\text{T}^{-1} = 62.5\,\text{mV}\,\text{T}^{-1}.$$

Q4.19 The Hall voltage is given by

$$V_H = \frac{iB}{nqt}$$

so, rearranging,

$$B = \frac{V_H nqt}{i}.$$

The minimum value of Hall voltage that can be detected is $V_H = 10\,\mu\text{V} = 10^{-5}\,\text{V}$. Other values are

$$i = 1\,\text{mA} = 10^{-3}\,\text{A}$$
$$q = |e| = 1.6 \times 10^{-19}\,\text{C}$$
$$n = 10^{23}\,\text{m}^{-3}$$
$$t = 0.1\,\text{mm} = 10^{-4}\,\text{m}.$$

Substituting into the expression for B gives the minimum value of magnetic field that can be detected:

$$B_{min} = \frac{10^{-5}\,\text{V} \times 10^{23}\,\text{m}^{-3} \times 1.6 \times 10^{-19}\,\text{C} \times 10^{-4}\,\text{m}}{10^{-3}\,\text{A}}$$
$$= 0.016\,\text{T}.$$

Q4.20 (a) Application of the right-hand grip rule shows that the current must be flowing clockwise in the coils to generate a magnetic field into the page.

(b) The upper plate must be positively charged with respect to the lower in order to produce an electric field pointing downwards.

(c) The resultant force on the particle is

$$\boldsymbol{F} = q(\boldsymbol{\mathscr{E}} + \boldsymbol{v} \times \boldsymbol{B}_1).$$

For the particle to continue to travel at the same speed in a straight line, the resultant force, $\boldsymbol{F}$, must be zero. This condition can only be met if

$$q\boldsymbol{\mathscr{E}} = -q(\boldsymbol{v} \times \boldsymbol{B}_1).$$

Assuming the ion is positively charged, then the electric force $q\boldsymbol{\mathscr{E}}$ is in the $-y$-direction and the magnetic force $q(\boldsymbol{v} \times \boldsymbol{B}_1)$ is in the $+y$-direction. (If the ion is negatively charged, both of these directions will be reversed and the argument will be unaffected.) Thus, we can simply equate the magnitudes, i.e. we require

$$|\boldsymbol{\mathscr{E}}| = |\boldsymbol{v} \times \boldsymbol{B}_1|.$$

Since $\boldsymbol{v}$ and $\boldsymbol{B}_1$ are perpendicular, this reduces to

$$\mathscr{E} = vB_1$$

or $\quad v = \mathscr{E}/B_1.$

Q4.21 (a) From the answer to Q4.20, we know that the ions passing through the selector will have speed $v = \mathscr{E}/B_1$. So, for ions to emerge with the required speed, the electric field must be of magnitude

$$\mathscr{E} = vB_1 = 2 \times 10^5\,\text{m}\,\text{s}^{-1} \times 0.5\,\text{T} = 10^5\,\text{V}\,\text{m}^{-1}.$$

This field depends on the plate separation d and the potential difference V according to $\mathscr{E} = V/d$ so the required potential difference is

$$V = \mathscr{E}d = 10^5\,\text{V}\,\text{m}^{-1} \times 0.1\,\text{m} = 10^4\,\text{V}.$$

(b) The ions travel in a circle with radius given by

$$R_C = \frac{mv}{qB_2}.$$

Since they travel in a semicircle they will strike the plate a distance $2R_C$ from the point at which they enter the chamber, i.e. the diameter of their cyclotron orbit. Thus $2R_C = 16.2$ cm and $R_C = 8.1$ cm. We can now solve for m to find

$$m = \frac{R_C q B_2}{v}.$$

Substituting the relevant values gives

$$m = \frac{8.1\times10^{-2}\,\mathrm{m}\times1.6\times10^{-19}\,\mathrm{C}\times1\,\mathrm{T}}{2\times10^{5}\,\mathrm{m\,s^{-1}}}$$

$$= 6.48\times10^{-26}\,\mathrm{kg}.$$

To find the relative atomic mass, we divide by the mass of the proton/neutron

$$\text{relative atomic mass} = \frac{6.48\times10^{-26}\,\mathrm{kg}}{1.66\times10^{-27}\,\mathrm{kg}} = 39.0$$

which identifies the likely candidate ion as potassium.

Q4.22 (a) The cyclotron frequency is given by Equation 4.9:

$$f_C = \frac{1}{2\pi}\frac{qB}{m} = \frac{1}{2\pi}\frac{1.6\times10^{-19}\,\mathrm{C}\times0.5\,\mathrm{T}}{1.67\times10^{-27}\,\mathrm{kg}}$$

$$= 7.6\times10^{6}\,\mathrm{Hz} = 7.6\,\mathrm{MHz}.$$

The accelerating potential must be reversed every half cycle, so the reversals will take place at twice f_C, i.e at a frequency of 15.2 MHz.

(b) The energy transferred to a charge q on 'falling' through a potential difference V is equal to qV. In each orbit the proton receives two 'kicks' and increases its energy by 1000 eV on each 'kick'. Thus on each orbit it receives 2000 eV of energy. To reach 25 MeV will therefore take

$$N = \frac{25\times10^{6}\,\mathrm{eV}}{2\times10^{3}\,\mathrm{eV\,orbit^{-1}}} = 12\,500 \text{ orbits}.$$

Q4.23 The magnitude of the force on the cable is given by Equation 4.12: $F = Bil$. Substituting the given values for B, i and l, we find

$$F = 10^{-4}\,\mathrm{T}\times10^{4}\,\mathrm{A}\times100\,\mathrm{m} = 100\,\mathrm{N}.$$

Q4.24 (a) Figure 4.83 shows the various steps in the argument. The electrons in wire 2 move in the opposite direction to i_2, with velocity $\boldsymbol{v}_2$. The magnetic field $\boldsymbol{B}_1$ due to wire 1 points downwards (perpendicular to $\boldsymbol{v}_2$). Since the particle involved is a negatively charged electron, the force $\boldsymbol{F}_2$ on wire 2 acts in the opposite direction to $(\boldsymbol{v}_2\times\boldsymbol{B}_1)$. $\boldsymbol{F}_2$ acts *towards* wire 1.

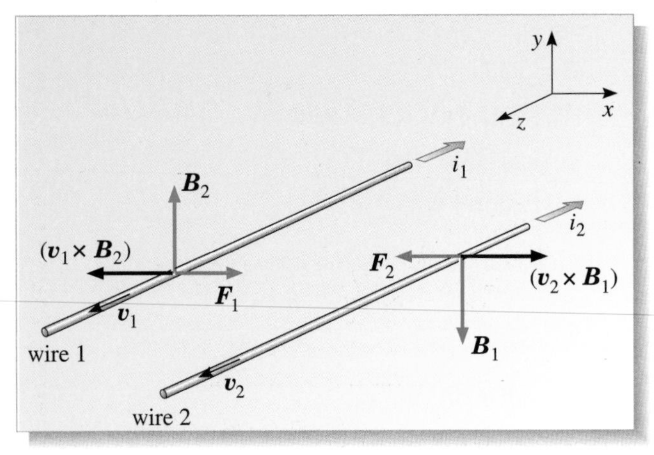

Figure 4.83 Answer to Q4.24.

Considering electrons flowing in wire 1, the direction of electron flow is the same ($\boldsymbol{v}_1$ is parallel to $\boldsymbol{v}_2$), but $\boldsymbol{B}_2$ is opposite to $\boldsymbol{B}_1$. The force $\boldsymbol{F}_1$ on wire 1 is therefore in the opposite direction to that on wire 2: $\boldsymbol{F}_1$ acts *towards* wire 2.

(b) A direct application of Equation 4.11 shows that, using the right-hand rule, $\boldsymbol{F}_2$ is in the direction of the vector product $\boldsymbol{i}_2\times\boldsymbol{B}_1$ which is in the direction shown in Figure 4.83. Similarly, $\boldsymbol{F}_1$ is in the direction of $\boldsymbol{i}_1\times\boldsymbol{B}_2$.

Q4.25 (a) Equating the units on either side of the equation $F = |q|vB$ gives:

$$\mathrm{N} = \mathrm{C}\times\mathrm{m\,s^{-1}}\times\mathrm{T}.$$

Using the definition of the coulomb:

$$1\,\mathrm{C\,s^{-1}} = 1\,\mathrm{A}$$

we find $\mathrm{N} = \mathrm{A\,m\,T}$, i.e. $\mathrm{A} = \mathrm{N\,m^{-1}\,T^{-1}}$.

(b) According to Equation 4.13, when a current of 1 A flows in each wire,

$$(2\times10^{-7})\,\mathrm{N\,m^{-1}} = \frac{\mu_0\times1\,\mathrm{A}\times1\,\mathrm{A}}{2\pi\times1\,\mathrm{m}}.$$

Substituting once for the units of the amp from part (a),

$$(2\times10^{-7})\,\mathrm{N\,m^{-1}} = \frac{\mu_0\times1\,\mathrm{A}\times1\,\mathrm{N\,m^{-1}\,T^{-1}}}{2\pi\times1\,\mathrm{m}}$$

i.e. $\mu_0 = 4\pi\times10^{-7}\,\mathrm{T\,m\,A^{-1}}$.

Q4.26 **Preparation** The sketch shows all the forces acting on the cylinder. $\boldsymbol{F}_{\text{fric}}$ represents the frictional force, which acts along the plane, $\boldsymbol{F}_{\text{m}}$ the magnetic force due to the current-carrying wires along the length of the cylinder and $\boldsymbol{R}$ the reaction force. $\boldsymbol{F}_{\text{m}}$ must be perpendicular to $\boldsymbol{B}$, i.e. horizontal. Known values are: $m = 0.15\,\text{kg}$; $B = 0.3\,\text{T}$; $l = 0.2\,\text{m}$, and $N = 25$ turns. We are not given r (the radius of the cylinder) or θ (the angle of inclination of the plane).

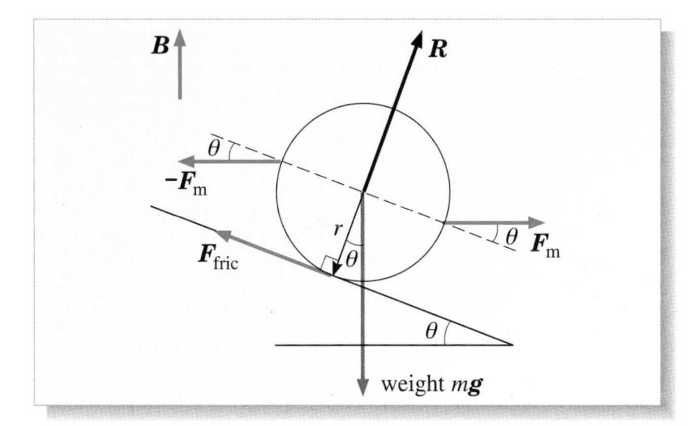

Figure 4.84 Sketch for Q4.26.

Working We need to find expressions for $\boldsymbol{F}_{\text{fric}}$ and $\boldsymbol{F}_{\text{m}}$ in terms of known values and the current i in the coil. If i is just sufficient to sustain static equilibrium, then the clockwise torque about the centre of mass due to friction will be equal to the anticlockwise torque due to the magnetic forces.

For the torques to balance, the magnetic forces must be outwards, as shown in Figure 4.84. Hence, by the right-hand rule, *electrons* must be moving into the page on the left of the cylinder and out of the page on the right — i.e. *conventional current* is flowing anticlockwise as seen from above.

Torque due to friction = $\boldsymbol{r} \times \boldsymbol{F}_{\text{fric}}$. But $\boldsymbol{F}_{\text{fric}}$ and $\boldsymbol{r}$ are perpendicular, so the magnitude of the torque due to friction

$$\Gamma = rF_{\text{fric}}$$

$$\Gamma = rmg\sin\theta$$

because the equilibrium of components along the plane requires $F_{\text{fric}} = mg\sin\theta$ (since there are equal but opposite contributions due to $\boldsymbol{F}_{\text{m}}$ along the plane).

Each side of the coil exerts a torque of magnitude $rF_{\text{m}}\sin\theta$, where $F_{\text{m}} = NBil$. Equating clockwise and anticlockwise torques, $rmg\sin\theta = 2rNBil\sin\theta$, i.e.

$$i = \frac{mg}{2NBl} = \frac{0.15\times 10}{2\times 25\times 0.3\times 0.2}\,\text{A} = 0.5\,\text{A}.$$

Checking The unit of i should be amperes.

mg has the unit $\text{kg}\times\text{m}\,\text{s}^{-2} = \text{N}$;

Bl has the unit $\text{T}\,\text{m} = \text{N}\,\text{s}\,\text{m}^{-1}\,\text{C}^{-1}\times\text{m} = \text{N}\,\text{s}\,\text{C}^{-1}$.

Therefore $\dfrac{mg}{Bl}$ has the unit $\dfrac{\text{N}}{\text{N}\,\text{s}\,\text{C}^{-1}} = \text{C}\,\text{s}^{-1} = \text{A}$, as required.

Q4.27 (a) Treating the twist as a circle, the radius is related to the field strength by Equation 4.8b:

$$R = \frac{m\,|\,v_{\text{perp}}\,|}{|\,q\,|\,B(\boldsymbol{r})}.$$

If ω is the magnitude of the angular velocity around the field line, then $v_{\text{perp}} = R\omega$.

Combining these two equations gives $\omega = |\,q\,|\,B(\boldsymbol{r})/m$.

If T is the time taken for one twist (i.e. to cover 2π radians), then $\omega = 2\pi/T$.

So $$T = \frac{2\pi m}{|\,q\,|\,B(\boldsymbol{r})}.$$

For the proton $$T_{\text{p}} = \frac{2\pi\times 1.7\times 10^{-27}\,\text{kg}}{1.6\times 10^{-19}\,\text{C}\times 1.5\times 10^{-5}\,\text{T}} = 4.5\times 10^{-3}\,\text{s}$$

The electron has a smaller mass than the proton, but the same magnitude of charge, so

$$T_{\text{e}} = T_{\text{p}}m_{\text{e}}/m_{\text{p}} = \frac{4.5\times 10^{-3}\,\text{s}\times 9.1\times 10^{-31}\,\text{kg}}{1.7\times 10^{-27}\,\text{kg}} = 2.4\times 10^{-6}\,\text{s}.$$

Note that the time needed for a single twist is determined only by the charge to mass ratio of the particle and the strength of the field.

(b) If the kinetic energy of the proton is 1 MeV, then $\frac{1}{2}m_{\text{p}}v^2 = 1\,\text{MeV} = 10^6\times 1.6\times 10^{-19}\,\text{J}$.

So $$v = \sqrt{\frac{2\times 1.6\times 10^{-13}\,\text{J}}{1.7\times 10^{-27}\,\text{kg}}} \approx 1.4\times 10^{7}\,\text{m}\,\text{s}^{-1}.$$

Therefore,
$$R = \frac{m_p v}{eB(\boldsymbol{r})}$$
$$= \frac{1.7 \times 10^{-27} \times 1.4 \times 10^{7}}{1.6 \times 10^{-19} \times 1.5 \times 10^{-5}}\,\text{m}$$
$$= 9.9 \times 10^{3}\,\text{m}.$$

Comment: *It gives some feeling for the speeds involved to realize that the proton executes a twist of diameter roughly 20 km in about 5 milliseconds! In fact, the proton only takes about 2 seconds to bounce from pole to pole, and the electron can make the same trip in about a tenth of a second.*

Q4.28 See Figure 4.85. The magnitude B of the field is given by
$$B = \sqrt{B_{\text{horiz}}^2 + B_{\text{vert}}^2}\,.$$
Therefore, in London
$$B = \sqrt{(19\,\mu\text{T})^2 + (-44\,\mu\text{T})^2}$$
$$= \sqrt{2297}\,\mu\text{T}$$
$$= 48\,\mu\text{T}.$$
The angle with respect to the horizontal is given by $\tan^{-1}(B_{\text{vert}}/B_{\text{horiz}})$ and so in London this angle is given by $\tan^{-1}(-44/19) = \tan^{-1}(-2.32) = -67°$ where the minus sign indicates that the field points below the horizon downward into the Earth.

In Bangui,
$$B = \sqrt{B_{\text{horiz}}^2 + B_{\text{vert}}^2}$$
$$= \sqrt{(32\,\mu\text{T})^2 + (9.3\,\mu\text{T})^2}$$
$$= \sqrt{1110}\,\mu\text{T}$$
$$= 33\,\mu\text{T}.$$
The angle with respect to the horizontal is given by $\tan^{-1}(B_{\text{vert}}/B_{\text{horiz}})$ and so in Bangui this angle is given by $\tan^{-1}(9.3/32) = \tan^{-1}(0.29) = +16°$, where the plus sign indicates that the field points above the horizon.

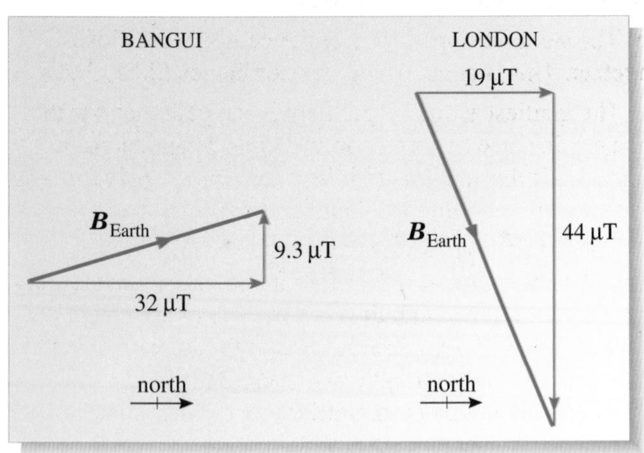

Figure 4.85 Answer to Q4.28.

Q4.29 Because the proton is moving at right angles to the magnetic field, we can use Equation 4.3 and have $F_m = qvB$. The magnetic field strength B is given by Equation 4.4, that is $B = \mu_0 i/(2\pi r)$, so
$$F_m = \frac{qv\mu_0 i}{2\pi r}.$$
Rearranging this and substituting the appropriate values gives
$$i = \frac{6.4 \times 10^{-17} \times 2\pi \times 0.1}{1.6 \times 10^{-19} \times 2 \times 10^{7} \times 4\pi \times 10^{-7}}\,\text{A} = 10\,\text{A}.$$

Q4.30 We can rearrange Equation 4.6 to give
$$\frac{N}{l} = \frac{B}{\mu_0 i}.$$
Substituting $B = 10\,\text{T}$ and $i = 100\,\text{A}$, we find
$$\frac{N}{l} = \frac{10}{4\pi \times 10^{-7} \times 100} \approx 80 \times 10^{3}\ \text{turns metre}^{-1}.$$
If each wire is 1mm in diameter then we can fit 1000 turns metre^{-1} on each layer and so we require 80 layers.

Q4.31 A tangent to a magnetic field line indicates the direction of the magnetic field at a given point. Since at each point, the magnetic field has a unique direction (it cannot point in two directions at once!), it is not possible for two field lines to pass through a point with different orientations. Hence there can be no points at which field lines cross one another.

Q4.32 Figure 4.86a shows the current-carrying wire and one of the concentric magnetic field lines surrounding it. Application of the right-hand grip rule demonstrates that at the location of the charged particle, on the $+x$-axis, the magnetic field $\boldsymbol{B}$ points in the $+y$-direction. Since the particle is positively charged, the direction of the magnetic force $\boldsymbol{F}_m$ on it will be the same as the direction of the vector product $(\boldsymbol{v} \times \boldsymbol{B})$, which can be found by the right-hand rule.

(a) $|\boldsymbol{v}|$ is zero, and therefore there is no magnetic force.

(b) $(\boldsymbol{v} \times \boldsymbol{B})$ points in the $+z$-direction, as illustrated in Figure 4.86b.

(c) $(\boldsymbol{v} \times \boldsymbol{B})$ points in the $-x$-direction, as illustrated in Figure 4.86c.

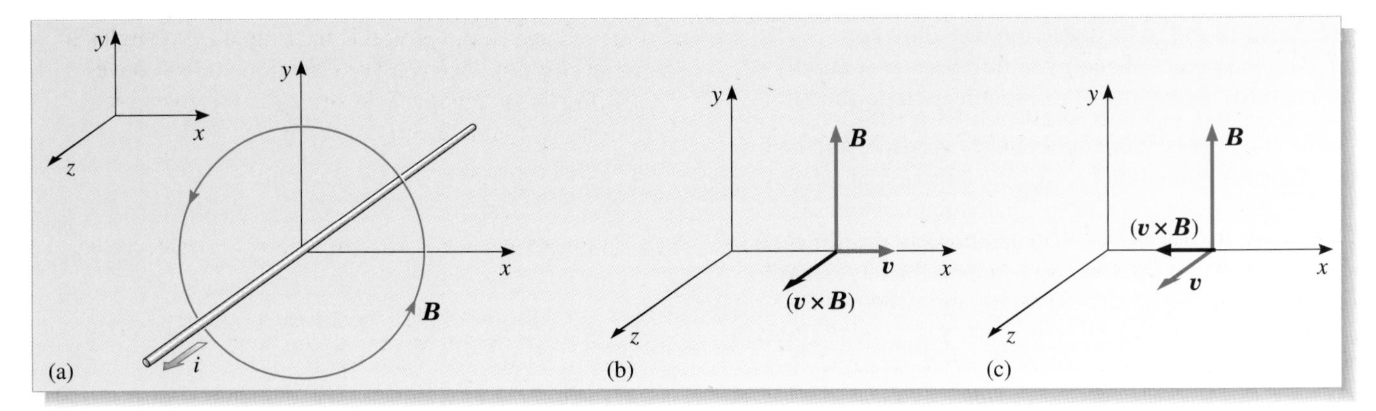

Figure 4.86 Answer to Q4.32.

Q4.33 The magnetic field lines point north. If the weight of the wire is balanced by the field, the force on the wire must point upwards (in the $+y$-direction in Figure 4.87, where the xz-plane is horizontal). If $\boldsymbol{F}_{\mathrm{m}}$ on a (negatively charged) electron points upwards, $\boldsymbol{v} \times \boldsymbol{B}$ points downwards. Application of the right-hand rule shows that $\boldsymbol{v}$ must point from east to west, so conventional current must flow from west to east, as illustrated in Figure 4.87.

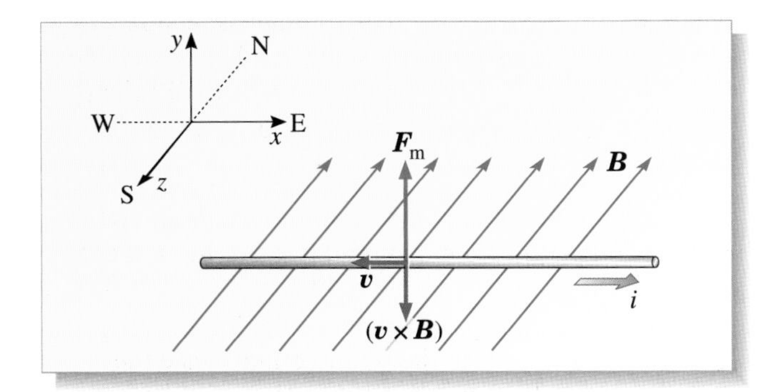

Figure 4.87 Answer to Q4.33.

The magnetic force is of magnitude $|\boldsymbol{F}_{\mathrm{up}}| = Bil$ and is equal but opposite to the downward gravitational force $\boldsymbol{F}_{\mathrm{down}} = m\boldsymbol{g}$.

Hence $Bil = mg$ and $i = mg/(Bl)$

so
$$i = \frac{0.1\,\mathrm{kg} \times 10\,\mathrm{m\,s^{-2}}}{5 \times 10^{-5}\,\mathrm{T} \times 2\,\mathrm{m}}$$
$$= 10^4\ \mathrm{A}.$$

(Note: ordinary household wire could not carry such a high current without melting, but superconducting wires will sustain these kinds of current.)

Q5.1 The attractive gravitational force between the electron and the proton has magnitude $F_{\mathrm{grav}} = Gm_{\mathrm{p}}m_{\mathrm{e}}/r^2$ where m_{p} is the mass of the proton, m_{e} the mass of the electron, and r the distance between them. The attractive electrostatic force between the charges has magnitude $F_{\mathrm{el}} = |-e^2|/(4\pi\varepsilon_0 r^2)$ where e is the charge on the proton. The ratio of these forces is given by

$$\frac{F_{\mathrm{el}}}{F_{\mathrm{grav}}} = \frac{e^2}{4\pi\varepsilon_0 G m_{\mathrm{p}} m_{\mathrm{e}}}.$$

Notice that the r^2 terms have cancelled. The ratio of the forces is independent of r since both forces are inverse square law forces, so the ratio would be unchanged if the radius were increased by a factor of 10.

Q5.2 (a) The coordinates are obtained from Figure 5.3 by reading the value from the east-west axis (along the bottom) and entering this before the comma, then reading the value from the north-south axis (along the left-hand side) and entering this after the comma. There are three contours within the 100 m contour for the summit located at (2.65, 1.95); there is only one contour within the 100 m contour for the summit at (2.50, 4.03). Therefore the height at (2.65, 1.95) exceeds 130 m, and is higher than (2.50, 4.03), which is lower than 120 m.

(b) The steepest slope is where the contours are closest together. This is at approximate coordinates (2.55, 3.95).

(c) The gentlest available gradient is the path that has the greatest distance between contour lines. From coordinates (2.65, 1.95) this is clearly in the direction towards the southwest (the bottom left-hand corner of the map).

(d) The gravitational potential energy per unit mass is proportional to altitude, therefore it decreases most rapidly in the direction in which the contours are closest together. From location (2.50, 2.25) this is in direction northwest (imagine drawing a straight line from (2.50, 2.25) and pivoting it about this point until it crosses the most contours in the shortest distance).

(e) A ball will roll in the direction that allows its gravitational potential energy to decrease most rapidly so, from part (d) above, it will roll in direction northwest.

(f) The contours here are quite widely spaced compared with their curvature, but locally, the direction of steepest slope will be perpendicular to a particular contour. At (2.25, 4.00) we are on the 90 m contour and the direction perpendicular to it at this point is approximately due west (due east is also perpendicular but points uphill).

Q5.3 Treating both the bodies as point masses

$$F_{\text{grav}} = Gm_1m_2/r^2$$

$$= \frac{6.67\times10^{-11}\,\text{N m}^2\,\text{kg}^{-2}\times1.00\,\text{kg}\times1.00\times10^{-3}\,\text{kg}}{1.00\,\text{m}^2}.$$

The forces exerted by the objects are equal in magnitude but opposite in direction. So, in each case,

$$F_{\text{grav}} = 6.67\times10^{-14}\,\text{N}.$$

Q5.4 We can apply the principle of superposition to calculate the net force on q by considering the forces due to each of the 10 fixed charges, resolving each in the x- and y-directions, and summing:

$$\boldsymbol{F}(\text{on } q) = \sum_{10\text{ charges}} F_x\hat{\boldsymbol{x}} + \sum_{10\text{ charges}} F_y\hat{\boldsymbol{y}}$$

where $\hat{\boldsymbol{x}}$ and $\hat{\boldsymbol{y}}$ are unit vectors in the x- and y-directions, respectively. Before embarking on this lengthy procedure, however, we notice that the charge at (2, 2) m will produce an equal and opposite force to that produced by the charge at (−2, −2) m. Similarly for the pair of charges located at (1, 2) m and (−1, −2) m. In fact eight of the ten charges produce equal and opposite fields at the origin, and we only need to consider the net force due to the charges located at (1, 3) m and (1, −3) m (Figure 5.9).

By symmetry the y-components of these two forces exactly cancel, and the net force on q is simply the sum of the two x-components:

$$F_{1x} = \frac{q}{4\pi\varepsilon_0}\left[\frac{1\,\text{C}}{(1^2+3^2)\,\text{m}^2}\right]\frac{1}{\sqrt{10}}$$

$$F_{2x} = \frac{q}{4\pi\varepsilon_0}\left[\frac{1\,\text{C}}{(1^2+3^2)\,\text{m}^2}\right]\frac{1}{\sqrt{10}}.$$

Note that in each case the factor of $1/\sqrt{10}$ represents the cosine of the angle between the direction of the force on the particle and the x-axis.

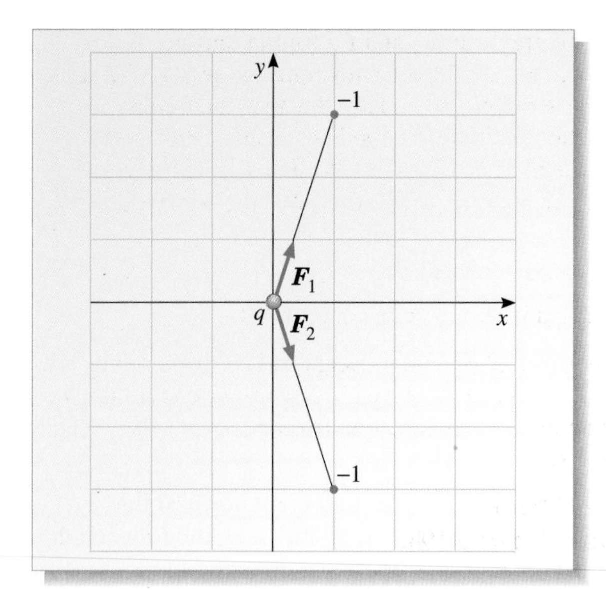

Figure 5.9 The forces on the charge at the origin due to the two charges placed at coordinates (1, 3) m and (1, −3) m are shown. When we resolve these forces into their x- and y-components, it is easy to see that the y-components are equal and opposite.

Also note that the forces in the x-direction will be positive if q is positive, so we have the signs correct. Hence the net force on q is

$$\boldsymbol{F}(\text{on } q) = \left(\frac{q(2\,\text{C})}{4\pi\varepsilon_0(10\,\text{m})\sqrt{10}}\right)\hat{\boldsymbol{x}}.$$

Q5.5 The change in potential energy when the satellite is put into orbit is

$$E_{\text{pot}} = E_{\text{grav}}(R_\text{s}) - E_{\text{grav}}(R_\text{E})$$

where $E_{\text{grav}}(r) = -\dfrac{GM_\text{E}m_\text{s}}{r}.$

So

$$\Delta E_{\text{pot}} = GM_\text{E}m_\text{s}\left[\frac{1}{R_\text{E}} - \frac{1}{R_\text{s}}\right]$$

$$= (6.67\times10^{-11}\,\text{N m}^2\,\text{kg}^{-2})\times(6.00\times10^{24}\,\text{kg})$$

$$\times(4.00\times10^3\,\text{kg})\times\left[\frac{1}{(6.40\times10^6\,\text{m})} - \frac{1}{(4.20\times10^7\,\text{m})}\right]$$

$$= 2.12\times10^{11}\,\text{N m}$$

$$= 2.12\times10^{11}\,\text{J}.$$

Q5.6 The graphs of F_x and E_{el} for the uniform field illustrated in Figure 5.5 are shown in Figure 5.10. We take $x = 0$ to be at the positive plate. Because the field is uniform, the force on a charge is the same everywhere between the plates and is equal to $\boldsymbol{F} = q\mathscr{E}$. Because the field is in the positive x-direction so is the force, so we have

$$F_x = q\mathscr{E}_x$$

and the graph is as shown in Figure 5.10a.

Taking the zero of potential energy to be at $x = 0$, the electrostatic potential energy E_{el} of q is equal to minus the work done by the conservative field on moving the charge to point x, so

$$E_{el} = -q\mathscr{E}_x x$$

as shown in Figure 5.10b.

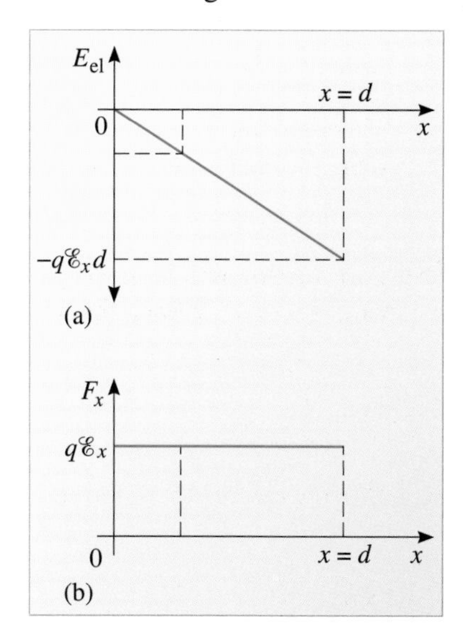

Figure 5.10 Graphs of electric force, F_x, and electrostatic potential energy, E_{el}, for the configuration shown in Figure 5.5.

Q5.7 If loose rocks are not to fly off, we require the gravitational force towards the centre of the moon to be at least as great as the centripetal force to maintain their circular motion, i.e.

$$mg_m = mR_m\,\omega^2$$

where m is the mass of a rock, R_m is the radius of the moon, ω is the angular speed in radians per second, and g_m is the magnitude of the acceleration due to gravity on the surface of the moon, which is given by:

$$g_m = \frac{GM_m}{R_m^2}$$

where M_m represents the mass of the moon.

Hence we require

$$\omega^2 \leq \frac{g_m}{R_m} \leq \frac{GM_m}{R_m^3}$$

$$\omega \leq \sqrt{\frac{6.67 \times 10^{-11}\ \text{N m}^2\ \text{kg}^{-2} \times 3.8 \times 10^{19}\ \text{kg}}{(2.5 \times 10^5\ \text{m})^3}}.$$

Maximum angular speed $\omega = 4.0 \times 10^{-4}\ \text{rad s}^{-1}$.

Q5.8 (a) The effective resistance between A and D is given by

$$\frac{1}{R_{eff}} = \left(\frac{1}{2} + \frac{1}{1}\right)\Omega^{-1}$$

$$\frac{1}{R_{eff}} = \frac{3}{2}\Omega^{-1}$$

$$R_{eff} = \frac{2}{3}\Omega.$$

(b) The current at point A can be calculated from Ohm's law:

$$i_A = \frac{V}{R_{eff}} = \frac{15 \times 3}{2}\ \text{A} = 22.5\ \text{A}$$

(c) The current at point B can be worked out using Ohm's law for the 1 Ω resistor:

$$i_B = \frac{V}{R} = \frac{15}{1}\ \text{A} = 15\ \text{A}.$$

The current at point C, again obtained using Ohm's law, is

$$i_C = \frac{V}{R} = \frac{15}{2} = 7.5\ \text{A}.$$

If the working is correct, Kirchhoff's second law tells us $i_A = i_B + i_C$ should be satisfied: 22.5 A = (15 + 7.5) A, so we can be confident in our result.

Kirchhoff's second law tells us that $i_D = i_A$, so the current at D is 22.5 A.

Q5.9 (a) Because the voltmeter reads 0.0 V, there is zero potential difference between points B and D. Hence no current will be driven along this part of the circuit, and we can redraw it as in Figure 5.11.

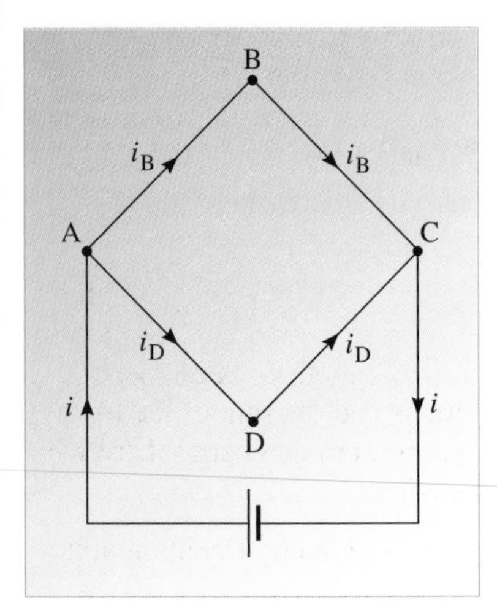

Figure 5.11 Current will flow around the Wheatstone bridge circuit as shown. No current flows directly from B to D when the voltmeter reads 0; in this case the circuit is 'balanced'.

(b) Along path ADC the total resistance is $1\ \Omega + 4\ \Omega$ (resistors in series) and applying Ohm's law we have

$$i_D = \frac{V}{R_{eff}} = \frac{10\ V}{5\ \Omega} = 2\ A.$$

(c) To calculate the current through the $20\ \Omega$ resistor, we note B and D are at the same potential, hence the potential difference between D and C is equal to the potential difference between B and C.

Ohm's law tells us that

$$V_{DC} = i_D \times 4\ \Omega = 8\ V$$

So $\quad V_{BC} = V_{DC} = 8$ V.

(d) Ohm's law applied to the $20\ \Omega$ resistor now tells us that

$$i_B = \frac{V_{BC}}{20\ \Omega} = \frac{8}{20}\ A = 0.4\ A.$$

(e) Kirchhoff's first law tells us that $V_{AB} = 10\ V - 8\ V = 2\ V$, so Ohm's law applied to the unknown resistance demands

$$R = \frac{V_{AB}}{i_B} = \frac{2}{0.4}\ \Omega = 5\ \Omega.$$

(f) If the $20\ \Omega$ resistor is replaced by a $10\ \Omega$ resistor then the currents will change. Assuming the voltmeter has infinite resistance, there will still be only two branches of current flow.

Applying Ohm's law to the path ABC we have
$i_B \times (5 + 10)\ \Omega = 10$ V

i.e. $\quad i_B = \frac{10}{15}\ A = \frac{2}{3}\ A.$

Applying Ohm's law to the $5\ \Omega$ resistor, we have

$$V_{AB} = i_B \times 5\ \Omega$$

$$= \frac{2}{3} \times 5\ V = 3.33\ V$$

Similarly, $i_D = \frac{10}{5}\ A = 2\ A$

and $\quad V_{AD} = i_D \times 1\ \Omega = 2$ V.

Thus the potential difference between points B and D is $(3.33 - 2)$ V = 1.33 V (point B is at the higher potential).

Q5.10 (a) Silver is the best conductor, so it would produce the minimum resistance in the wires.

(b) If copper were used instead, the resistance would increase due to the higher resistivity. Equation 3.4 relates the resistance, R, to the resistivity, ρ, and the length, L, and cross-sectional area, A, of the wire:

$$R = \frac{\rho L}{A}.$$

So we can write

$$R_{copper} = \frac{\rho_{copper} L}{A_{copper}}$$

and $$R_{silver} = \frac{\rho_{silver} L}{A_{silver}}.$$

If $R_{copper} = R_{silver}$ as required, then

$$\frac{\rho_{copper} L}{A_{copper}} = \frac{\rho_{silver} L}{A_{silver}}.$$

The Ls cancel, giving

$$\frac{A_{copper}}{A_{silver}} = \frac{\rho_{copper}}{\rho_{silver}}.$$

But $A \propto d^2$, so

$$\frac{d^2_{copper}}{d^2_{silver}} = \frac{\rho_{copper}}{\rho_{silver}}.$$

Hence, taking the square root of both sides,

$$\frac{d_{copper}}{d_{silver}} = \left(\frac{\rho_{copper}}{\rho_{silver}}\right)^{1/2}$$

$$= \left(\frac{1.7 \times 10^{-8}}{1.6 \times 10^{-8}}\right)^{1/2} = 1.03.$$

Thus the diameter of the copper wires needs to be only 3% bigger than that of the silver wires — hence copper is generally used.

Q5.11 (a) $P(t) = |\, i(t) \,||\, V(t) \,|$.

(b) If $q(t)$ represents the charge remaining on the plates at time t, then

$$V(t) = \frac{q(t)}{C}.$$

If $i(t)$ represents the current through the resistor at time t, then $V(t) = i(t)R$, so

$$i(t) = \frac{q(t)}{RC}.$$

But because the loss of charge from the capacitor accounts for the current through the resistor, it follows that

$$i(t) = -\mathrm{d}q(t)/\mathrm{d}t.$$

Combining the last two results to eliminate $i(t)$, we obtain the differential equation

$$\frac{\mathrm{d}q}{\mathrm{d}t} = -\frac{q}{RC}$$

which has the solution

$$q(t) = q_0 \mathrm{e}^{-t/RC}.$$

So $$V(t) = \frac{q_0}{C} e^{-t/RC}$$

and $$i(t) = -\frac{1}{RC} q_0 \mathrm{e}^{-t/RC}$$

implying that $$P(t) = \frac{q_0^2}{RC^2} \mathrm{e}^{-2t/RC}.$$

(c) The graph of power as a function of time (Figure 5.12) illustrates an exponential decay, with time constant $RC/2$.

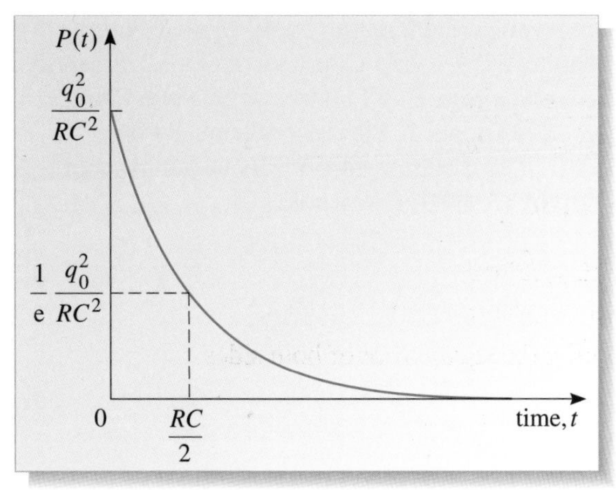

Figure 5.12 The graph that you sketched for Question 5.11c should look like this exponential decay (you were not asked to include the quantitative indications of the rate of the decay).

Q5.12 At point A the magnitude of the magnetic field is, according to Equation 4.4,

$$B = \frac{4\pi \times 10^{-7} \times 10}{2\pi \times 2}\,\mathrm{T} = 1.0 \times 10^{-6}\,\mathrm{T}.$$

From the right-hand grip rule, the field at point A is directed *into* the page. At point B the field strength is 4.0×10^{-7} T and the field points *out* of the page. At point C the field strength is 1.0×10^{-6} T and the field points *out* of the page.

Q5.13 Each one of the moving charges with a component of velocity perpendicular to the magnetic field experiences a force. But the charges are in random motion, so there will on average be equal numbers of charges travelling in opposite directions and the sum of the forces on all the charge carriers will be zero and hence there will be no resultant force on the wire.

Q5.14 The cyclotron radius is given by Equation 4.8:

$$R_\mathrm{c} = \frac{mv}{|\,q\,|\,B}.$$

So $$R_\mathrm{c} = \frac{1.67 \times 10^{-27} \times 2.0 \times 10^{6}}{1.60 \times 10^{-19} \times 2.1}\,\mathrm{m}.$$

Note that the speed has been converted into metres per second, so that all the quantities on the right-hand side are in SI units, hence the radius is in metres. Evaluating the expression, we obtain

$$R_\mathrm{c} = 9.9 \times 10^{-3}\,\mathrm{m}.$$

Acknowledgements

Grateful acknowledgement is made to the following sources for permission to reproduce material in this book:

Front cover - Science Photo Library

Figure 1.3 1980 UKATC Royal Observatory, Edinburgh; *Figures 1.4, 1.6 and 1.40* Science and Society Picture Library/Science Museum; *Figure 1.5* British Library; *Figure 1.8* Ann Ronan/Image Select; *Figure 1.22* Versailles et Trianon Lecomte-Vernet, portrait du physicien Charles-Auguste de Coulomb. © Photo Réunion des Musées Nationaux; *Figure 1.26* Science Photo Library; *Figure 1.31* Courtesy of Hunterian Museum and Art Gallery, Glasgow University;

Figure 2.1 Akira Fujii; *p.57 Earth/Moon photo* - JPL; *Figures 2.9 and 2.28* Mary Evans Picture Library; *Figure 2.10* Simon Fraser/Science Photo Library; *Figure 2.19* Image Select/Ann Ronan; *Figure 2.20a* Science Museum/Science and Society Picture Library; *Figure 2.26* Dennis Hardley Scottish Photo Library; *Figure 2.29* Anglo Australian Telescope Board Photograph by David Malin; *Figure 2.30* NASA/Chandra Science Center; *Figure 2.31* Photo Aerospatiale DS21170 & Matra Marconi Space; *Figure 2.34* Michael Gilbert/Science Photo Library;

Figure 3.1 upper left Keith Kent/Science Photo Library; *Figure 3.1 upper right* NASA/Science Photo Library; *Figure 3.1 lower left and 3.31* Science Photo Library; *Figure 3.7b* NASA; *Figure 3.13* Mary Evans Picture Library; *Figure 3.27* Reproduced with permission from *Traceable Temperatures* by Nicholas & White. Copyright John Wiley & Sons Ltd. *Figure 3.36* David Frazier/Science Photo Library; *Figure 3.37* Courtesy of Sven Bräutigam, Open University Department of Physics and Astronomy;

p. 135 (bottom) and Figure 4.23 Mary Evans Picture Library; *Figure 4.10 Electromagnet* - Alex Bartel/Science Photo Library; *Sunspot* - Courtesy of Calvin J. Hamilton National Solar Observatory/Sacramento Peak; *Crab Nebula* - National Optical Astronomy Observations; *Figure 4.35* Courtesy of Open University Biomagnetism Group; *Figure 4.51* Lawrence Berkeley Laboratory/Science Photo Library; *Figure 4.56* Terraplus; *Figure 4.74* Lionel F. Stevenson/Science Photo Library; *Figure 4.75* Worpole, I., (1989) 'The dynamic aurora', Syun-Ichi Akasofu, *Scientific American*, May 1989. Copyright © 1989, I. Worpole.

Index

Entries and page numbers in **bold type** refer to key words which are printed in **bold** in the text and which are defined in the Glossary.